Statistics for Industry and Technology

Series Editor

Statistics for Industry and Technology

For further volumes:
http://www.springer.com/series/4982

N. Unnikrishnan Nair • P.G. Sankaran
N. Balakrishnan

Quantile-Based Reliability Analysis

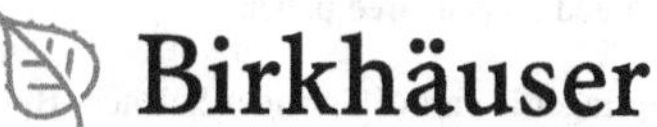

N. Unnikrishnan Nair
Department of Statistics
Cochin University of Science
and Technology
Cochin, Kerala, India

P.G. Sankaran
Department of Statistics
Cochin University of Science
and Technology
Cochin, Kerala, India

N. Balakrishnan
Department of Mathematics and Statistics
McMasters University
Hamilton, ON, Canada

ISBN 978-1-4939-5167-3 ISBN 978-0-8176-8361-0 (eBook)
DOI 10.1007/978-0-8176-8361-0
Springer New York Heidelberg Dordrecht London

Mathematics Subject Classification (2010): 62E-XX, 62N05, 62P30

Printed on acid-free paper

Springer is part of Springer Science+Business Media (www.springer.com)

To Jamuna, Anoop and Arun
NUN
To Sandhya, Revathi and Nandana
PGS
To Julia, Sarah and Colleen
NB

Foreword

Quantile functions are a fundamental, and often the most natural, way of representing probability distributions and data samples. In much of environmental science, finance, and risk management, there is a need to know the event magnitude corresponding to a given return period or exceedance probability, and the quantile function provides the most direct expression of the solution. The book *Statistical Modelling with Quantile Functions* by Warren Gilchrist provides an extensive survey of quantile-based methods of inference for complete distributions.

Reliability analysis is another field in which quantile-based methods are particularly useful. In many applications one must deal with censored data and truncated distributions, and concepts such as hazard rate and residual life become important. Quantile-based methods need some extensions to deal with these issues, and the present book goes beyond Prof. Gilchrist's and provides a thorough grounding in the relevant theory and practical methods for reliability analysis.

I have been fortunate in having been able for over 20 years to develop a theory of L-moments, statistics that are simple and effective inferences about probability distributions. Although not restricted to quantile-based inference, L-moments lend themselves well to quantile methods since many of the key results regarding L-moments are most conveniently expressed in terms of the quantile function. As with quantile methods generally, an L-moment-based approach to reliability analysis has not been developed in detail. In this book the authors have made significant progress, and in particular the development of L-moment methods for residual life analysis is a major step forward. I congratulate the authors on their achievements, and I invite the readers of this book to enjoy a survey of material that is rich in both theoretical depth and practical utility.

Yorktown Heights, NY J.R.M. Hosking

Preface

Reliability theory has taken rapid strides in the last four decades to become an independent discipline that influences our daily lives and schedules through our dependence on good and reliable functioning of devices and systems that we constantly use. The extensive literature on reliability theory, along with its applications, is scattered over various disciplines including statistics, engineering, applied probability, demography, economics, medicine, survival analysis, insurance and public policy. Life distributions specified by their distribution functions and various concepts and characteristics derived from it occupy a big portion of reliability analysis. Although quantile functions also represent life distributions and would facilitate one to carry out all the principal functions enjoyed by distribution functions in the existing theory and practice, this feature is neither fully appreciated nor exploited. The objective of this book is to attempt a systematic study of various aspects of reliability analysis with the aid of quantile functions, so as to provide alternative methodologies, new models and inferential results that are sometimes difficult to accomplish through the conventional approach.

Due to the stated objective, the material presented in this book is loaded with a quantile flavour. However, all through the discussion, we first present a concept or methodology in terms of the conventional approach and only introduce the quantile-based counterpart. This will enable the reader to transfer the methodology from one form to the other and to choose the one that fits his/her taste and need. Being an introductory text in quantile-based reliability methods, there is scope for further improvements and extensions of the results discussed here.

The book is biased towards the mathematical theory, with examples intended to clarify various notions and applications to real data being limited to demonstrate the utility of quantile functions. For those with interest in practical aspects of quantile-based model building, relevant tools and descriptive data analysis, the book by Gilchrist would provide a valuable guidance.

This book is organized into nine chapters. Chapter 1 deals with the definition, properties and various descriptive measures based on the quantile functions. Various reliability concepts like hazard rate, and mean residual life, in the conventional form as well as their quantile equivalents, are discussed in Chap. 2. This is followed, in Chap. 3, by a detailed presentation of the distributional and reliability aspects of quantile function models along with some applications to real data. Different ageing concepts in quantile versions are described in Chap. 4. Total time on test transforms, an essentially quantile-based notion, is detailed in Chap. 5. As alternatives to the conventional moments, the L-moments and partial moments in relation to residual life are presented. In Chap. 6, the definitions, properties and characterizations of these concepts are explained along with their use in inferential methods. Bathtub hazard models are considered in Chap. 7 along with their quantile counterparts and some new quantile functions that exhibit nonmonotone hazard quantile functions. The definitions and properties of various stochastic orders encountered in reliability theory are described in Chap. 8. Finally, Chap. 9 deals with various methods of estimation and modelling problems. A more detailed account of the contents of each chapter is provided in the Abstract at the beginning of each chapter.

Within the space available for this book, it has not been possible to include all the topics pertinent to reliability analysis. Likewise, the work of many authors who have contributed to these topics, as well as to those in the text, could not be included in the book. Our sincere apologies for these shortcomings. Any suggestion for the improvement in the contents and/or indication of possible errors in the book are wholeheartedly welcomed.

Finally, we wish to thank Sanjai Varma, Vinesh Kumar and K. P. Sasidharan for their contributions to the cause of this work. We are grateful to the colleagues in the Department of Statistics and Administration of the Cochin University of Science and Technology. Our sincere thanks also go to Ms. Debbie Iscoe for her help with the final typesetting and production of the volume.

Finally, we would like to state formally that this project was catalysed and supported by the Department of Science and Technology, Government of India, under its Utilisation of Scientific Expertise of Retired Scientist Scheme.

Cochin, India — N. Unnikrishnan Nair
Cochin, India — P.G. Sankaran
Hamilton, ON, Canada — N. Balakrishnan

Contents

List of Figures

Acronyms

X, Y, Z	Continuous random variables
$F(x)$	Distribution function of X
$Q(u)$	Quantile function corresponding to $F(x)$
$f(x)$	Probability density function of X
$q(u)$	Quantile density function
X_t	Residual life at age t
$\bar{F}(x)$	Survival function of X
$\bar{F}_t(x)$	Survival function of X_t
M	Median
S	Galton's coefficient of skewness
T	Moors' measure of kurtosis
Δ	Gini's mean difference
$X_{r:n}$	rth order statistic in a sample of size n
L_r	rth L-moment
τ_2	L-coefficient of variation
τ_3	L-coefficient of skewness
τ_4	L-coefficient of kurtosis
$h(x)$	Hazard rate of X
$H(u)$	Hazard quantile function of X
$m(x)$	Mean residual life function
$M(u)$	Mean residual quantile function
$\sigma^2(x)$	Variance residual life function
$V(u)$	Variance residual quantile function
$p_\alpha(x)$	Percentile residual life function
$P_\alpha(u)$	Percentile residual quantile function
$\lambda(x)$	Reversed hazard rate function
$\Lambda(u)$	Reversed hazard quantile function
$r(x)$	Reversed mean residual life
$R(u)$	Reversed mean residual quantile function
$v(x)$	Reversed variance residual life function

$D(u)$	Reversed variance quantile function
$q_\alpha(x)$	Reversed percentile residual life function
σ^2	Variance of X
μ_r	rth central moment of X
l_r	rth sample Λ-moment
$J(u)$	Score function
$\tau(\cdot)$	Total time on test statistic
$F_n(x)$	Empirical distribution function
$T(u)$	Total time on test transform
$\phi(u)$	Scaled total time on test transform
$L(u)$	Lorenz curve
G	Gini index
$B(u)$	Bonferroni curve
$K(u)$	Leimkuhler curve
$T_n(u)$	Total time on test transform of order n
$L_r(t)$	rth L-moment of $X\|(X>t)$
$l_r(u)$	Quantile version of $L_r(t)$
$B_r(t)$	rth L-moment of $X\|(X\leq t)$
$\theta_r(u)$	Quantile version of $B_r(t)$
$P_r(u)$	Upper partial moment of order r
$P_r^*(u)$	Lower partial moment of order r
$Q_n(p)$	Empirical quantile function
ξ_p	Sample quantile
$M_{p,rs}$	Probability weighted moments
$\leq_{\text{st}}$	Usual stochastic order
$\leq_{\text{hr}}$	Hazard rate order
$\leq_{\text{mrl}}$	Mean residual life order
$\leq_{\text{rmrl}}$	Renewal mean residual life order
$\leq_{\text{hrmrl}}$	Harmonic renewal mean residual life order
$\leq_{\text{vrl}}$	Variance residual life order
$\leq_{\text{prl}-\alpha}$	Percentile residual life order
$\leq_{\text{rh}}$	Reversed hazard rate order
$\leq_{\text{MIT}}$	Mean inactivity time order
$\leq_{\text{VIT}}$	Variance inactivity time order
$\leq_{\text{TTT}}$	Total time on test transform order
$\leq_{\text{c}}$	Convex transform order
$\leq_{*}$	Star order
$\leq_{\text{su}}$	Super additive order
$\leq_{\text{NBUHR}}$	New better than used in hazard rate order
$\leq_{\text{DMRL}}$	Decreasing mean residual life order
$\leq_{\text{NBUE}}$	New better than used in expectation order
$\leq_{\text{MTTF}}$	Mean time to failure order
BT (UBT)	Bathtub-shaped hazard rate (upside-down bathtub-shaped hazard rate)

DCSS	Decreasing cumulative survival class
DHRA	Decreasing hazard rate average
DMERL	Decreasing median residual life
DMRL	Decreasing mean residual life
DMRLHA	Decreasing mean residual life in harmonic average
DMTTF	Decreasing mean time to failure
DPRL	Decreasing percentile residual life
DRMRL	Decreasing renewal mean residual life
DVRL	Decreasing variance residual life
GIMRL	Generalized increasing mean residual life
HNBUE	Harmonically new better than used
HNWUE	Harmonically new worse than used
HRNBUE	Harmonically renewal new better than used in expectation
HUBAE	Harmonically used better than aged
ICSS	Increasing cumulative survival class
IHR (2)	Increasing hazard rate of order 2
IHR (DHR)	Increasing hazard rate (decreasing hazard rate)
IHRA	Increasing hazard rate average
IMERL	Increasing median residual life
IMIT	Increasing mean inactivity time
IMRL	Increasing mean residual life
IPRL	Increasing percentile residual life
IQR	Interquartile range
IRMRL	Increasing renewal mean residual life
IVRL	Increasing variance residual life
NBRU	New better than renewal used
NBU	New better than used
NBUC	New better than used in convex order
NBUCA	New better than used in convex average
NBUE	New better than used in expectation
NBUHR	New better than used in hazard rate
NBUHRA	New better than used in hazard rate average
NBUL	New better than used in Laplace order
NDMRL	Net decreasing mean residual life
NWRU	New worse than renewal used
NWU	New worse than used
NWUC	New worse than used in convex order
NWUE	New worse than used in expectation
NWUHR	New worse than used in hazard rate
NWUHRA	New worse than used in hazard rate average
NWUL	New worse than used in Laplace order
RNBRU	Renewal new is better than renewal used
RNBRUE	Renewal new is better than renewal used in expectation
RNBU	Renewal new is better than used
RNBUE	Renewal new is better than used in expectation

RNWU	Renewal new is worse than used
SIHR	Stochastically increasing hazard rate
SNBU	Stochastically new better than used
TTT	Total time on test transform
UBA	Used better than aged
UBAE	Used better than aged in expectation
UWA	Used worse than aged
UWAE	Used worse than aged in expectation

Chapter 1
Quantile Functions

Abstract A probability distribution can be specified either in terms of the distribution function or by the quantile function. This chapter addresses the problem of describing the various characteristics of a distribution through its quantile function. We give a brief summary of the important milestones in the development of this area of research. The definition and properties of the quantile function with examples are presented. In Table 1.1, quantile functions of various life distributions, representing different data situations, are included. Descriptive measures of the distributions such as location, dispersion and skewness are traditionally expressed in terms of the moments. The limitations of such measures are pointed out and some alternative quantile-based measures are discussed. Order statistics play an important role in statistical analysis. Distributions of order statistics in quantile forms, their properties and role in reliability analysis form the next topic in the chapter. There are many problems associated with the use of conventional moments in modelling and analysis. Exploring these, and as an alternative, the definition, properties and application of *L*-moments in describing a distribution are presented. Finally, the role of certain graphical representations like the *Q*-*Q* plot, box-plot and leaf-plot are shown to be useful tools for a preliminary analysis of the data.

1.1 Introduction

As mentioned earlier, a probability distribution can be specified either in terms of the distribution function or by the quantile function. Although both convey the same information about the distribution, with different interpretations, the concepts and methodologies based on distribution functions are traditionally employed in most forms of statistical theory and practice. One reason for this is that quantile-based studies were carried out mostly when the traditional approach either is difficult or fails to provide desired results. Except in a few isolated areas, there have been no systematic parallel developments aimed at replacing distribution functions in modelling and analysis by quantile functions. However, the feeling that through

N.U. Nair et al., *Quantile-Based Reliability Analysis*, Statistics for Industry and Technology, DOI 10.1007/978-0-8176-8361-0_1,

an appropriate choice of the domain of observations, a better understanding of a chance phenomenon can be achieved by the use of quantile functions, is fast gaining acceptance.

Historically, many facts about the potential of quantiles in data analysis were known even before the nineteenth century. It appears that the Belgian sociologist Quetelet [499] initiated the use of quantiles in statistical analysis in the form of the present day inter-quantile range. A formal representation of a distribution through a quantile function was introduced by Galton (1822–1911) [206] who also initiated the ordering of observations along with the concepts of median, quartiles and interquartile range. Subsequently, the research on quantiles was directed towards estimation problems with the aid of sample quantiles, their large sample behaviour and limiting distributions (Galton [207, 208]). A major development in portraying quantile functions to represent distributions is the work of Hastings et al. [264], who introduced a family of distributions by a quantile function. This was refined later by Tukey [568]. The symmetric distribution of Tukey [568] and his articulation of exploratory data analysis sparked considerable interest in quantile functional forms that continues till date. Various aspects of the Tukey family and generalizations thereto were studied by a number of authors including Hogben [273], Shapiro and Wilk [536], Filliben [197], Joiner and Rosenblatt [304], Ramberg and Schmeiser [504], Ramberg [501], Ramberg et al. [502], MacGillivray [407], Freimer et al. [203], Gilchrist [215] and Tarsitano [563]. We will discuss all these models in Chap. 3. Another turning point in the development of quantile functions is the seminal paper by Parzen [484], in which he emphasized the description of a distribution in terms of the quantile function and its role in data modelling. Parzen [485–487] exhibits a sequential development of the theory and application of quantile functions in different areas and also as a tool in unification of various approaches.

Quantile functions have several interesting properties that are not shared by distributions, which makes it more convenient for analysis. For example, the sum of two quantile functions is again a quantile function. There are explicit general distribution forms for the quantile function of order statistics. In Sect. 1.2, we mention these and some other properties. Moreover, random numbers from any distribution can be generated using appropriate quantile functions, a purpose for which lambda distributions were originally conceived. The moments in different forms such as raw, central, absolute and factorial have been used effectively in specifying the model, describing the basic characteristics of distributions, and in inferential procedures. Some of the methods of estimation like least squares, maximum likelihood and method of moments often provide estimators and/or their standard errors in terms of moments. Outliers have a significant effect on the estimates so derived. For example, in the case of samples from the normal distribution, all the above methods give sample mean as the estimate of the population mean, whose values change significantly in the presence of an outlying observation. Asymptotic efficiency of the sample moments is rather poor for heavy tailed distributions since the asymptotic variances are mainly in terms of higher order moments that tend to be large in this case. In reliability analysis, a single long-term survivor can have a marked effect on mean life, especially in the case of heavy tailed models which are commonly encountered for lifetime data. In such cases, quantile-based estimates are generally found to be more precise and robust against outliers. Another advantage in choosing quantiles is that in life testing experiments, one need not wait until the failure of all the items on test, but just a portion of them for proposing useful estimates. Thus, there is a case for adopting quantile functions as models of lifetime and base their analysis with the aid of functions derived from them. Many other facets of the quantile approach will be more explicit in the sequel in the form of alternative methodology, new opportunities and unique cases where there are no corresponding results if one adopts the distribution function approach.

1.2 Definitions and Properties

In this section, we define the quantile function and discuss some of its general properties. The random variable considered here has the real line as its support, but the results are valid for lifetime random variables which take on only non-negative values.

Definition 1.1. Let X be a real valued continuous random variable with distribution function $F(x)$ which is continuous from the right. Then, the quantile function $Q(u)$ of X is defined as

$$Q(u) = F^{-1}(u) = \inf\{x : F(x) \geq u\}, \quad 0 \leq u \leq 1. \tag{1.1}$$

For every $-\infty < x < \infty$ and $0 < u < 1$, we have

$$F(x) \geq u \text{ if and only if } Q(u) \leq x.$$

Thus, if there exists an x such that $F(x) = u$, then $F(Q(u)) = u$ and $Q(u)$ is the smallest value of x satisfying $F(x) = u$. Further, if $F(x)$ is continuous and strictly increasing, $Q(u)$ is the unique value x such that $F(x) = u$, and so by solving the equation $F(x) = u$, we can find x in terms of u which is the quantile function of X. Most of the distributions we consider in this work are of this form and nature.

Definition 1.2. If $f(x)$ is the probability density function of X, then $f(Q(u))$ is called the density quantile function. The derivative of $Q(u)$, i.e.,

$$q(u) = Q'(u),$$

is known as the quantile density function of X. By differentiating $F(Q(u)) = u$, we find

$$q(u)f(Q(u)) = 1. \tag{1.2}$$

Some important properties of quantile functions required in the sequel are as follows.

1. From the definition of $Q(u)$ for a general distribution function, we see that
   (a) $Q(u)$ is non-decreasing on $(0,1)$ with $Q(F(x)) \le x$ for all $-\infty < x < \infty$ for which $0 < F(x) < 1$;
   (b) $F(Q(u)) \ge u$ for any $0 < u < 1$;
   (c) $Q(u)$ is continuous from the left or $Q(u-) = Q(u)$;
   (d) $Q(u+) = \inf\{x : F(x) > u\}$ so that $Q(u)$ has limits from above;
   (e) Any jumps of $F(x)$ are flat points of $Q(u)$ and flat points of $F(x)$ are jumps of $Q(u)$.

2. If U is a uniform random variable over $[0,1]$, then $X = Q(U)$ has its distribution function as $F(x)$. This follows from the fact that

$$P(Q(U) \le x) = P(U \le F(x)) = F(x).$$

   This property enables us to conceive a given data set as arising from the uniform distribution transformed by the quantile function $Q(u)$.
3. If $T(x)$ is a non-decreasing function of x, then $T(Q(u))$ is a quantile function. Gilchrist [215] refers to this as the Q-transformation rule. On the other hand, if $T(x)$ is non-increasing, then $T(Q(1-u))$ is also a quantile function.

   Example 1.1. Let X be a random variable with Pareto type II (also called Lomax) distribution with

$$F(x) = 1 - \alpha^c (x+\alpha)^{-c}, \quad x > 0;\ \alpha, c > 0.$$

   Since $F(x)$ is strictly increasing, setting $F(x) = u$ and solving for x, we obtain

$$x = Q(u) = \alpha[(1-u)^{-\frac{1}{c}} - 1].$$

   Taking $T(X) = X^\beta$, $\beta > 0$, we have a non-decreasing transformation which results in

$$T(Q(u)) = \alpha^\beta [(1-u)^{-\frac{1}{c}} - 1]^\beta.$$

   When $T(Q(u)) = y$, we obtain, on solving for u,

$$u = G(y) = 1 - \left(1 + \frac{y^{\frac{1}{\beta}}}{\alpha}\right)^{-c}$$

   which is a Burr type XII distribution with $T(Q(u))$ being the corresponding quantile function.

Example 1.2. Assume X has Pareto type I distribution with

$$F(x) = 1 - \left(\frac{x}{\sigma}\right)^{\alpha}, \quad x > \sigma;\ \alpha > 0,\ \sigma > 0.$$

Then, working as in the previous example, we see that

$$Q(u) = \sigma(1-u)^{-\frac{1}{\alpha}}.$$

Apply the transformation $T(X) = Y = X^{-1}$, which is non-increasing, we have

$$T(Q(1-u)) = \sigma^{-1}u^{\frac{1}{\alpha}}$$

and equating this to y and solving, we get

$$G(y) = (y\sigma)^{\alpha}, \quad 0 \le y \le \frac{1}{\sigma}.$$

$G(y)$ is the distribution function of a power distribution with $T(Q(1-u))$ being the corresponding quantile function.

4. If $Q(u)$ is the quantile function of X with continuous distribution function $F(x)$ and $T(u)$ is a non-decreasing function satisfying the boundary conditions $T(0) = 0$ and $T(1) = 1$, then $Q(T(u))$ is a quantile function of a random variable with the same support as X.

   Example 1.3. Consider a non-negative random variable with continuous distribution function $F(x)$ and quantile function $Q(u)$. Taking $T(u) = u^{\frac{1}{\theta}}$, for $\theta > 0$, we have $T(0) = 0$ and $T(1) = 1$. Then,

$$Q_1(u) = Q(T(u)) = Q(u^{\frac{1}{\theta}}).$$

   Further, if $y = Q_1(u)$, $u^{\frac{1}{\theta}} = y$ and so the distribution function corresponding to $Q_1(u)$ is

$$G(x) = F^{\theta}(x).$$

   The random variable Y with distribution function $G(x)$ is called the proportional reversed hazards model of X. There is considerable literature on such models in reliability and survival analysis. If we take X to be exponential with

$$F(x) = 1 - e^{-\lambda x}, \quad x > 0;\ \lambda > 0,$$

   so that

$$Q(u) = \lambda^{-1}(-\log(1-u)),$$

then

$$Q_1(u) = \lambda^{-1}(-\log(1-u^{\frac{1}{\theta}}))$$

provides

$$G(x) = (1-e^{-\lambda x})^{\theta},$$

the generalized or exponentiated exponential law (Gupta and Kundu [250]). In a similar manner, Mudholkar and Srivastava [429] take the baseline distribution as Weibull. For some recent results and survey of such models, we refer the readers to Gupta and Gupta [240]. In Chap. 3, we will come across several quantile functions that represent families of distributions containing some life distributions as special cases. They are highly flexible empirical models capable of approximating many continuous distributions. The above transformation on these models generates new proportional reversed hazards models of a general form. The analysis of lifetime data employing such models seems to be an open issue.

Remark 1.1. From the form of $G(x)$ above, it is clear that for positive integral values of θ, it is simply the distribution function of the maximum of a random sample of size θ from the exponential population with distribution function $F(x)$ above. Thus, $G(x)$ may be simply regarded as the distribution function of the maximum from a random sample of real size θ (instead of an integer). This viewpoint was discussed by Stigler [547] under the general idea of 'fractional order statistics'; see also Rohatgi and Saleh [509].

Remark 1.2. Just as $G(x)$ can be regarded as the distribution function of the maximum from a random sample of (real) size θ from the population with distribution function $F(x)$, we can consider $G^*(x) = 1-(1-F(x))^{\theta}$ as a generalized form corresponding to the minimum of a random sample of (real) size θ. The model $G^*(x)$ is, of course, the familiar proportional hazards model. It is important to mention here that these two models are precisely the ones introduced by Lehmann [382], as early as in 1953, as stochastically ordered alternatives for nonparametric tests of equality of distributions.

Remark 1.3. It is useful to bear in mind that for distributions closed under minima such as exponential and Weibull (i.e., the distributions for which the minima have the same form of the distribution but with different parameters), the distribution function $G(x)$ would provide a natural generalization while, for distributions closed under maxima such as power and inverse Weibull (i.e., the distributions for which the maxima have the same form of the distribution but with different parameters), the distribution function $G^*(x)$ would provide a natural generalization.

5. The sum of two quantile functions is again a quantile function. Likewise, two quantile density functions, when added, produce another quantile density function.

6. The product of two positive quantile functions is a quantile function. In this case, the condition of positivity cannot be relaxed, as in general, there may be negative quantile functions that affect the increasing nature of the product. Since we are dealing primarily with lifetimes, the required condition will be automatically satisfied.
7. If X has quantile function $Q(u)$, then $\frac{1}{X}$ has quantile function $1/Q(1-u)$.

   Remark 1.4. Property 7 is illustrated in Example 1.2. Chapter 3 contains some examples wherein quantile functions are generated as sums and products of quantile functions of known distributions. It becomes evident from Properties 3–7 that they can be used to produce new distributions from the existing ones. Thus, in our approach, a few basic forms are sufficient to begin with since new forms can always be evolved from them that match our requirements and specifications. This is in direct contrast to the abundance of probability density functions built up, each to satisfy a particular data form in the distribution function approach. In data analysis, the crucial advantage is that if one quantile function is not an appropriate model, the features that produce lack of fit can be ascertained and rectification can be made to the original model itself. This avoids the question of choice of an altogether new model and the repetition of all inferential procedures for the new one as is done in most conventional analyses.

8. The concept of residual life is of special interest in reliability theory. It represents the lifetime remaining in a unit after it has attained age t. Thus, if X is the original lifetime with quantile function $Q(u)$, the associated residual life is the random variable $X_t = (X - t|X > t)$. Using the definition of conditional probability, the survival function of X_t is

$$\bar{F}_t(x) = P(X_t > x) = \frac{\bar{F}(x+t)}{\bar{F}(t)},$$

where $\bar{F}(x) = P(X > x) = 1 - F(x)$ is the survival function. Thus, we have

$$F_t(x) = \frac{F(x+t) - F(t)}{1 - F(t)}. \tag{1.3}$$

Let $F(t) = u_0$, $F(x+t) = v$ and $F_t(x) = u$. Then, with

$$x + t = Q(v), \quad x = Q_1(u), \text{ say,}$$

we have

$$Q_1(u) = Q(v) - Q(u_0)$$

and consequently from (1.3),

$$u(1 - u_0) = v - u_0$$

or

$$v = u_0 + (1 - u_0)u.$$

Thus, the quantile function of the residual life X_t becomes

$$Q_1(u) = Q(u_0 + (1 - u_0)u) - Q(u_0). \tag{1.4}$$

Equation (1.4) will be made use of later in defining mean residual quantile function in Chap. 2.

9. In some reliability and quality control situations, truncated forms of lifetime models arise naturally, and the truncation may be on the right or on the left or on both sides. Suppose $F(x)$ is the underlying distribution function and $Q(u)$ is the corresponding quantile function. Then, if the distribution is truncated on the right at $x = U$ (i.e., the observations beyond U cannot be observed), then the corresponding distribution function is

$$F_{RT}(x) = \frac{F(x)}{F(U)}, \quad 0 \leq x \leq U,$$

and its quantile function is

$$Q_{RT}(x) = Q(uQ^{-1}(U)).$$

Similarly, if the distribution is truncated on the left at $x = L$ (i.e., the observations below L cannot be observed), then the corresponding distribution function is

$$F_{LT}(x) = \frac{F(x) - F(L)}{1 - F(L)}, \quad x \geq L,$$

and its quantile function is

$$Q_{LT}(u) = Q(u + (1 - u)Q^{-1}(L)).$$

Finally, if the distribution is truncated on the left at $x = L$ and also on the right at $x = U$, then the corresponding distribution function is

$$F_{DT}(x) = \frac{F(x) - F(L)}{F(U) - F(L)}, \quad L \leq x \leq U,$$

and its quantile function is

$$Q_{DT}(u) = Q(uQ^{-1}(U) + (1 - u)Q^{-1}(L)).$$

Example 1.4. Suppose the underlying distribution is logistic with distribution function $F(x) = 1/(1+e^{-x})$ on the whole real line $\mathbb{R}$. It is easily seen that the corresponding quantile function is $Q(u) = \log\left(\frac{u}{1-u}\right)$. Further, suppose we consider the distribution truncated on the left at 0, i.e., $L = 0$, for proposing a lifetime model. Then, from the expression above and the fact that $Q^{-1}(0) = \frac{1}{2}$, we arrive at the quantile function

$$Q_{LT}(u) = Q\left(u + (1-u)\frac{1}{2}\right) = \log\left(\frac{u + \frac{1}{2}(1-u)}{1 - u - \frac{1}{2}(1-u)}\right) = \log\left(\frac{1+u}{1-u}\right)$$

corresponding to the half-logistic distribution of Balakrishnan [47, 48]; see Table 1.1.

1.3 Quantile Functions of Life Distributions

As mentioned earlier, we concentrate here on distributions of non-negative random variables representing the lifetime of a component or unit. The distribution function of such random variables is such that $F(0-) = 0$. Often, it is more convenient to work with

$$\bar{F}(x) = 1 - F(x) = P(X > x),$$

which is the probability that the unit survives time (referred to as the age of the unit) x. It is also called the reliability or survival function since it expresses the probability that the unit is still reliable at age x.

In the previous section, some examples of quantile functions and a few methods of obtaining them were explained. We now present in Table 1.1 quantile functions of many distributions considered in the literature as lifetime models. The properties of these distributions are discussed in the references cited below each of them. Models like gamma, lognormal and inverse Gaussian do not find a place in the list as their quantile functions are not in a tractable form. However, in the next chapter, we will see quantile forms that provide good approximations to them.

1.4 Descriptive Quantile Measures

The advent of the Pearson family of distributions was a major turning point in data modelling using distribution functions. The fact that members of the family can be characterized by the first four moments gave an impetus to the extensive use of moments in describing the properties of distributions and their fitting to observed data. A familiar pattern of summary measures took the form of mean

Table 1.1 Quantile functions of some lifetime distributions

No.	Distribution	$\bar{F}(x)$	$Q(u)$
1	Exponential (Marshall and Olkin [412])	$\exp[-\lambda x]$ $x>0;\ \lambda>0$	$\lambda^{-1}(-\log(1-u))$
2	Weibull (Murthy et al. [434], Hahn and Shapiro [257])	$\exp[-(\frac{x}{\sigma})^{\lambda}]$ $x>0;\ \lambda,\sigma>0$	$\sigma(-\log(1-u))^{\frac{1}{\lambda}}$
3	Pareto II (Marshall and Olkin [412])	$\alpha^{c}(x+\alpha)^{-c}$ $x>0;\ \alpha,c>0$	$\alpha[(1-u))^{-\frac{1}{c}}-1]$
4	Rescaled beta (Marshall and Olkin [412])	$(1-\frac{x}{R})^{c}$ $0\leq x\leq R;\ c,R>0$	$R[1-(1-u))^{\frac{1}{c}}]$
5	Half-logistic (Balakrishnan [47,48], Balakrishnan and Wong [61])	$2\left[1+\exp\left(\frac{x}{\sigma}\right)\right]^{-1}$ $x>0;\ \sigma>0$	$\sigma\log\left(\frac{1+u}{1-u}\right)$
6	Power (Marshall and Olkin [412])	$1-(\frac{x}{\alpha})^{\beta}$ $0\leq x\leq\alpha;\ \alpha,\beta>0$	$\alpha u^{\frac{1}{\beta}}$
7	Pareto I (Marshall and Olkin [412])	$(\frac{x}{\sigma})^{-\alpha}$ $x>\sigma>0;\ \alpha,\sigma>0$	$\sigma(1-u)^{-\frac{1}{\alpha}}$
8	Burr type XII (Zimmer et al. [604], Fry [204])	$(1+x^{c})^{-k}$ $x>0;\ c,k>0$	$[(1-u)^{\frac{1}{k}}-1]^{\frac{1}{c}}$
9	Gompertz (Lai and Xie [368])	$\exp[\frac{-B(C^{x}-1)}{\log C}]$ $x>0;\ B,C>0$	$\frac{1}{\log C}[1-\frac{\log C\log(1-u)}{B}]$
10	Greenwich [225]	$(1+\frac{x^{2}}{b^{2}})^{-\frac{a}{2}}$ $x\geq 0;\ a,b>0$	$b[(1-u)^{\frac{2}{a}}-1]^{\frac{1}{2}}$
11	Kus [364]	$\frac{1-e^{\lambda e^{-\beta x}}}{1-e^{\lambda}}$ $x>0;\ \lambda,\beta>0$	$-\frac{1}{\beta}\log[\lambda^{-1}\log\{1-(1-u)$ $(1-e^{-\lambda})\}]$
12	Logistic exponential (Lan and Leemis [372])	$\frac{1+(e^{\lambda\theta}-1)^{k}}{1+(e^{\lambda(x+\theta)}-1)^{k}}$ $x\geq 0;\ \lambda>0,$ $k>0,\ \theta\geq 0$	$\frac{1}{\lambda}\log[1+\{\frac{(e^{\lambda\theta}-1)^{k}+u}{1-u}\}^{\frac{1}{k}}]$
13	Dimitrakopoulou et al. [178]	$\exp[1-(1+\lambda x^{\beta})^{\alpha}]$ $x>0;\ \alpha,\beta,\lambda>0$	$\lambda^{-1}[\{1-\log(1-u)\}^{\frac{1}{\alpha}}-1]^{\frac{1}{\beta}}$
14	Log Weibull (Avinadav and Raz [41])	$\exp[-(\log(1+\rho x))^{k}]$ $x>0;\ \rho,k>0$	$\rho^{-1}[\exp(-\log(1-u))^{\frac{1}{k}}-1]$
15	Modified Weibull extension (Xie et al. [595])	$\exp[-\alpha\sigma(e^{(\frac{x}{\sigma})^{\lambda}}-1)]$ $x>0;\ \alpha,\sigma,\lambda>0$	$\sigma[\log(1+\frac{\log(1-u)}{\alpha\sigma})]^{\frac{1}{\lambda}}$
16	Exponential power (Paranjpe et al. [482])	$\exp[e^{-(\lambda t)^{\alpha}}-1]$ $x>0;\ \lambda,\alpha>0$	$\frac{1}{\lambda}[-\log(1+\log(1-u))]^{\frac{1}{\alpha}}$
17	Generalized Pareto (Lai and Xie [368])	$(1+\frac{ax}{b})^{-\frac{a+1}{a}}$ $x>0,b>0,a>-1$	$\frac{b}{a}[(1-u)^{-\frac{a}{a+1}}-1]$
18	Inverse Weibull (Erto [188])	$1-\exp[-(\frac{\sigma}{x})^{\lambda}]$ $x>0;\ \sigma,\lambda>0$	$\sigma(-\log u)^{-\frac{1}{\lambda}}$

(continued)

Table 1.1 (continued)

No.	Distribution	$\bar{F}(x)$	$Q(u)$
19	Extended Weibull	$\frac{\theta\exp[-(\frac{x}{\sigma})^{\lambda}]}{1-(1-\theta)\exp[-(\frac{x}{\sigma})^{\lambda}]}$	$\sigma[\log\frac{\theta+(1-\theta)(1-u)}{1-u}]^{\frac{1}{\lambda}}$
	(Marshall and Olkin [411])	$x>0;\ \theta,\lambda,\sigma>0$	
20	Generalized exponential	$1-[1-\exp(-\frac{x}{\sigma})]^{\theta}$	$\sigma[-\log(1-u^{\frac{1}{\theta}})]$
	(Gupta et al. [239])	$x>0;\ \sigma,\theta>0$	
21	Exponentiated Weibull	$1-[1-\exp(-\frac{x}{\sigma})^{\lambda}]^{\theta}$	$\sigma[-\log(1-u^{\frac{1}{\theta}})]^{\frac{1}{\lambda}}$
	(Mudholkar et al. [427])	$x>0;\ \sigma,\theta,\lambda>0$	
22	Generalized Weibull	$[1-\lambda(\frac{x}{\beta})^{\alpha}]^{[\frac{1}{\lambda}]}$	$\beta[\frac{1-(1-u)^{\lambda}}{\lambda}]^{\frac{1}{\alpha}},\ \lambda\neq 0$
	(Mudholkar and Kollia [426])	$x>0$ for $\lambda\leq 0$	
		$0<x<\frac{\beta}{\lambda^{\frac{1}{\alpha}}},\ \lambda>0$	$\beta[-\log(1-u)]^{\frac{1}{\alpha}},\ \lambda=0$
		$\alpha,\beta>0$	
23	Exponential geometric	$(1-p)e^{-\lambda x}(1-pe^{-\lambda x})^{-1}$	$\frac{1}{\lambda}\log(\frac{1-pu}{1-u})$
	(Adamidis and Loukas [18])	$x>0,\ \lambda>0,\ 0<p<1$	
24	Log logistic	$(1+(\alpha x)^{\beta})^{-1}$	$\alpha^{-1}(\frac{u}{1-u})^{\frac{1}{\beta}}$
	(Gupta et al. [237])	$x>0,\ \alpha,\beta>0$	
25	Generalized half-logistic	$\frac{2(1-kx)^{1/k}}{1+(1-kx)^{1/k}}$	$\frac{1}{k}\left\{1-\left(\frac{1-u}{1+u}\right)^{k}\right\}$
	(Balakrishnan and Sandhu [59], Balakrishnan and Aggarwala [49])	$0\leq x\leq\frac{1}{k},\ k\geq 0$	

for location, variance for dispersion, and the Pearson's coefficients $\beta_1=\frac{\mu_3^2}{\mu_2^3}$ for skewness and $\beta_2=\frac{\mu_4}{\mu_2^2}$ for kurtosis. While the mean and variance claimed universal acceptance, several limitations of β_1 and β_2 were subsequently exposed. Some of the concerns with regard to β_1 are: (1) it becomes arbitrarily large or even infinite making it difficult for comparison and interpretation as relatively small changes in parameters produce abrupt changes, (2) it does not reflect the sign of the difference (mean-median) which is a traditional basis for defining skewness, (3) there exist asymmetric distributions with $\beta_1=0$ and (4) instability of the sample estimate of β_1 while matching with the population value. Similarly, for a standardized variable X, the relationship

$$E(X^4)=1+V(X^2) \tag{1.5}$$

would mean that the interpretation of kurtosis depends on the concentration of the probabilities near $\mu\pm\sigma$ as well as in the tails of the distribution.

The specification of a distribution through its quantile function takes away the need to describe a distribution through its moments. Alternative measures in terms of quantiles that reduce the shortcomings of the moment-based ones can be thought of. A measure of location is the median defined by

$$M = Q(0.5). \tag{1.6}$$

Dispersion is measured by the interquartile range

$$\text{IQR} = Q_3 - Q_1, \tag{1.7}$$

where $Q_3 = Q(0.75)$ and $Q_1 = Q(0.25)$.

Skewness is measured by Galton's coefficient

$$S = \frac{Q_1 + Q_3 - 2M}{Q_3 - Q_1}. \tag{1.8}$$

Note that in the case of extreme positive skewness, $Q_1 \to M$ while in the case of extreme negative skewness $Q_3 \to M$ so that S lies between -1 and $+1$. When the distribution is symmetric, $M = \frac{Q_1+Q_3}{2}$ and hence $S = 0$. Due to the relation in (1.5), kurtosis can be large when the probability mass is concentrated near the mean or in the tails. For this reason, Moors [421] proposed the measure

$$T = [Q(0.875) - Q(0.625) + Q(0.375) - Q(0.125)]/\text{IQR} \tag{1.9}$$

as a measure of kurtosis. As an index, T is justified on the grounds that the differences $Q(0.875) - Q(0.625)$ and $Q(0.375) - Q(0.125)$ become large (small) if relatively small (large) probability mass is concentrated around Q_3 and Q_1 corresponding to large (small) dispersion in the vicinity of $\mu \pm \sigma$.

Given the form of $Q(u)$, the calculations of all the coefficients are very simple, as we need to only substitute the appropriate fractions for u. On the other hand, calculation of moments given the distribution function involves integration, which occasionally may not even yield closed-form expressions.

Example 1.5. Let X follow the Weibull distribution with (see Table 1.1)

$$Q(u) = \sigma(-\log(1-u))^{\frac{1}{\lambda}}.$$

Then, we have

$$M = Q\left(\frac{1}{2}\right) = \sigma(\log 2)^{\frac{1}{\lambda}},$$

$$S = \frac{(\log 4)^{\frac{1}{\lambda}} + (\log \frac{4}{3})^{\frac{1}{\lambda}} - 2(\log 2)^{\frac{1}{\lambda}}}{(\log 4)^{\frac{1}{\lambda}} - (\log \frac{4}{3})^{\frac{1}{\lambda}}},$$

$$\text{IQR} = \sigma\left[(\log 4)^{\frac{1}{\lambda}} - \left(\log \frac{4}{3}\right)^{\frac{1}{\lambda}}\right],$$

and

$$T = \frac{(\log 8)^{\frac{1}{\lambda}} - (\log \frac{8}{3})^{\frac{1}{\lambda}} + (\log \frac{8}{5})^{\frac{1}{\lambda}} - (\log \frac{8}{7})^{\frac{1}{\lambda}}}{(\log 4)^{\frac{1}{\lambda}} - (\log \frac{4}{3})^{\frac{1}{\lambda}}}.$$

The effect of a change of origin and scale on $Q(u)$ and the above four measures are of interest in later studies. Let X and Y be two random variables such that $Y = aX + b$. Then,

$$F_Y(y) = P(Y \le y) = P\left(X \le \frac{y-b}{a}\right) = F_X\left(\frac{y-a}{b}\right).$$

If $Q_X(u)$ and $Q_Y(u)$ denote the quantile functions of X and Y, respectively,

$$F_X\left(\frac{y-a}{b}\right) = u \Rightarrow Q_X(u) = \frac{y-b}{a} = \frac{Q_Y(u)-b}{a}$$

or

$$Q_Y(u) = aQ_X(u) + b.$$

So, we simply have

$$M_Y = Q_Y(0.5) = aQ_X(0.5) + b = aM_X + b.$$

Similar calculations using (1.7), (1.8) and (1.9) yield

$$\text{IQR}_Y = a\text{IQR}_X,\ S_Y = S_X \text{ and } T_Y = T_X.$$

Other quantile-based measures have also been suggested for quantifying spread, skewness and kurtosis. One measure of spread, similar to mean deviation in the distribution function approach, is the median of absolute deviation from the median, viz.,

$$A = \text{Med}\,(|X - M|). \tag{1.10}$$

For further details and justifications for (1.10), we refer to Falk [194]. A second popular measure that has received wide attention in economics is Gini's mean difference defined as

$$\begin{aligned} \Delta &= \int_{-\infty}^{\infty}\int_{-\infty}^{\infty} |x-y| f(x) f(y)\,dx\,dy \\ &= 2\int_{-\infty}^{\infty} F(x)(1-F(x))\,dx, \end{aligned} \tag{1.11}$$

where $f(x)$ is the probability density function of X. Setting $F(x) = u$ in (1.11), we have

$$\Delta = 2\int_0^1 u(1-u)q(u)du \tag{1.12}$$

$$= 2\int_0^1 (2u-1)Q(u)du. \tag{1.13}$$

The expression in (1.13) follows from (1.12) by integration by parts. One may use (1.12) or (1.13) depending on whether $q(u)$ or $Q(u)$ is specified. Gini's mean difference will be further discussed in the context of reliability in Chap. 4.

Example 1.6. The generalized Pareto distribution with (see Table 1.1)

$$Q(u) = \frac{b}{a}\left\{(1-u)^{-\frac{a}{a+1}} - 1\right\}$$

has its quantile density function as

$$q(u) = \frac{b}{a+1}(1-u)^{-\frac{a}{a+1}-1}.$$

Then, from (1.12), we obtain

$$\Delta = \frac{2b}{a+1}\int_0^1 u(1-u)^{-\frac{a}{a+1}}du = \frac{2b}{a+1}B\left(2, \frac{1}{a+1}\right),$$

where $B(m,n) = \int_0^1 t^{m-1}(1-t)^{n-1}dt$ is the complete beta function. Thus, we obtain the simplified expression

$$\Delta = \frac{2b(a+1)}{a+2}.$$

Hinkley [271] proposed a generalization of Galton's measure of skewness of the form

$$S(u) = \frac{Q(u) + Q(1-u) - 2Q(0.5)}{Q(u) - Q(1-u)}. \tag{1.14}$$

Obviously, (1.14) reduces to Galton's measure when $u = 0.75$. Since (1.14) is a function of u and u is arbitrary, an overall measure of skewness can be provided as

$$S_2 = \sup_{\frac{1}{2} \le u \le 1} S(u).$$

Groeneveld and Meeden [227] suggested that the numerator and denominator in (1.14) be integrated with respect to u to arrive at the measure

$$S_3 = \frac{\int_{\frac{1}{2}}^{1}\{Q(u)+Q(1-u)-2Q(0.5)\}du}{\int_{\frac{1}{2}}^{1}\{Q(u)+Q(1-u)\}du}.$$

Now, in terms of expectations, we have

$$\int_{\frac{1}{2}}^{1} Q(u)du = \int_{M}^{x} xf(x)dx,$$

$$\int_{\frac{1}{2}}^{1} Q(1-u)du = \int_{0}^{\frac{1}{2}} Q(u)du = \int_{0}^{M} xf(x)dx,$$

$$\int_{\frac{1}{2}}^{1} Q(0.5)du = \frac{1}{2}M,$$

and thus

$$S_3 = \frac{E(X)-M}{\int_M^\infty xf(x)dx - \int_0^M xf(x)dx} = \frac{\mu - M}{E(|X-M|)}. \tag{1.15}$$

The numerator of (1.15) is the traditional term (being the difference between the mean and the median) indicating skewness and the denominator is a measure of spread used for standardizing S_3. Hence, (1.15) can be thought of as an index of skewness in the usual sense. If we replace the denominator by the standard deviation σ of X, the classical measure of skewness will result.

Example 1.7. Consider the half-logistic distribution with (see Table 1.1)

$$Q(u) = \sigma \log\left(\frac{1+u}{1-u}\right),$$

$$\mu = \int_0^1 Q(u)du = \sigma \log 4,$$

$$\int_{\frac{1}{2}}^{1} Q(u)du = \sigma\left(\log 16 - \frac{3}{2}\log 3\right),$$

$$\int_0^{\frac{1}{2}} Q(u)du = \sigma\left(\frac{3}{2}\log 3 - 2\log 2\right),$$

and hence $S_3 = \log(\frac{4}{3})/\log(\frac{64}{27})$.

Instead of using quantiles, one can also use percentiles to define skewness. Galton [206] in fact used the middle 50 % of observations, while Kelly's measure takes 90 % of observations to propose the measure

$$S_4 = \frac{Q(0.90)+Q(0.10)-2M}{Q(0.90)-Q(0.10)}.$$

For further discussion of alternative measures of skewness and kurtosis, a review of the literature and comparative studies, we refer to Balanda and MacGillivray [63], Tajuddin [559], Joannes and Gill [299], Suleswki [552] and Kotz and Seier [355].

1.5 Order Statistics

In life testing experiments, a number of units, say n, are placed on test and the quantity of interest is their failure times which are assumed to follow a distribution $F(x)$. The failure times $X_1, X_2, \ldots, X_n$ of the n units constitute a random sample of size n from the population with distribution function $F(x)$, if $X_1, X_2, \ldots, X_n$ are independent and identically distributed as $F(x)$. Suppose the realization of X_i in an experiment is denoted by x_i. Then, the order statistics of the random sample $(X_1, X_2, \ldots, X_n)$ are the sample values placed in ascending order of magnitude denoted by $X_{1:n} \leq X_{2:n} \leq \cdots \leq X_{n:n}$, so that $X_{1:n} = \min_{1 \leq i \leq n} X_i$ and $X_{n:n} = \max_{1 \leq i \leq n} X_i$. The sample median, denoted by m, is the value for which approximately 50 % of the observations are less than m and 50 % are more than m. Thus

$$m = \begin{cases} X_{\frac{n+1}{2}:n} & \text{if } n \text{ is odd} \\ \frac{1}{2}(X_{\frac{n}{2}:n} + X_{\frac{n}{2}+1:n}) & \text{if } n \text{ is even.} \end{cases} \tag{1.16}$$

Generalizing, we have the percentiles. The $100p$-th percentile, denoted by x_p, in the sample corresponds to the value for which approximately np observations are smaller than this value and $n(1-p)$ observations are larger. In terms of order statistics we have

$$x_p = \begin{cases} X_{[np]:n} & \text{if } \frac{1}{2n} < p < 0.5 \\ X_{(n+1)-[n(1-p)]} & \text{if } 0.5 < p < 1 - \frac{1}{2n} \end{cases}, \tag{1.17}$$

where the symbol $[t]$ is defined as $[t] = r$ whenever $r - 0.5 \leq t < r + 0.5$, for all positive integers r. We note that in the above definition, if x_p is the ith smallest observation, then the ith largest observation is x_{1-p}. Obviously, the median m is the 50th percentile and the lower quartile q_1 and the upper quartile q_3 of the sample are, respectively, the 25th and 75th percentiles. The sample interquartile range is

$$iqr = q_3 - q_1. \tag{1.18}$$

All the sample descriptive measures are defined in terms of the sample median, quartiles and percentiles analogous to the population measures introduced in Sect. 1.4. Thus, iqr in (1.18) describes the spread, while

$$s = \frac{q_3 + q_1 - 2m}{q_3 - q_1} \tag{1.19}$$

and

$$t = \frac{e_7 - e_5 + e_3 - e_1}{iqr}, \tag{1.20}$$

where $e_i = \frac{i}{8}$, $i = 1,3,5,7$, describes the skewness and kurtosis.

Parzen [484] introduced the empirical quantile function

$$\bar{Q}(u) = F_n^{-1}(u) = \inf(x : F_n(x) \geq u),$$

where $F_n(x)$ is the proportion of $X_1, X_2, \ldots, X_n$ that is at most x. In other words,

$$\bar{Q}(u) = X_{r:n} \quad \text{for } \frac{r-1}{n} < u < \frac{r}{n}, \quad r = 1,2,\ldots,n, \tag{1.21}$$

which is a step function with jump $\frac{1}{n}$. For $u = 0$, $\bar{Q}(u)$ is taken as $X_{1:n}$ or a natural minimum if one is available. In the case of lifetime variables, this becomes $\bar{Q}(0)$. When a smooth function is required for $\bar{Q}(u)$, Parzen [484] suggested the use of

$$\bar{Q}_1(u) = n\left(\frac{r}{n} - u\right)X_{r-1:n} + n\left(u - \frac{r-1}{n}\right)X_{r:n}$$

for $\frac{r-1}{n} \leq u \leq \frac{r}{n}$, $r = 1,2,\ldots,n$. The corresponding empirical quantile density function is

$$\bar{q}_1(u) = \frac{d}{du}\bar{Q}_1(u) = n(X_{r:n} - X_{r-1:n}), \quad \text{for } \frac{r-1}{n} < u < \frac{r}{n}.$$

In this set-up, we have $q_i = \bar{Q}(\frac{i}{4})$, $i = 1,3$ and $e_i = \bar{Q}(\frac{i}{8})$, $i = 1,3,7,8$.

It is well known that the distribution of the rth order statistic $X_{r:n}$ is given by Arnold et al. [37]

$$F_r(x) = P(X_{r:n} \leq x) = \sum_{k=r}^{n} \binom{n}{k} F^k(x)(1 - F(x))^{n-k}. \tag{1.22}$$

In particular, $X_{n:n}$ and $X_{1:n}$ have their distributions as

$$F_n(x) = F^n(x) \tag{1.23}$$

and

$$F_1(x) = 1 - (1 - F(x))^n. \tag{1.24}$$

Recalling the definitions of the beta function

$$B(m,n) = \int_0^1 t^{m-1}(1-t)^{n-1}dt, \quad m,n > 0,$$

and the incomplete beta function ratio

$$I_x(m,n) = \frac{B_x(m,n)}{B(m,n)},$$

where

$$B_x(m,n) = \int_0^x t^{m-1}(1-t)^{n-1}dt,$$

we have the upper tail of the binomial distribution and the incomplete beta function ratio to be related as (Abramowitz and Stegun [15])

$$\sum_{k=r}^{n} \binom{n}{k} p^k (1-p)^{n-k} = I_p(r, n-r+1). \tag{1.25}$$

Comparing (1.22) and (1.25) we see that, if a sample of n observations from a distribution with quantile function $Q(u)$ is ordered, then the quantile function of the rth order statistic is given by

$$Q_r(u_r) = Q(I_{u_r}^{-1}(r, n-r+1)), \tag{1.26}$$

where

$$u_r = I_u(r, n-r+1) \tag{1.27}$$

and I^{-1} is the inverse of the beta function ratio I. Thus, the quantile function of the rth order statistic has an explicit distributional form, unlike the expression for distribution function in (1.22). However, the expression for $Q_r(u_r)$ is not explicit in terms of $Q(u)$. This is not a serious handicap as the $I_u(\cdot,\cdot)$ function is tabulated for various values of n and r (Pearson [489]) and also available in all statistical softwares for easy computation. The distributions of $X_{n:n}$ and $X_{1:n}$ have simple quantile forms

$$Q_n(u_n) = Q\left(u_n^{\frac{1}{n}}\right)$$

and

$$Q_1(u_1) = Q[1-(1-u_1)^{\frac{1}{n}}].$$

The probability density function of $X_{r:n}$ becomes

$$f_r(x) = \frac{n!}{(r-1)!\,(n-r)!} F^{r-1}(x)(1-F(x))^{n-r} f(x)$$

and so

$$\mu_{r:n} = E(X_{r:n}) = \int x f_r(x) dx$$
$$= \frac{n!}{(r-1)!(n-1)!} \int_0^1 u^{r-1}(1-u)^{n-r} Q(u) du. \tag{1.28}$$

This mean value is referred to as the rth mean rankit of X. For reasons explained earlier with reference to the use of moments, often the median rankit

$$M_{r:n} = Q(I_{0.5}^{-1}(r, n-r+1)), \tag{1.29}$$

which is robust, is preferred over the mean rankit.

The importance and role of order statistics in the study of quantile function become clear from the discussions in this section. Added to this, there are several topics in reliability analysis in which order statistics appear quite naturally. One of them is system reliability. We consider a system consisting of n components whose lifetimes $X_1, X_2, \ldots, X_n$ are independent and identically distributed. The system is said to have a series structure if it functions only when all the components are functioning, and the lifetime of this system is the smallest among the X_i's or $X_{1:n}$. In the parallel structure, on the other hand, the system functions if and only if at least one of the components work, so that the system life is $X_{n:n}$. These two structures are embedded in what is called a k-out-of-n system, which functions if and only if at least k of the components function. The lifetime of such a system is obviously $X_{n-k+1:n}$.

In life testing experiments, when n identical units are put on test to ascertain their lengths of life, there are schemes of sampling wherein the experimenter need not have to wait until all units fail. The experimenter may choose to observe only a prefixed number of failures of, say, $n-r$ units and terminate the experiment as soon as the $(n-r)$th unit fails. Thus, the lifetimes of r units that are still working get censored. This sampling scheme is known as type II censoring. The data consists of realizations of $X_{1:n}, X_{2:n}, \ldots, X_{n-r:n}$. Another sampling scheme is to prefix a time T^* and observe only those failures that occur up to time T^*. This scheme is known as type I censoring, and in this case the number of failures to be observed is random. One may refer to Balakrishnan and Cohen [51] and Cohen [154] for various methods of inference for type I and type II censored samples from a wide array of lifetime distributions. Yet another sampling scheme is to prefix the number of failures at $n-r$ and also a time T^*. If $(n-r)$ failures occur before time T^*, then the experiment is terminated; otherwise, observe all failures until time T^*. Thus, the time of truncation of the experiment is now $\min(T, X_{n-r:n})$. This is referred to as type I hybrid censoring; see Balakrishnan and Kundu [53] for an overview of various developments on this and many other forms of hybrid censoring schemes. A third important application of order statistics is in the construction of tests regrading the nature of ageing of a device; see Lai and Xie [368]. For an encyclopedic treatment on the theory, methods and applications of order satistics, one may refer to Balakrishnan and Rao [56, 57].

1.6 Moments

The emphasis given to quantiles in describing the basic properties of a distribution does not in any way minimize the importance of moments in model specification and inferential problems. In this section, we look at various types of moments through quantile functions. The conventional moments

$$\mu_r' = E(X^r) = \int_0^\infty x^r f(x)dx$$

are readily expressible in terms of quantile functions, by the substitution $x = Q(u)$, as

$$\mu_r' = \int_0^1 \{Q(u)\}^r du. \tag{1.30}$$

In particular, as already seen, the mean is

$$\mu = \int_0^1 Q(u)du = \int_0^1 (1-u)q(u)du. \tag{1.31}$$

The central moments and other quantities based on it are obtained through the well-known relationships they have with the raw moments μ_r' in (1.30).

Some of the difficulties experienced while employing the moments in descriptive measures as well as in inferential problems have been mentioned in the previous sections. The L-moments to be considered next can provide a competing alternative to the conventional moments. Firstly, by definition, they are expected values of linear functions of order statistics. They have generally lower sampling variances and are also robust against outliers. Like the conventional moments, L-moments can be used as summary measures (statistics) of probability distributions (samples), to identify distributions and to fit models to data. The origin of L-moments can be traced back to the work on linear combination of order statistics in Sillito [537] and Greenwood et al. [226]. It was Hosking [276] who presented a unified theory on L-moments and made a systematic study of their properties and role in statistical analysis. See also Hosking [277, 279, 280] and Hosking and Wallis [282] for more elaborate details on this topic.

The rth L-moment is defined as

$$L_r = \frac{1}{r}\sum_{k=0}^{r-1}(-1)^k\binom{r-1}{k}E(X_{r-k:r}), \quad r = 1,2,3,\ldots \tag{1.32}$$

Using (1.28), we can write

$$L_r = \frac{1}{r}\sum_{k=0}^{r-1}(-1)^k\binom{r-1}{k}\frac{r!}{(r-k-1)!k!}\int_0^1 u^{r-k-1}(1-u)^k Q(u)du.$$

Expanding $(1-u)^k$ in powers of u using binomial theorem and combining powers of u, we get

$$L_r = \int_0^1 \sum_{k=0}^{r-1}(-1)^{r-1-k}\binom{r-1}{k}\binom{r-1+k}{k}u^k Q(u)du. \tag{1.33}$$

Jones [306] has given an alternative method of establishing the last relationship. In particular, we obtain:

$$L_1 = \int_0^1 Q(u)du = \mu, \tag{1.34}$$

$$L_2 = \int_0^1 (2u-1)Q(u)du, \tag{1.35}$$

$$L_3 = \int_0^1 (6u^2-6u+1)Q(u)du, \tag{1.36}$$

$$L_4 = \int_0^1 (20u^3-30u^2+12u-1)Q(u)du. \tag{1.37}$$

Sometimes, it is convenient (to avoid integration by parts while computing the integrals in (1.34)–(1.37)) to work with the equivalent formulas

$$L_1 = \int_0^1 (1-u)q(u)du, \tag{1.38}$$

$$L_2 = \int_0^1 (u-u^2)q(u)du, \tag{1.39}$$

$$L_3 = \int_0^1 (3u^2-2u^3-u)q(u)du, \tag{1.40}$$

$$L_4 = \int_0^1 (u-6u^2+10u^3-5u^4)q(u)du. \tag{1.41}$$

Example 1.8. For the exponential distribution with parameter λ, we have

$$Q(u) = -\lambda^{-1}\log(1-u) \quad \text{and} \quad q(u) = \frac{1}{\lambda(1-u)}.$$

Hence, using (1.38)–(1.41), we obtain

$$L_1 = \int_0^1 \frac{1}{\lambda} du = \lambda^{-1},$$

$$L_2 = \int_0^1 u(1-u)q(u)du = \int_0^1 \frac{u}{\lambda} du = (2\lambda)^{-1},$$

$$L_3 = \int_0^1 u(1-u)(2u-1)q(u)du = (6\lambda)^{-1},$$

$$L_4 = \int_0^1 u(1-u)(1-5u+5u^2)q(u)du = (12\lambda)^{-1}.$$

More examples are presented in Chap. 3 when properties of various distributions are studied.

The L-moments have the following properties that distinguish themselves from the usual moments:

1. The L-moments exist whenever $E(X)$ is finite, while additional restrictions may be required for the conventional moments to be finite for many distributions;
2. A distribution whose mean exists is characterized by $(L_r : r = 1,2,\ldots)$. This result can be compared with the moment problem discussed in probability theory. However, any set that contains all L-moments except one is not sufficient to characterize a distribution. For details, see Hosking [279, 280];
3. From (1.12), we see that $L_2 = \frac{1}{2}\Delta$, and so L_2 is a measure of spread. Thus, the first (being the mean) and second L-moments provide measures of location and spread. In a recent comparative study of the relative merits of the variance and the mean difference Δ, Yitzhaki [596] noted that the mean difference is more informative than the variance in deriving properties of distributions that depart from normality. He also compared the algebraic structure of variance and Δ and examined the relative superiority of the latter from the stochastic dominance, exchangability and stratification viewpoints. For further comments on these aspects and some others in the reliability context, see Chap. 7;
4. Forming the ratios $\tau_r = \frac{L_r}{L_2}$, $r = 3,4,\ldots$, for any non-degenerate X with $\mu < \infty$, the result $|\tau_r| < 1$ holds. Hence, the quantities τ_r's are dimensionless and bounded;
5. The skewness and kurtosis of a distribution can be ascertained through the moment ratios. The L-coefficient of skewness is

$$\tau_3 = \frac{L_3}{L_2} \tag{1.42}$$

and the L-coefficient of kurtosis is

$$\tau_4 = \frac{L_4}{L_2}. \tag{1.43}$$

These two measures satisfy the criteria presented for coefficients of skewness and kurtosis in terms of order relations. The range of τ_3 is $(-1,1)$ while that of τ_4 is $\frac{1}{4}(5\tau_3^2 - 1) \le \tau_4 < 1$. These results are proved in Hosking [279] and

Jones [306] using different approaches. It may be observed that both τ_3 and τ_4 are bounded and do not assume arbitrarily large values as β_1 (for example, in the case of $F(x) = 1 - x^{-3}, x > 1$);

6. The ratio

$$\tau_2 = \frac{L_2}{L_1} \tag{1.44}$$

is called L-coefficient of variation. Since X is non-negative in our case, $L_1 > 0$, $L_2 > 0$ and further

$$L_2 = \int_0^1 u(1-u)q(u)du < \int_0^1 (1-u)q(u)du = L_1$$

so that $0 < \tau_2 < 1$.

The above properties of L-moments have made them popular in diverse applications, especially in hydrology, civil engineering and meteorology. Several empirical studies (as the one by Sankarasubramonian and Sreenivasan [517]) comparing L-moments and the usual moments reveal that estimates based on the former are less sensitive to outliers. Just as matching the population and sample moments for the estimation of parameters, the same method (method of L-moments) can be applied with L-moments as well. Asymptotic approximations to sampling distributions are better achieved with L-moments. An added advantage is that standard errors of sample L-moments exist whenever the underlying distribution has a finite variance, whereas for the usual moments this may not be enough in many cases.

When dealing with the conventional moments, the (β_1, β_2) plot is used as a preliminary tool to discriminate between candidate distributions for the data. For example, if one wishes to choose a distribution from the Pearson family as a model, (β_1, β_2) provide exclusive classification of the members of this family. Distributions with no shape parameters are represented by points in the β_1-β_2 plane, those with a single shape parameter have their (β_1, β_2) values lie on the line $2\beta_2 - 3\beta_1 - 6 = 0$, while two shape parameters in the distribution ensure that for them, (β_1, β_2) falls in a region between the lines $2\beta_2 - 3\beta_1 - 6 = 0$ and $\beta_2 - \beta_1 - 1 = 0$. These cases are, respectively, illustrated by the exponential distribution (which has $(\beta_1, \beta_2) = (4, 9)$ as a point), the gamma family and the beta family; see Johnson et al. [302] for details. In a similar manner, one can construct (τ_2, τ_3)-plots or (τ_3, τ_4)-plots for distribution functions or quantile functions to give a visual identification of which distribution can be expected to fit a given set of observations. Vogel and Fennessey [574] articulate the need for such diagrams and provide several examples on how to construct them. Some refinements of the L-moments are also studied in the name of trimmed L-moments (Elamir and Seheult [187], Hosking [281]) and LQ-moments (Mudholkar and Hutson [424]).

Example 1.9. The L-moments of the exponential distribution were calculated earlier in Example 1.8. Applying the formulas for τ_2, τ_3 and τ_4 in (1.44), (1.42) and (1.43), we have

$$\tau_2 = \frac{1}{2}, \; \tau_3 = \frac{1}{3}, \; \tau_4 = \frac{1}{6}.$$

Thus, $(\tau_2, \tau_3) = (\frac{1}{2}, \frac{1}{3})$ and $(\tau_3, \tau_4) = (\frac{1}{3}, \frac{1}{6})$ are points in the τ_2-τ_3 and τ_3-τ_4 planes, respectively.

Example 1.10. The random variable X has generalized Pareto distribution with

$$Q(u) = \frac{b}{a}\{(1-u)^{-\frac{a}{a+1}} - 1\}, \quad a > -1, b > 0.$$

Then, straightforward calculations yield

$$L_1 = b, \qquad L_2 = b(a+1)(a+2)^{-1},$$

$$L_3 = b(a+1)(2a+1)[(2a+3)(a+2)]^{-1},$$

$$L_4 = b(a+1)(6a^2+7a+2)[(a+2)(2a+3)(3a+4)]^{-1},$$

so that

$$\tau_2 = \frac{a+1}{a+2}, \quad \tau_3 = \frac{2a+1}{2a+3} \quad \text{and} \quad \tau_4 = \frac{6a^2+7a+2}{6a^2+17a+12}.$$

Then, eliminating a between τ_2 and τ_3, we obtain

$$\tau_3 = \frac{3\tau_2 - 1}{\tau_2 + 1}.$$

Thus, the plot of (τ_2, τ_3) for all values of a and b lies on the curve $(\tau_2+1)(3-\tau_3) = 4$. Note that the exponential plot is $(\frac{1}{2}, \frac{1}{3})$ which lies on the curve when $a \to 0$. The estimation and other related inferential problems are discussed in Chap. 7.

We now present probability weighted moments (PWM) which is a forerunner to the concept of L-moments. Introduced by Greenwood et al. [226], PWMs are of considerable interest when the distribution is expressed in quantile form. The PWMs are defined as

$$M_{p,r,s} = E[X^p F^r(X) \bar{F}^s(X)], \tag{1.45}$$

where p, r, s are non-negative real numbers and $E|X|^p < \infty$. Two special cases of (1.45) in general use are

$$\begin{aligned} \beta_{p,r} &= E(X^p F^r(X)) \\ &= \int x^p F^r(x) f(x) dx \\ &= \int_0^1 (Q(u))^p u^r du \end{aligned} \tag{1.46}$$

and

$$\alpha_{p,s} = E(X^p \bar{F}^s(X))$$
$$= \int_0^1 (Q(u))^p (1-u)^s du. \tag{1.47}$$

Like L-moments, PWMs are more robust to outliers in the data. They have less bias in estimation even for small samples and converge rapidly to asymptotic normality.

Example 1.11. The PWMs of the Pareto distribution with (see Table1.1)

$$Q(u) = \sigma(1-u)^{-\frac{1}{\alpha}}, \quad \sigma, \alpha > 0,$$

are

$$\alpha_{p,s} = \sigma \int_0^1 (1-u)^{-\frac{p}{\alpha}+s} du = \frac{\sigma\alpha}{\alpha(s+1)-p}, \quad \alpha(s+1) > p.$$

Similarly, for the power distribution with (see Table 1.1)

$$Q(u) = \alpha u^{\frac{1}{\beta}}, \quad \alpha, \beta > 0,$$

we have

$$\beta_{p,r} = \alpha \int_0^1 u^{-\frac{p}{\beta}+r} du = \frac{\alpha\beta}{1+\beta(r+1)}.$$

Further specializing (1.46) for $p = 1$, we see that the L-moments are linear combination of the PW moments. The relationships are

$$L_1 = \beta_{1,0},$$
$$L_2 = 2\beta_{1,1} - \beta_{1,0},$$
$$L_3 = 6\beta_{1,2} - 6\beta_{1,1} + \beta_{1,0},$$
$$L_4 = 20\beta_{1,3} - 30\beta_{1,2} + 12\beta_{1,1} - \beta_{1,0}$$

in the first four cases. Generally, we have the relationship

$$L_{r+1} = \sum_{k=0}^{r} \frac{(-1)^{r-k}(r+k)!}{(k!)^2(r-k)!} \beta_{1,k}.$$

The conventional moments can also be deduced as $M_{p,0,0}$ or $\beta_{p,0}$ or $\alpha_{p,0}$. The role of PW moments in reliability analysis will be taken up in the subsequent chapters. In spite of its advantages, Chen and Balakrishnan [140] have pointed out some infeasibility problems in estimation. While estimating the parameters of some distributions like the generalized forms of Pareto, the estimated distributions have an upper or lower bound and one or more of the data values lie outside this bound.

1.7 Diagrammatic Representations

In this section, we demonstrate a few graphical methods other than the conventional ones. The primary goal is fixed as the choice of model for the data represented by a quantile function. An important tool in this category is the *Q*-*Q* plot. The *Q*-*Q* plot is the plot of points $(Q(u_r), x_{r:n})$, $r = 1,2,\ldots,n$, where $u_r = \frac{r-0.5}{n}$.[1] For application purposes, we may replace $Q(u_r)$ by the fitted quantile function. One use of this plot is to ascertain whether the sample could have arisen from the target population $Q(u)$. In the ideal case, the graph should show a straight line that bisects the axes, since we are plotting the sample and population quantiles. However, since the sample is random and the fitted values of $Q(u)$ are used, the points lying approximately around the line is indicative of the model being adequate. The points in the *Q*-*Q* plot are always non-decreasing when viewed from left to right.

The *Q*-*Q* plot can also be used for comparing two competing models by plotting the *r*th quantile of one against the *r*th quantile of the other. When the two distributions are similar, the points on the graph should show approximately the straight line $y = x$. A general trend in the plot, like steeper (flatter) than $y = x$, will mean that the distribution plotted on the y-axis (x-axis) is more dispersed. On the other hand, *S*-shaped plots often suggest that one of the distributions exhibits more skewness or tail-heaviness. It should also be noted that the relationship in quantile plot can be linear when the constituent distributions are linearly related. This procedure is direct when the data sets from two distributions contain the same number of observations. Otherwise, it is necessary to use interpolated quantile estimates in the shorter set to equal the number in the larger sets. Often, *Q*-*Q* plots are found to be more powerful and informative than histogram comparisons.

Example 1.12. The times to failure of a set of 10 units are given as 16, 34, 53, 75, 93, 120, 150, 191, 240 and 390 h (Kececioglu [322]). A Weibull distribution with quantile function

$$Q(u) = \sigma(-\log(1-u))^{1/\lambda}$$

is proposed for the data. The parameters of the model were estimated by the method of maximum likelihood as $\hat{\sigma} = 146.2445$ and $\hat{\lambda} = 1.973$. The *Q*-*Q* plot pertaining to the model is presented in Fig. 1.1. From the figure, it is seen that the above model seems to be adequate.

---

[1] There are different choices for these plotting points and recently Balakrishnan et al. [52] discussed the determination of optimal plotting points by the use of Pitman closeness criterion.

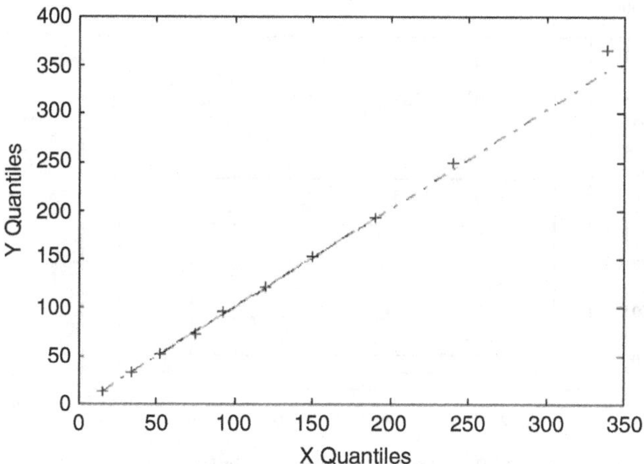

Fig. 1.1 *Q-Q* plot for Example 1.12

A second useful graphical representation is the box plot introduced by Tukey [569]. It depicts graph of the numerical data through a five-figure summary in the form of extremes, quartiles and the median. The steps required for constructing a box plot are (Parzen [484])

(i) compute the median $m = \bar{Q}(0.50)$, the lower quartile $q_1 = \bar{Q}(0.25)$ and the upper quartile $q_3 = \bar{Q}(0.75)$;
(ii) draw a vertical box of arbitrary width and length equal to $q_3 - q_1$;
(iii) a solid line is marked within the box at a distance $m - q_1$ above the lower end of the box. Dashed lines are extended from the lower and upper ends of the box at distances equal to $x_{n:n} - q_3$ and $x_{1:n} - q_1$. This constitutes the H-plot, H standing for hinges or quartiles. Instead, one can use $\bar{Q}(0.125) = e_1$ and $\bar{Q}(0.875) = e_7$ resulting in E-box plots. Similarly, the quantiles $\bar{Q}(0.0625)$ and $\bar{Q}(0.9375)$ constitute the D-box plots;
(iv) A quantile box plot consists of the graph of $\bar{Q}(u)$ on $[0,1]$ along with the three boxes in (iii), superimposed on it.

Parzen [484] proposed the following information to be drawn from the plot. By drawing a perpendicular line to the median line at its midpoint and of length $\pm n^{-\frac{1}{2}} - (q_3 - q_1)$, a confidence interval for the median can be obtained. The graph $x = \bar{Q}(u)$ exhibiting sharp rises is likely to have a density with more than one mode. If such points lie inside the H-box, the presence of several distinct populations generating the data is to be suspected, while, if they are outside the D-box, presence of outliers is indicated. Horizontal segments in the graph may be the results of the discrete distributions. By calculating

$$\frac{\bar{Q}(\frac{1}{2}) - \frac{1}{2}[\bar{Q}(u) + \bar{Q}(1-u)]}{\bar{Q}(1-u) - \bar{Q}(u)}$$

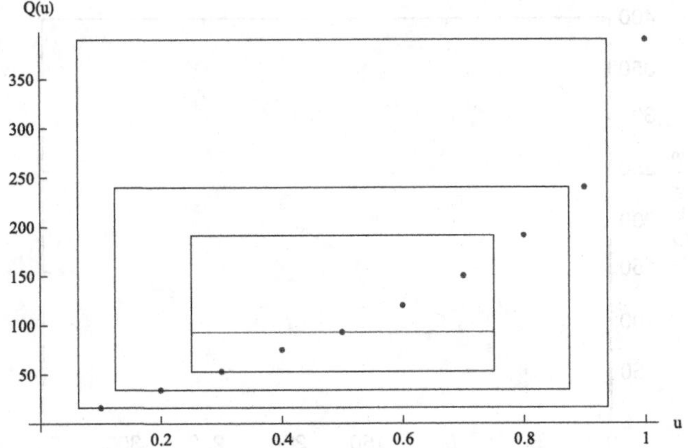

Fig. 1.2 Box plot for the data given in Example 1.12

for u values, one can get a feel for skewness with a value near zero suggesting symmetry. Parzen [484] also suggested some measures of tail classification.

Example 1.13. The box plot corresponding to the data in Example 1.12 is exhibited in Fig. 1.2. It may be noticed that the observation 390 is a likely outlier.

A stem-leaf plot can also be informative about some meaningful characteristics of the data. To obtain such a plot, we first arrange the observations in ascending order. The leaf is the last digit in a number. The stem contains all other digits (When the data consists of very large numbers, rounded values to a particular place, like hundred or thousand, are used a stem and leaves). In the leaf plot, there are two columns, first representing the stem, separated by a line from the second column representing the leaves. Each stem is listed only once and the leaves are entered in a row. The plot helps to understand the relative density of the observations as well as the shape. The mode is easily displayed along with the potential outliers. Finally, the descriptive statistics can be easily worked out from the diagram.

Example 1.14. We illustrate the stem-leaf plot for a small data set: 36, 57, 52, 44, 47, 51, 46, 63, 59, 68, 66, 68, 72, 73, 75, 81, 84, 106, 76, 88, 91, 41, 84, 68, 34, 38, 54.

Stem	Leaves
3	4 6 8
4	1, 4, 6, 7
5	1 2 4 7 9
6	3 6 8 8 8
7	2 3 5 6
8	1, 4, 4, 8
9	1
10	6

Chapter 2
Quantile-Based Reliability Concepts

Abstract There are several functions in reliability theory used to describe the patterns of failure in different mechanisms or systems as a function of age. The functional forms of many of these concepts characterize the life distribution and therefore enable the identification of the appropriate model. In this chapter, we discuss these basic concepts, first using the distribution function approach and then introduce their analogues in terms of quantile functions. Various important concepts introduced here include the hazard rate, mean residual life, variance residual life, percentile residual life, coefficient of variation of residual life, and their counterparts in reversed time. The expressions for all these functions for standard life distributions are given in the form of tables to facilitate easy reference. Formulas for the determination of the distribution from these functions, their characteristic properties and characterization theorems for different life distributions by relationships between various functions are reviewed. Many of the quantile functions in the literature do not have closed-form expressions for their distributions, and they have to be evaluated numerically. This renders analytic manipulation of these reliability functions based on the distribution function rather difficult. Accordingly, we introduce equivalent definitions and properties of the traditional concepts in terms of quantile functions. This leads to hazard quantile function, mean residual quantile function and so on. The interrelationships between these functions are presented along with characterizations. Various examples given in the sequel illustrate how the quantile based reliability functions can be found directly from the quantile functions of life distributions. Expressions of such functions for standard life distributions can also be read from the tables provided in each case.

2.1 Concepts Based on Distribution Functions

The notion of reliability, in the statistical sense, is the probability that an equipment or unit will perform the required function, under conditions specified for its operation, for a given period of time. In Sect. 1.3, we defined life distributions and

N.U. Nair et al., *Quantile-Based Reliability Analysis*, Statistics for Industry and Technology, DOI 10.1007/978-0-8176-8361-0_2,

gave several examples of such distributions used in the literature under different contexts. When a unit does not perform its intended function, we say that it has failed. This can happen in different forms such as mechanical breakdown, decrease in performance below an assigned level, defective performance, and so on. The primary concern in reliability theory is to understand the patterns in which failures occur, for different mechanisms and under varying operating environments, as a function of age. Accordingly, several concepts have been developed that help in evaluating the effect of age, based on the distribution function of the lifetime random variable X and its residual life X_t introduced earlier in Sect. 1.2. In this section, we present some key concepts and their properties as background material for later discussions using the quantile functions as the basic fabric.

2.1.1 *Hazard Rate Function*

The hazard rate of X is defined as

$$h(x) = \lim_{\delta \downarrow 0} \frac{P(x \le X < x+\delta | X > x)}{\delta} \tag{2.1}$$

so that $\delta h(x)$ is approximately the conditional probability that a unit will fail in the next small interval of time δ, given that the unit has survived age x. When $F(x)$ is absolutely continuous with probability density function $f(x)$, (2.1) reduces to

$$h(x) = \frac{f(x)}{\bar{F}(x)} = -\frac{d \log \bar{F}(x)}{dx} \tag{2.2}$$

for all x for which $\bar{F}(x) > 0$. Treated as a function of age x, the hazard rate function is also referred to as the failure rate function, instantaneous death rate, force of mortality, and intensity function in other areas of study like survival analysis, actuarial science, biosciences, demography and extreme value theory. The origin of hazard rate can be traced back to the 'force of mortality' used in connection with the construction of life tables as models of human mortality.

Integrating (2.2) over $(0,x)$ and using $F(0) = 0$, we get

$$\bar{F}(x) = \exp\left\{ -\int_0^x h(t)dt \right\}. \tag{2.3}$$

The inversion formula in (2.3) is often used to characterize life distributions in terms of the functional form of $h(x)$, that could be postulated from the physical properties of the failure rate patterns. While postulating the form of $h(x)$, the following theorem is helpful in the choice of $h(x)$.

Theorem 2.1 (Marshall and Olkin [412]). *A necessary and sufficient condition that a function $h(x)$ is the hazard rate of a distribution is that*

(i) $h(x) \geq 0$;
(ii) $\int_0^x h(t)dt < \infty$ *for some* $x > 0$;
(iii) $\int_0^\infty h(t)dt = \infty$;
(iv) $\int_0^x h(t)dt = \infty$ *implies* $h(y) = \infty$ *for every* $y > x$.

Example 2.1. The exponential power model has survival function

$$\bar{F}(x) = \exp[-(e^{(\lambda x)^\alpha} - 1)], \quad x > 0.$$

The probability density function is

$$f(x) = -\frac{d\bar{F}(x)}{dx} = \lambda^\alpha \alpha x^{\alpha-1} \exp[-(e^{(\lambda x)^\alpha} - 1)] e^{(\lambda x)^\alpha},$$

and so

$$h(x) = \frac{f(x)}{\bar{F}(x)} = \alpha \lambda^\alpha x^{\alpha-1} \exp[(\lambda x)^\alpha].$$

Example 2.2. Let $X_1, X_2, \ldots, X_n$ be independent random variables and $Z = \min(X_1, X_2, \ldots, X_n)$. Then,

$$P(Z > x) = P(X_1 > x, X_2 > x, \ldots, X_n > x)$$

or

$$\bar{F}_Z(x) = \bar{F}_{X_1}(x) \ldots \bar{F}_{X_n}(x).$$

Logarithmic differentiation leads to

$$h_z(x) = h_{X_1}(x) + \cdots + h_{X_n}(x). \tag{2.4}$$

The above model constitutes a series system with n independent components having life distributions $F_{X_1}, \ldots, F_{X_n}$. This is more general than the system illustrated in Sect. 1.5, since the components here are not identically distributed. Formula (2.4) could be employed to construct new life distributions from standard ones.

Hjorth [272] chose X_1 to be Pareto II (see Table 1.1) and X_2 to be Rayleigh with

$$\bar{F}_{X_2}(x) = \exp\left(-\frac{1}{2}\alpha x^2\right), \quad x > 0,$$

to obtain the model

$$\bar{F}(x) = e^{-\frac{\alpha x^2}{2}} (1 + \theta x)^{-\beta}.$$

The resulting hazard rate is

$$h(x) = \alpha x + \beta\theta(1+\theta x)^{-1}.$$

In a similar manner, Jaisingh et al. [291] considered a three-component model consisting of exponential, Pareto II and Weibull (see Table 1.1) to produce the model

$$\bar{F}(x) = \beta^{\theta}(x+\beta)^{-\theta}\exp[-\alpha x - \lambda^{-1}\delta x^{\lambda}], \quad x > 0,$$

with corresponding hazard rate

$$h(x) = \alpha + \theta(x+\beta)^{-1} + \delta\lambda x^{\lambda-1}.$$

Further examples of similar models can be seen in Wang [577] and Jiang and Murthy [294]. The linear failure rate distribution and quadratic failure rate distribution (Gore et al. [223]) with respective hazard functions

$$h_1(x) = a + bx$$

and

$$h_2(x) = a + bx + cx^2$$

can also be interpreted in the same manner, although they have been derived independently without such assumptions. Some of these distributions also figure in the context of additive hazards models considered in Nair and Sankaran [446].

The hazard rate functions of various distributions, their analysis with extensive references are given in Lai and Xie [368]. We have presented in Table 2.1 the expressions for $h(x)$ of the life distributions given in Table 1.1. It may be noticed that all the distributions in Example 2.2 do not have closed-form expressions for $Q(u)$ and therefore do not form part of Table 1.1

2.1.2 Mean Residual Life Function

Another important notion is based on the residual life introduced earlier in Sect. 1.2. For a unit which has survived until x, the lifetime remaining to it is $(X - x|X > x)$ with survival function (1.3)

$$\bar{F}_x(t) = \frac{\bar{F}(x+t)}{\bar{F}(x)}.$$

Table 2.1 Hazard rate functions of distributions in Table 1.1

No.	Distribution	Hazard rate
1	Exponential	λ
2	Weibull	$\lambda\sigma^{-\lambda}x^{\lambda-1}$
3	Pareto II	$c(x+\alpha)^{-1}$
4	Rescaled beta	$c(R-x)^{-1}$
5	Half-logistic	$e^{\frac{x}{\sigma}}[\sigma(1+e^{\frac{x}{\sigma}})]^{-1}$
6	Power	$\beta x^{\beta-1}(\alpha^{\beta}-x^{\beta})^{-1}$
7	Pareto I	αx^{-1}
8	Burr XII	$kcx^{c-1}(1+x^{c})^{-1}$
9	Gompertz	BC^{x}
10	Log logistic	$\beta\alpha^{\beta}x^{\beta-1}(1+\alpha^{\beta}x^{\beta})$
11	Exponential geometric	$\lambda(1-pe^{-\lambda x})^{-1}$
12	Generalized Weibull	$\alpha x^{\alpha-1}(\beta^{\alpha}-\lambda x^{\alpha})^{-1}$
13	Exponentiated Weibull	$\dfrac{\lambda\theta(\frac{x}{\sigma})^{\lambda-1}[1-\exp(-\frac{x}{\sigma})^{\lambda}]^{\theta-1}\exp[-(\frac{x}{\sigma})^{\lambda}]}{1-[1-\exp(-\frac{x}{\sigma})^{\lambda}]^{\theta}}$
14	Generalized exponential	$\dfrac{\theta(1-e^{-\frac{x}{\sigma}})^{\theta-1}e^{-\frac{x}{\sigma}}}{\sigma[1-(1-e^{-\frac{x}{\sigma}})^{\theta}]}$
15	Extended Weibull	$\lambda\sigma^{-\lambda}x^{\lambda-1}[1-(1-\theta)e^{-(\frac{x}{\sigma})^{\lambda}}]^{-1}$
16	Inverse Weibull	$\dfrac{\lambda\sigma^{\lambda}x^{-\lambda-1}e^{-(\frac{x}{\sigma})^{\lambda}}}{[1-e^{-(\frac{\sigma}{x})^{\lambda}}]}$
17	Generalized Pareto	$\dfrac{a+1}{ax+b}$
18	Exponential power	$\alpha\lambda^{\alpha}x^{\alpha-1}\exp[-(\lambda x)^{\alpha}]$
19	Modified Weibull	$\alpha\lambda(\frac{x}{\sigma})^{\lambda-1}\exp[\frac{x}{\sigma}]^{\lambda}$
20	Log Weibull	$\frac{k\rho}{1+\rho x}[\log(1+\rho x)]^{k-1}$
21	Dimitrakopoulou et al.	$\alpha\lambda\beta x^{\beta-1}(1+\lambda x^{\beta})^{\alpha-1}$
22	Logistic exponential	$\dfrac{ke^{\lambda(x+\theta)}(e^{\lambda(x+\theta)}-1)^{k-1}}{1+(e^{\lambda(x+\theta)}-1)^{k}}$
23	Kus	$\dfrac{\beta\lambda e^{-\beta x}\exp[\lambda e^{-\beta x}]}{1-\exp[\lambda e^{-\beta x}]}$
24	Greenwich	$ax(b^{2}+x^{2})^{-1}$
25	Generalized half-logistic	$\frac{(1-kx)^{-1}}{1+(1-kx)^{1/k}}$

The expected value of the distribution $\bar{F}_x(t)$ is called the mean residual life function and is denoted by $m(x)$. Thus, when $E(X) < \infty$,

$$m(x) = \int_0^\infty \frac{\bar{F}(t+x)}{\bar{F}(x)} dt = \frac{1}{\bar{F}(x)} \int_x^\infty \bar{F}(t)dt \tag{2.5}$$

for all x for which $\bar{F}(x) > 0$. When $\bar{F}(x)$ has a density $f(x)$, we have

$$m(x) = \frac{1}{\bar{F}(x)} \int_x^\infty (t-x)f(t)dt. \tag{2.6}$$

Like the hazard rate function, the origin of mean residual life also traces back to the life table function 'expectation of life', used by actuaries. It is called the mean excess function in actuarial science.

Differentiating (2.5) with respect to x and rearranging the terms, the identity

$$h(x) = \frac{1+m'(x)}{m(x)} \tag{2.7}$$

results. The function $m(x)$ determines the distribution of X uniquely by virtue of the formula

$$\bar{F}(x) = \frac{\mu}{m(x)} \exp\left\{-\int_0^x \frac{dt}{m(t)}\right\}. \tag{2.8}$$

Both the hazard function and the mean residual life function are conditional on the survival until x. The former provides information in an infinitesimal interval after x, while the latter contributes to the entire interval $[x,\infty)$. For further comparison of the two measures, we refer to Muth [436]. Guess and Proschan [228] and Nanda et al. [458] have both reviewed the basic results and various applications of the mean residual life function and associated orderings and properties.

Example 2.3. Consider the Weibull distribution with survival function

$$\bar{F}(x) = \exp\left\{-\left(\frac{x}{\sigma}\right)^{\frac{1}{2}}\right\}, \quad x > 0.$$

Then,

$$\begin{aligned} m(x) &= \frac{1}{\bar{F}(x)} \int_x^\infty e^{-(\frac{t}{\sigma})^{\frac{1}{2}}} dt \\ &= 2e^{(\frac{x}{\sigma})^{\frac{1}{2}}} \int_{x^{\frac{1}{2}}}^\infty y e^{-\sigma^{-\frac{1}{2}} y} dy \quad (\text{with } y = x^{\frac{1}{2}}) \\ &= 2\sigma^{\frac{1}{2}}(\sigma^{\frac{1}{2}} + x^{\frac{1}{2}}). \end{aligned}$$

Table 2.2 Mean residual life functions of some distributions

Distribution	$m(x)$
Exponential	λ^{-1}
Power	$\dfrac{(\beta+1)\alpha^{\beta}(1-x)+x^{\beta+1}-1}{(\beta+1)(\alpha^{\beta}-x^{\beta})}$
Pareto II	$\dfrac{x+\alpha}{c-1}$
Rescaled beta	$\dfrac{R-x}{R+1}$
Pareto	$(\alpha-1)^{-1}x$
Half-logistic	$\sigma(1+e^{\frac{x}{\sigma}})\log(1+e^{-\frac{x}{\sigma}})$
Exponential geometric	$-(\lambda p)^{-1}e^{\lambda x}(1-pe^{-\lambda x})\log(1-pe^{-\lambda x})$
Exponential geometric extension	$-\{\beta(1-\theta)\}^{-1}e^{\beta x}[1-(1-\theta)e^{-\beta x}]$ $\log[1-(1-\theta)e^{-\beta x}]$
Adamidis et al. [17]	$(\bar{F}(x)=\theta e^{-\beta x}[1-(1-\theta)e^{-\beta x}]^{-1},\ x>0)$

Example 2.4. Let X be distributed as exponential geometric with

$$\bar{F}(x)=(1-p)e^{-\lambda x}(1-pe^{-\lambda x})^{-1}.$$

Then,

$$\begin{aligned}m(x)&=\frac{1-pe^{-\lambda x}}{(1-p)e^{-\lambda x}}\int_x^{\infty}\frac{(1-p)e^{-\lambda x}}{1-pe^{-\lambda x}}dx\\&=-(\lambda p)^{-1}e^{-\lambda x}(1-pe^{-\lambda x})\log(1-pe^{-\lambda x}).\end{aligned}$$

Further examples of mean residual life functions are presented in Table 2.2.

Not every function can be the mean residual life function of a life distribution. The following theorem helps to conclude whether a given function can represent a mean residual life.

Theorem 2.2 (Guess and Proschan [228] and Nanda et al. [458]). *A necessary and sufficient condition for $m(x)$ to be a mean residual life function is that*

(i) $m(x)$ has range $[0,\infty)$ for all $x\ge 0$;
(ii) $m(0)=\mu>0$;
(iii) $m(x)$ is right continuous;
(iv) $m(x)+x$ is increasing;
(v) when there exists an x_0 satisfying $\lim_{x\downarrow x_0}m(x)=0$, then $m(x)=0$ holds for x in $[x_0,\infty)$. If there is no x_0 for which the above limit is 0, then $\int_0^{\infty}\frac{dx}{m(x)}=\infty$.

The formula in (2.7) makes it easy to find the hazard function when $m(x)$ is given. However, the problem is with the converse. When $h(x)$ is known, the differential equation resulting from (2.7) in $m(x)$ is difficult to solve for most distributions. Hence, efforts were put in to finding simpler relationships between $m(x)$ and $h(x)$ satisfied by distributions. The price paid for simplicity in such cases is the limitation to the range of applicability. Starting with individual distributions like gamma and negative binomial (Osaki and Li [475]), the work in this direction progressed to characterization of Pearson family (Nair and Sankaran [442]), the exponential family (Consul [155]), mixtures of distributions (Abraham and Nair [12]), generalized Pearson system (Sankaran et al. [516]) and other generalizations (Gupta and Bradley [238]). The general relationship

$$E(C(X)|X > x) = \mu_C + \sigma_C h(x) g(x), \tag{2.9}$$

for some $g(x)$ and a measurable function $C(x)$ with $\mu_C = EC(X)$ and $\sigma_C^2 = V(C(X))$, is seen to hold for the class of distributions satisfying

$$\frac{f'(x)}{f(x)} = \frac{\mu_C - C(x) - g'(x)}{\sigma_C g(x)}$$

and conversely (Nair and Sudheesh [449]). The special case $C(X) = X$ gives the necessary relationship in terms of $m(x)$ and $h(x)$. Apart from providing such relationships for a wider class of distributions, (2.9) was employed to develop lower bound to the variance that compares favourably with the well-known Cramer–Rao and Chapman–Robbins inequalities. Details can be seen in Nair and Sudheesh [449, 450] and also in the references in Nair and Sankaran [442]. In an alternative approach, Nair and Sankaran [443] viewed the mean residual life function as the expectation of the conditional distribution of residual life given age, arising from the joint distribution of age and residual life in renewal theory.

2.1.3 Variance Residual Life Function

For a lifetime random variable X with $E(X^2) < \infty$, the variance residual life function is defined as

$$\begin{aligned} \sigma^2(x) &= V(X - x|X > x) = V(X|X > x) \\ &= E(X - x^2|X > x) - m^2(x) \\ &= \frac{1}{\bar{F}(x)} \int_x^\infty (t-x)^2 f(t) dt - m^2(x). \end{aligned} \tag{2.10}$$

The integral on the right side can be simplified by integration by parts as

$$\int_x^\infty (t-x)^2 f(t) dt = 2\int_x^\infty (t-x)\bar{F}(t) dt = 2\int_x^\infty \int_u^\infty \bar{F}(t)\, dt\, du$$

and therefore

$$\sigma^2(x) = 2[\bar{F}(x)]^{-1} \int_x^\infty \int_u^\infty \bar{F}(t)\, dt\, du - m^2(x). \tag{2.11}$$

Apart from the usual meaning as a measure of spread of the residual life distribution and its role in finding the variance of the sample mean residual life, $\sigma^2(x)$ has some other important applications in reliability analysis. Launer [377], who introduced this concept, used it to distinguish life distributions based on its monotonic properties while Gupta and Kirmani [243, 245] considered characterizations using $\sigma^2(x)$ (see also Gupta [234] and Gupta et al. [246]). Gupta [234] established that

$$\frac{d}{dx}\sigma^2(x) = h(x)(\sigma^2(x) - m^2(x)). \tag{2.12}$$

It follows from (2.3) that (Abouammoh et al. [8])

$$\bar{F}(x) = \exp\left[-\int_0^x \frac{\frac{d\sigma^2(t)}{dt}}{\sigma^2(t) - m^2(t)}\right]. \tag{2.13}$$

Equation (2.13) makes it clear that both $\sigma^2(x)$ and $m(x)$ are required to retrieve $F(x)$. With the variance and mean in place, the coefficient of variation of residual life becomes

$$C(x) = \frac{\sigma(x)}{m(x)}. \tag{2.14}$$

Gupta [234] showed that

$$\frac{d}{dx}\sigma^2(x) = m(x)(1 + m'(x))(C^2(x) - 1)$$

which arises from (2.12), (2.14) and (2.7). In a later work, Gupta and Kirmani [243] found

$$m(x) = (1 + C^2(x))^{-1}\left\{\int_0^x C^2(t)dt + m(1 + C^2(0)) - x\right\}. \tag{2.15}$$

Since $m(x)$ characterizes $F(x)$ and $m(x)$ is expressed uniquely in terms of $C(x)$ by (2.15), it is evident that $C(x)$ also determines $F(x)$ uniquely. They also showed that if two life distributions $F(x)$ and $G(x)$ have the same means and equal residual coefficient of variation for all x, then $F = G$. This was further strengthened in Gupta and Kirmani [243] by the conditions $m_F(x_0) = m_G(x_0)$ for some $x_0 \geq 0$ and $\sigma_F^2(x) = \sigma_G^2(x)$, for the equality of $F(x)$ and $G(x)$ for all x.

Unlike the hazard and mean residual life functions, there is no direct formula that expresses $F(x)$ in terms of $\sigma^2(x)$ only. This brings in the importance of characterizing specific distributions or families by the functional form of $\sigma^2(x)$. The works of Dallas [166], Adatia et al. [19], Koicheva [351], Ghitany et al. [213], Navarro et al. [465] and El-Arishi [184] all belong to this category. Most of these results are subsumed in the general formula given in Nair and Sudheesh [451], which states that if (2.9) holds, then

$$V(C(X)|X>x) = \sigma_C E[C'(X)g(X)|X>x] + (\mu_C - M(x))(M(x) - C(x)), \quad (2.16)$$

where $M(x) = E(C(X)|X > x)$. Conversely, if there exists a measurable function $C(x)$ for which $C'(x) \neq 0$ for all $x > 0$, satisfying (2.15), then (2.9) holds and

$$\frac{f'(x)}{f(x)} = \frac{\mu_C - C(x) - g'(x)}{\sigma_C g(x)}.$$

When $C(X) = X$, the implication to $\sigma^2(x)$ from (2.16) is obvious.

Example 2.5. The generalized Pareto distribution has survival function

$$\bar{F}(x) = \left(1 + \frac{ax}{b}\right)^{-\frac{a+1}{a}}, \; x > 0; b > 0, a > -1. \quad (2.17)$$

The form of this distribution is quite amenable to deriving several characterizations based on reliability functions. It consists of three distributions, viz., the exponential ($a \to 0$), the Pareto II when $a = (C^{-1} - 1)$ and $b = a\alpha$, and the rescaled beta when $a = -(1 + C^{-1})$ and $b = Ra$; see Table 1.1 for details. All the three constituent distributions are important models in reliability on their own accord. For the model in (2.17), we have

$$m(x) = \left(1 + \frac{ax}{b}\right)^{\frac{a+1}{a}} \int_x^\infty \left(1 + \frac{at}{b}\right)^{-\frac{a+1}{a}} dt = ax + b.$$

Hence, a linear mean residual life function characterizes the generalized Pareto model of which, the exponential has $m(x) = b$ ($a = 0$), Pareto II has $a > 0$ so that $m(x)$ is increasing, and rescaled beta has $-1 < a < 0$ giving a decreasing mean residual life.

Notice also that the hazard rate of (2.16) is

$$h(x) = \frac{(a+1)}{(ax+b)}.$$

Hence,

$$m(x)h(x) = \text{constant},$$

a relationship that affords another characterization; see Mukherjee and Roy [431] for further details.

Again, we have

$$\sigma^2(x) = 2\left(1+\frac{ax}{b}\right)^{\frac{a+1}{a}} \int_x^\infty \int_u^\infty \left(1+\frac{at}{b}\right)^{-\frac{a+1}{a}} dt\,du$$

from (2.11). After simplification, we obtain

$$\sigma^2(x) = \frac{a+1}{1-a} b^2 (ax+b)^2.$$

Thus, we have the identity

$$\sigma^2(x) = Km^2(x). \tag{2.18}$$

Conversely, if we assume (2.18), upon substituting it in (2.13), we get

$$m'(x) = \frac{k-1}{k+1}$$

which implies that $m(x)$ is linear and X is distributed as generalized Pareto.

2.1.4 Percentile Residual Life Function

The mean and variance of residual life are popular measures in lifelength analysis with potential applications in other fields of study. However, there are instances like censored data, or observations from heavily skewed distributions in which the empirical counterparts of the two functions are difficult to compute. Moreover, the other limitations that were described in Chap. 1 in connection with the use of conventional moments are also true for $m(x)$ and $\sigma^2(x)$. An alternative in such cases is the percentile residual life function first studied by Haines and Singpurwalla [258]; see also Launer [376].

For any $0 < \alpha < 1$, the αth percentile residual life function is the αth percentile of the residual life distribution of X. Thus, recalling from (2.4) the expression for the survival function of the residual life, the αth percentile residual life function, denoted by $p_\alpha(x)$, is

$$\begin{aligned} p_\alpha(x) &= F_x^{-1}(\alpha) \\ &= \inf\{x|F_x(t) \geq \alpha\} \\ &= \inf\left\{x|1 - \frac{\bar{F}(x+t)}{\bar{F}(x)} \geq \alpha\right\} \end{aligned}$$

$$= \inf\{y|\bar{F}(y) \le (1-\alpha)\bar{F}(x)\} - x$$
$$= F^{-1}(1-(1-\alpha)\bar{F}(x)) - x. \tag{2.19}$$

Thus, $p_\alpha(x)$ can be expressed in terms of the baseline distribution function $F(x)$. From (2.19), it is clear that $p_\alpha(x)$ is a solution of the functional equation

$$F(p_\alpha(x)+x) = 1-(1-\alpha)\bar{F}(x) = \alpha + (1-\alpha)F(x). \tag{2.20}$$

We interpret $p_\alpha(x)$ as the age that will be survived, on the average, by $100(1-\alpha)\%$ of units that have lived beyond age x.

Example 2.6. The exponential distribution $\bar{F}(x) = e^{-x}, x > 0$, has $p_\alpha(x)$ defined by (2.20) as

$$1 - e^{-(x+p_\alpha)} = 1-(1-\alpha)e^{-x}$$

which simplifies to

$$p_\alpha = -\log(1-\alpha)$$

which is a constant, independent of x for any choice of α in $(0,1)$. On the other hand, choosing (Song and Cho [545]) as

$$\bar{F}(x) = e^{-x}(1+\theta \sin x),\ x \ge 0, |\theta| < 2^{-\frac{1}{2}},$$

and $\alpha = 1 - e^{-2\pi}$, (2.20) yields

$$\bar{F}(p_\alpha(x)+x) = 1 - e^{-2\pi}e^{-x}(1+\theta\sin x)$$
$$= 1 - e^{-(2\pi+x)}(1+\theta\sin(x+2\pi))$$
$$= \bar{F}(2\pi + x).$$

Since F is continuous and strictly increasing, we get

$$p_\alpha(x) = 2\pi$$

which is the same as that of the exponential when $\alpha = 1 - e^{-2\pi}$.

It is clear from the above example that the percentile residual life function does not determine $F(x)$ uniquely. Thus, the problem of searching conditions for characterizing distributions in terms of $p_\alpha(x)$ has received the attention of many researchers like Schmittlein and Morrison [523], Arnold and Brockett [38], Gupta and Langford [247], Joe [300], Song and Cho [545], Lillo [399] and Lin [402]. A comprehensive solution was offered by Gupta and Langford [247] (see also Joe [300]) who identified (2.20) as a particular case of the Schroder functional equation

$$R(\phi(t)) = uR(t), \quad 0 \le t < \infty, \tag{2.21}$$

where $0 < u < 1$ and $\phi(t)$ is a continuous and strictly increasing function on $[0,\infty]$ satisfying $\phi(t) > t$ for all t. The general solution of (2.22) is

$$R(t) = R_0(t)K(\log R_0(t)),$$

where $K(\cdot)$ is a periodic function with period $-\log u$ and $R_0(t)$ is a particular solution of (2.21) which is positive, continuous and strictly decreasing such that $R(0) = 1$. Thus, there is no unique solution to (2.20). Song and Cho [545] (correcting a result of Arnold and Brockett [38]) proved that if F is continuous and strictly increasing and if for $0 < \alpha_1 < \alpha_2 < 1$, $\frac{\log(1-\alpha_1)}{\log(1-\alpha_2)}$ is irrational, then F is uniquely determined by $p_{\alpha_1}(x)$ and $p_{\alpha_2}(x)$. More general results due to Lin [402] are the following:

1. If $F(x)$ and $G(x)$ are continuous distributions on $[0,\infty)$ such that $F(0) = G(0) = u_0$ in $[0,1)$, then for a fixed number α, $p_{\alpha,F}(x) = p_{\alpha,G}(x)$ if and only if

$$\bar{F}(x) = \bar{G}K_1(-\log \bar{G}(x)) \quad 0 \le x < r,$$
$$\bar{G}(x) = \bar{F}K_2(-\log \bar{F}(x)) \quad 0 \le x < r,$$

   where K_i, $i = 1,2$, are periodic functions with the same period $1-\alpha$ and r is the common right extremity of the supports of F and G;
2. For real numbers α_i in $(0,1)$ such that $\frac{\log(1-\alpha_1)}{\log(1-\alpha_2)}$ is irrational and $p_{\alpha_i,F}(x) = p_{\alpha_i,G}(x)$, $i = 1,2$, we have $F(x) = G(x)$.

2.2 Reliability Functions in Reversed Time

2.2.1 *Reversed Hazard Rate*

In this section, we consider functions similar to those explained in Sect. 2.1 but are conditioned on the event $X \le x$, that is, the unit is assumed to have a lifetime less than or equal to x. The primary notion in this connection is the reversed hazard rate $\lambda(x)$ given by

$$\lim_{\Delta \to 0} \frac{P(x-\Delta < X \le x | X \le x)}{\Delta} \tag{2.22}$$

and hence

$$\Delta\lambda(x) = P(x-\Delta < X < x | X \le x) + o(\Delta).$$

Thus, for all x for which $F(x) > 0$,

$$\lambda(x) = \frac{d}{dx}\log F(x) = \frac{f(x)}{F(x)}.$$

Hence, the probability that a unit with life X which has survived age $x-\Delta$ will fail in the next small interval of time Δ given that it will not survive age x is $\Delta\lambda(x)$. Introduced by Keilson and Sumita [323], the function $\lambda(x)$ has been used in various contexts such as estimation and modelling for left censored data, stochastic orderings, characterization of distributions, and in developing repair and maintenance strategies. Block et al. [111] have shown that there does not exist a non-negative random variable having increasing or constant reversed hazard rate function. If X has support (a,b), $-\infty \le a < b \le \infty$, then $\lambda_{-X}(x) = h_X(-x)$ where x, $-x \in (a,b)$ which is a duality property that justifies the adjective 'reversed' associated with $\lambda(x)$. Finkelstein [198] observed that for possible application of $\lambda(x)$ in reliability studies, the above duality property is not relevant. Moreover, with the upper extremity of the interval of support being usually infinity, the properties of the reversed hazard rate for non-negative random variables cannot be formally observed from the corresponding properties of $h(x)$. Like $h(x)$, we can use $\lambda(x)$ also to recover the distribution of X by means of the relation

$$F(x) = \exp\left\{-\int_x^\infty \lambda(t)dt\right\} \tag{2.23}$$

obtained by integrating (2.22) over (x,∞).

Example 2.7. The generalized exponential distribution with (see Table 1.1)

$$F(x) = (1-e^{-\lambda x})^\theta,\ x > 0;\ \lambda,\theta > 0,$$

has

$$f(x) = \lambda\theta(1-e^{-\lambda x})^{\theta-1}e^{-\lambda x},$$

and so

$$\lambda(x) = \theta\lambda(e^{\lambda x}-1)^{-1}.$$

Example 2.8. Let $X_1, X_2, \ldots, X_n$ be independent random variables and $W = \max(X_1, X_2, \ldots, X_n)$. Then,

$$P(W \le x) = P(X_1 \le x, X_2 \le x, \ldots, X_n \le x).$$

Logarithmic differentiation leads to

$$\lambda_W(x) = \lambda_{X_1}(x) + \lambda_{X_2}(x) + \cdots + \lambda_{X_n}(x).$$

Table 2.3 Reversed hazard rate functions of some life distributions

Distribution	$F(x)$	$\lambda(x)$
Power	$(\frac{x}{\alpha})^{\beta}, 0 \le x \le \alpha$	βx^{-1}
Reciprocal exponential	$\exp(-\frac{\lambda}{x}), x > 0, \lambda > 0$	λx^{-2}
Reciprocal Lomax	$(1+\frac{1}{\alpha x})^{-c}, x > 0$	$\frac{c}{x(1+\alpha x)}$
Reciprocal Weibull	$\exp[-(\frac{1}{\sigma x})^{\lambda}], x > 0$	$\frac{\lambda}{\sigma^{\lambda} x^{\lambda+1}}$
Reciprocal beta	$(1-\frac{1}{Rx})^{c}$, $\frac{1}{R} < x < \infty$	$\frac{C}{x(Rx-1)}$
Reciprocal Gompertz	$\exp[\frac{-B(C^{x^{-1}}-1)}{\log C}]$	$\frac{B}{x^2} C^{\frac{1}{x}}$
Generalized exponential	$(1-e^{-\lambda x})^{\theta}, x > 0$	$\frac{\theta\lambda}{e^{\lambda x}-1}$
Burr	$(1+x^{-C})^{-k}, x > 0$	$\frac{kc}{x(1+x^{C})}$
Generalized power	$(1-x^{-\beta})^{\theta}, x > 1$	$\frac{\beta\theta}{x(x^{\beta}-1)}$
Negative Weibull	$\exp[-\theta(x^{-\beta-1})]$	$\frac{\theta\beta}{x^{\beta+1}}$

This model constitutes a parallel system with n independent components with life distribution functions $F_{X_1}, \ldots, F_{X_n}$.

A review of the main results and applications of $\lambda(x)$ are given in Nair and Asha [439]. For many of the distributions in Table 1.1, the reversed hazard functions are complicated, though they can be obtained as in the above example. A useful result that enables one to get models with simple expressions for $\lambda(x)$ is the following.

Theorem 2.3. *For a non-negative random variable X with hazard rate $h(x)$, its reciprocal $\frac{1}{X}$ has reversed hazard rate $\lambda^*(x)$ that satisfies*

$$h(x) = \frac{1}{x^2}\lambda^*\left(\frac{1}{x}\right)$$

or

$$\lambda^*(x) = \frac{1}{x^2}h\left(\frac{1}{x}\right).$$

Table 2.3 contains some distributions belonging to the above category.

2.2.2 Reversed Mean Residual Life

The random variable $(x - X|X \le x)$ is called the inactivity time or reversed residual life of X. It represents the time elapsed since the failure of a unit given that its lifetime is at most x. We can write the distribution function of the reversed residual life as

$$F_x(t) = P((x-X) \le t | X \le x)$$
$$= \frac{F(x) - F(x-t)}{F(x)}$$

with corresponding density function

$$f_x(t) = \frac{f(x-t)}{F(x)}.$$

Accordingly, the mean inactivity time (reversed mean residual life) becomes

$$r(x) = \int_0^x \frac{t f(x-t)}{F(x)} dt = \frac{1}{F(x)} \int_0^x F(t) dt \qquad (2.24)$$

and

$$r(x)F(x) = \int_0^x F(t) dt.$$

Differentiating with respect to x and using the definition of $\lambda(x)$, we obtain

$$\lambda(x) = \frac{1 - r'(x)}{r(x)}. \qquad (2.25)$$

Hence, from (2.23), we get

$$F(x) = \exp\left\{ - \int_x^\infty \frac{1 - r'(t)}{r(t)} dt \right\}. \qquad (2.26)$$

As in the case of the mean residual life function, for a chosen function $r(x)$ to be a reversed mean residual life function, the following conditions have to be satisfied.

Theorem 2.4 (Finkelstein [198]). *A function $r(x)$ is a reversed mean residual life of a non-negative random variable X if and only if*

(i) $r(x) \ge 0$ *for all* $x > 0$, *with* $r(0) = 0$;

(ii) $r'(x) < 1$;

(iii) $\int_0^\infty \frac{1-r'(t)}{r(t)} dt = \infty$;

(iv) $\int_x^\infty \frac{1-r'(t)}{r(t)} dt < \infty$ *for* $x > 0$.

2.2.3 Some Other Functions

Kundu and Nanda [360] discussed the properties of the reversed variance residual life function

$$\begin{aligned} v(x) &= V(x-X|X \leq x) \\ &= E((x-X)^2|X \leq x) - r^2(x) \\ &= \frac{2}{F(x)} \int_0^x \int_0^u F(t)\,dt\,du - r^2(x) \end{aligned} \tag{2.27}$$

and also the corresponding coefficient of variation given by

$$a(x) = \frac{[v(x)]^{\frac{1}{2}}}{r(x)}.$$

They obtained the identity

$$\frac{dv(x)}{dx} = \lambda(x) r^2(x)[1 - a^2(x)]$$

and used it to characterize the distribution

$$F(x) = \frac{[\{(2b-\mu)C^2 - \mu\} + (1-C^2)x]^{\frac{2C^2}{1-C^2}}}{(b-\mu)(1+C^2)}, \quad \frac{\mu + (\mu - 2b)C^2}{1-C^2} < x < b,$$

by the property $a(x) = C$.

Example 2.9. In the case of the power distribution with

$$F(x) = \left(\frac{x}{\alpha}\right)^{\beta}, \; 0 \leq x \leq \alpha,$$

$$f(x) = \alpha^{-\beta} \beta x^{\beta - 1},$$

we have $\lambda(x) = \beta x^{-1}$. Again, $r(x) = (\beta+1)x$, and so

$$r(x)\lambda(x) = \beta(\beta+1)^{-1}, \text{ a constant.}$$

Upon using

$$\int_0^x \int_0^u F(t)dtdu = \frac{2}{\alpha^{\beta}} \frac{x^{\beta+2}}{(\beta+1)(\beta+2)}$$

and (2.27), we obtain

$$v(x) = \frac{\beta x^2}{(\beta+2)(\beta+1)^2} = \frac{\beta}{\beta+2} r^2(x).$$

One can prove that all these are characterizations, with the help of (2.23), (2.26), (2.25) and (2.27). Note the similarity between the above and those of the generalized Pareto distribution in Example 2.3.

The reversed percentile residual life $q_\alpha(x)$, for $0 < \alpha < 1$, is defined as (Nair and Vineshkumar [453])

$$\begin{aligned} q_\alpha(x) &= F_x^{-1}(\alpha) = \inf[t|F_x(t) \geq \alpha] \\ &= \inf[t|F(x-t) \leq (1-\alpha)F(x)] \\ &= x - F^{-1}[(1-\alpha)F(x)]. \end{aligned}$$

The functional equation that solves for $q_\alpha(x)$ is

$$F(x - q_\alpha(x)) = (1-\alpha)F(x). \tag{2.28}$$

By obtaining a solution of the form

$$F(x) = G(x)K(-\log(x)),$$

where $K(\cdot)$ is a periodic function with period $-\log(1-\alpha)$ and $G(x)$ is a particular solution, Nair and Vineshkumar [453] concluded that $F(x)$ is uniquely determined by two percentile functions $q_\alpha(x)$ and $q_\beta(x)$, with $\frac{\log(1-\alpha)}{\log(1-\beta)}$ being irrational. They also showed that $q_\alpha(x)$ and $\lambda(x)$ are related through

$$q'_\alpha(x) = 1 - \frac{\lambda(x)}{\lambda(x - q_\alpha(x))}.$$

2.3 Hazard Quantile Function

We have seen several distribution functions for which the corresponding quantile functions cannot be obtained in explicit algebraic form. In practice, the solution of $F(x) = u$ is obtained numerically. Similarly, there are quantile functions that do not permit closed-form expressions for $F(x)$. Hence, the reliability functions introduced in the last two sections and their properties are of limited use for algebraic manipulations and analysis. In view of this, we need translation of the definitions and properties in terms of quantile functions. This approach will facilitate all forms of analysis with the same scope and strength as in the distribution function approach. In addition, they offer new results and opportunities by way of models and methods of analysis. The main source of the discussions in the rest of this chapter is Nair and Sankaran [444]. We assume that $F(x)$ is continuous and strictly increasing so that all quantile related functions are well defined.

Setting $x = Q(u)$ in (2.2) and using the relationship

$$f(Q(u)) = [q(u)]^{-1},$$

we have the definition of the hazard quantile function as

$$H(u) = h(Q(u)) = [(1-u)q(u)]^{-1}. \tag{2.29}$$

In this definition, $H(u)$ is interpreted as the conditional probability of the failure of a unit in the next small interval of time given the survival of the unit at $100(1-u)\%$ point of the distribution. Gilchrist [215] refers to (2.29) as the p-hazard (with p taking the place of u in our notation) and points out some forms of hazard functions. From (2.29), we have

$$q(u) = [(1-u)H(u)]^{-1} \tag{2.30}$$

and so

$$Q(u) = \int_0^u \frac{dp}{(1-p)H(p)}. \tag{2.31}$$

The last two equations can be employed for the unique determination of the distribution of X as illustrated in the following examples.

Example 2.10. Taking

$$Q(u) = u^{\theta+1}(1+\theta(1-u)), \quad \theta > 0,$$

we have

$$q(u) = u^{\theta}[1+\theta(\theta+1)(1-u)]$$

and so

$$H(u) = [(1-u)u^{\theta}(1+\theta(\theta+1)(1-u))]^{-1}.$$

Note that there is no analytic solution for $x = Q(u)$ that gives $F(x)$ in terms of x.

Example 2.11. Given the hazard quantile function of a distribution as

$$H(u) = \frac{a+1}{b}(1-u)^{\frac{a}{a+1}},$$

from (2.30), we have

$$q(u) = \frac{b}{a+1}(1-u)^{-\frac{a}{a+1}-1},$$

and so from (2.31), we obtain

$$Q(u) = \frac{b}{a}[(1-u)^{-\frac{a}{a+1}} - 1],$$

the quantile function of the generalized Pareto distribution.

The hazard quantile functions that characterize the life distributions in Table 1.1 are presented in Table 2.4. More examples are available in Chap. 3 wherein we discuss new models.

Example 2.12. Suppose we are given the hazard quantile function of a distribution as

$$H(u) = \frac{1}{2}\left(\frac{1+u}{1-u}\right)^{k+1} \quad \text{for } k > 0.$$

Then, from (2.30), we have the quantile density function as

$$q(u) = \frac{1}{(1-u)H(u)} = 2\frac{(1-u)^k}{(1+u)^{k+1}}.$$

So, from (2.31), we obtain the quantile function of the distribution as

$$Q(u) = \int_0^u q(p)dp = 2\int_0^u \frac{(1-p)^k}{(1+p)^{k+1}}dp = \frac{1}{k}\left\{1 - \left(\frac{1-u}{1+u}\right)^k\right\},$$

which is the quantile function of the generalized half-logistic distribution as presented in Table 1.1.

The application of hazard quantile functions is not limited to the appraisal of the mechanism of failures in a specific failure time model. It can also provide the identification of the model in a given data situation by means of characterization theorems. The characterization problems discussed earlier and elsewhere in the distribution function approach automatically hold for quantile functions under the transformation $x = Q(u)$. Other than these, we can find new characterizations exclusively in the quantile set-up, which is illustrated in the following theorem.

Table 2.4 Hazard quantile functions of distributions in Table 1.1

Distribution	$H(u)$
Exponential	λ
Weibull	$\lambda\sigma^{-1}(-\log(1-u))^{1-\frac{1}{\lambda}}$
Pareto II	$c\alpha^{-1}(1-u)^{1/c}$
Rescaled beta	$cR^{-1}(1-u)^{-\frac{1}{c}}$
Half-logistic	$(2\sigma)^{-1}(1+u)$
Power	$\beta\alpha^{-1}(1-u)^{-1}u^{1-\frac{1}{\beta}}$
Pareto	$\alpha\sigma^{-1}(1-u)^{\frac{1}{\alpha}}$
Burr XII	$ck(1-u)^{\frac{1}{k}}[(1-u)^{-\frac{1}{k}}-1]^{1-\frac{1}{c}}$
Loglogistic	$\alpha\beta(1-u)^{\frac{1}{\beta}}u^{1-\frac{1}{\beta}}$
Exponential geometric	$\lambda(1-p)^{-1}(1-pu)$
Generalized Weibull	$\alpha\beta^{-1}(1-u)^{-\lambda}[1-\frac{(1-u)^{\lambda}}{\lambda}]^{1-\frac{1}{\alpha}}$
Exponentiated Weibull	$\lambda\theta(1-u^{\frac{1}{\theta}})(-\log(1-u^{\frac{1}{\theta}})^{1-\frac{1}{\lambda}}$ $(1-u)^{-1}u^{1-\frac{1}{\theta}}$
Generalized exponential	$\sigma^{-1}\theta(1-u^{\frac{1}{\theta}})(1-u)^{-1}u^{1-\frac{1}{\theta}}$
Exponential power	$\lambda\alpha[1+\log(1-u)]$ $[-\log(1+\log(1-u))]^{1-\frac{1}{\alpha}}$
Modified Weibull extension	$\lambda\sigma^{-1}[\alpha\sigma-\log(1-u)]$ $[\log(1-\frac{\log(1-u)}{\alpha\sigma})]^{1-\frac{1}{\lambda}}$
Log Weibull	$\rho k\exp[-(-\log(1-u))^{\frac{1}{k}}]$ $[-\log(1-u)]^{1-\frac{1}{k}}$
Greenwich	$ab^{-1}(1-u)^{-\frac{2}{a}}[(1-u)^{\frac{2}{a}}-1]^{\frac{1}{2}}$
Extended Weibull	$\frac{\lambda}{\theta\sigma}[\log\frac{\theta+(1-\theta)(1-u)}{1-u}]^{1-\frac{1}{\lambda}}$ $(\theta+(1-\theta)(1-u))$
Inverse Weibull	$\lambda\sigma^{-1}u(1-u)^{-1}(-\log u)^{\frac{1}{\lambda}+1}$
Generalized Pareto	$b^{-1}(a+1)(1-u)^{\frac{a}{a+1}}$
Generalized half-logistic	$\frac{1}{2}\left(\frac{1+u}{1-u}\right)^{k+1}$

Theorem 2.5. *Let X be a non-negative random variable with absolutely continuous distribution function $F(x)$ and quantile function $Q(u)$. Then, the hazard quantile function of X is of the linear form*

$$H(u) = a + bu,\ a > 0, \tag{2.32}$$

for all $0 < u < 1$, if and only if

$$Q(u) = \log \left(\frac{a+bu}{a(1-u)} \right)^{\frac{1}{a+b}}. \tag{2.33}$$

Proof. When X has its quantile function as in (2.33), we have

$$q(u) = [(1-u)(a+bu)]^{-1}$$

and so

$$H(u) = a + bu.$$

To prove the sufficiency, we obtain from (2.32) that

$$q(u) = [(1-u)(a+bu)]^{-1},$$

and upon integrating it from 0 to u, we get

$$Q(u) = (a+b)^{-1} \log \left(\frac{a+bu}{1-u} \right) + C.$$

Setting $u = 0$, the condition $Q(0) = 0$ readily gives $C = -\frac{\log a}{a+b}$, and consequently (2.33) holds. We notice that (2.33) represents a family of distributions with some well-known models as particular cases. For the special case when $b = 0$, $a = \lambda^{-1}$, $\lambda > 0$, we have the exponential distribution; for the case when $a = b = (2\sigma)^{-1} > 0$, we have the half-logistic distribution; the case $a = \lambda(1-p)^{-1} > 0$, $b = -\frac{p\lambda}{1-p} = -pa < 0$ corresponds to the exponential geometric, and finally $a = \frac{1}{\alpha} > 0$, $b = -\frac{1}{\alpha} = -1 < 0$ leads to Pareto II distribution with parameter $(\alpha, 1)$. □

The above characterization theorem can be used to identify the adequacy of the model for a given data in the following manner. Using (2.29), we obtain the empirical version of $H(u)$ as

$$\bar{H}(u) = [(1-u)\bar{q}(u)]^{-1}.$$

If the points $(u, \bar{H}(u))$ plotted on a graph for different values of u in $(0,1)$ lie approximately on a straight line, it suggests the distribution in (2.33). The specific member of the model is identified on the basis of the estimates of a and b derived from the plot, for example.

The family of distributions in (2.33) will be referred to in the sequel as the linear hazard quantile family. It is easy to invert (2.33) to obtain the distribution function as

$$F(x) = \frac{1-e^{-(a+b)x}}{1+\frac{b}{a}e^{-(a+b)x}} \quad x > 0,\ a > 0.$$

2.4 Mean Residual Quantile Function

Recall that the mean residual life function is defined as

$$m(x) = \frac{1}{\bar{F}(x)} \int_x^\infty t f(t) dt - x.$$

In terms of quantiles, the mean residual quantile function is thus given by

$$\begin{aligned} M(u) = m(Q(u)) &= \frac{1}{1-u} \int_u^1 Q(p) dp - Q(u) \\ &= (1-u)^{-1} \int_u^1 [Q(p) - Q(u)] dp. \end{aligned} \tag{2.34}$$

The same expression can also be obtained from (1.4). Also

$$M(u) = (1-u)^{-1} \int_u^1 (1-p) q(p) dp. \tag{2.35}$$

We interpret $M(u)$ as the average remaining life beyond the $100(1-u)\%$ of the distribution.

Equivalence of (2.34) and (2.35) is readily verified by integrating by parts the RHS of (2.35). From (2.35) and the definition of the hazard quantile function $H(u)$ in (2.29), we have

$$M(u) = (1-u)^{-1} \int_u^1 \frac{dp}{H(p)}. \tag{2.36}$$

Differentiating (2.35) with respect to u, we obtain

$$(1-u)q(u) = M(u) - (1-u)M'(u)$$

or

$$[H(u)]^{-1} = M(u) - (1-u)M'(u), \tag{2.37}$$

where $M'(u)$ is the derivative of $M(u)$ with respect to u. The last two equations determine $M(u)$ from $H(u)$ and vice versa. Finally, the distribution of X is recovered from $M(u)$ when (2.37) is inserted into (2.31) as

Table 2.5 Mean residual quantile functions

Distribution	$M(u)$
Exponential	λ^{-1}
Pareto II rescaled	$\frac{\alpha}{c-1}(1-u)^{-\frac{1}{c}}$
Beta	$\frac{R}{c+1}(1-u)^{\frac{1}{c}}$
Half-logistic	$\frac{2\sigma}{1-u}\log\frac{2}{1+u}$
Exponential geometric	$\frac{1-p}{\lambda p(1-u)}\log\frac{1-pu}{1-p}$
Power	$\frac{\alpha}{1-u}[1-u^{\frac{1}{\beta}}-(\beta+1)^{-1}(1-u^{1+\frac{1}{\beta}})]$
Generalized Pareto	$b(1-u)^{-\frac{a}{a+1}}$

$$
\begin{aligned}
Q(u) &= \int_0^u \frac{M(p)-(1-p)M'(p)}{1-p}dp \\
&= \int_0^u \frac{M(p)}{1-p}dp - M(u) + \mu \quad \text{since } M(0)=\mu. \qquad (2.38)
\end{aligned}
$$

Since the quantile density function also specifies the distribution, a simpler formula is

$$
q(u) = (1-u)^{-1}M(u) - M'(u). \qquad (2.39)
$$

Example 2.13. Let

$$
Q(u) = u^{\theta+1}[1+\theta(1-u)], \quad \theta > 0.
$$

Then, we have

$$
q(u) = u^{\theta}[1+\theta(\theta+1)(1-u)]
$$

and so

$$
\begin{aligned}
M(u) &= \int_u^1 (p^{\theta}-p^{\theta+1})[1+\theta(\theta+1)(1-p)]\,dp \\
&= \frac{3+\theta(2\theta+3)}{(\theta+1)(\theta+2)(\theta+3)} - \frac{u^{\theta+1}}{(\theta+1)(\theta+2)} \\
&\quad \times \Bigg[\{(1+(1-u)(1+\theta))(1+\theta(\theta+1)(1-u))\} \\
&\quad - \frac{\theta(\theta+1)}{\theta+3}u(2+(1-u)(1+\theta))\Bigg].
\end{aligned}
$$

The expressions for $M(u)$ for most distributions are quite complicated involving special functions. A few simple cases are presented in Table 2.5.

We now prove a characterization result by the functional form of $M(u)$.

Theorem 2.6. *A lifetime random variable X has the linear hazard quantile distribution in* (2.33) *if and only if, for all* $0 < u < 1$,

$$M(u) = \frac{1}{b(1-u)} \log\left(\frac{a+b}{a+bu}\right). \tag{2.40}$$

Proof. From Theorem 2.5, we use the expression for $q(u)$ to write

$$M(u) = \frac{1}{1-u} \int_u^1 (1-p)[(1-p)(a+bp)]^{-1} dp$$

which leads to (2.40) and the 'only if' part. Conversely, from (2.40), we have

$$M'u) = \frac{1}{b}\left[\frac{1}{(1-u)^2} \log \frac{a+b}{a+bu} - \frac{b}{(1-u)(a+bu)}\right]$$

and so (2.37) yields

$$H(u) = a + bu.$$

Hence, by Theorem 2.4, X has its quantile function as (2.33). The special cases are slightly different from Theorem 2.4 and so need enumeration. Firstly, for the Pareto II case with parameter $(\alpha, 1)$, the mean does not exist and so it is not a member of the class for which (2.40) is true.

Secondly, the case of the exponential distribution needs the evaluation of $M(u)$ as a limit when $b \to 0$ using L'Hospital rule. In fact,

$$\begin{aligned} M(u) &= \lim_{b \to 0} \frac{1}{b(1-u)} [\log(a+b) - \log(a+bu)] \\ &= \lim_{b \to 0} \frac{1}{1-u}\left[\frac{1}{a+b} - \frac{u}{a+bu}\right] \\ &= \frac{1}{a}. \end{aligned}$$

The other cases are as in Theorem 2.4 and this completes the proof. □

Another characterization is, as in the distribution function approach, from the relationship between $M(u)$ and $H(u)$.

Theorem 2.7. *The relationship*

$$M(u) = (1-u)^{-1}[A + B \log H(u)], \tag{2.41}$$

for all $0 < u < 1$, holds for a lifetime random variable X if and only if it has linear hazard quantile distribution in (2.33).

Proof. First, we assume X has distribution specified by (2.33). Then,

$$\begin{aligned} M(u) &= \frac{1}{b(1-u)}[\log(a+b) - \log(a+bu)] \\ &= (1-u)^{-1}[A + B\log H(u)], \end{aligned}$$

where $A = b^{-1}\log(a+b)$ and $B = -b^{-1}$. Conversely, let (2.41) hold. Then

$$\int_u^1 (1-p)q(p)\,dp = A + B\log H(u)$$

which yields, on differentiation,

$$-(1-u)q(u) = \frac{B}{H(u)}H'(u)$$

or

$$-\frac{1}{H(u)} = \frac{BH'(u)}{H(u)}$$

giving $H(u)$ as a linear function. This completes the proof. □

2.5 Residual Variance Quantile Function

From the definition of the variance residual life function in (2.10), the corresponding quantile-based function takes on the form

$$\begin{aligned} V(u) = \sigma^2(Q(u)) &= (1-u)^{-1}\int_u^1 Q^2(p)dp - [(1-u)^{-1}\int_u^1 Q(p)dp]^2 \\ &= (1-u)^{-1}\int_u^1 Q^2(p)dp - (M(u)+Q(u))^2. \end{aligned} \tag{2.42}$$

Since

$$(1-u)(M(u)+Q(u)) = \int_u^1 Q(p)dp, \tag{2.43}$$

upon differentiating (2.43), we obtain

$$(1-u)(M'(u)+Q'(u)) - (M(u)+Q(u)) = -Q(u)$$

giving

$$M(u) = (1-u)(M'(u)+Q'(u)). \tag{2.44}$$

Also, from (2.42), we have

$$(1-u)V(u) = \int_u^1 Q^2(p)dp - (1-u)(M(u)+Q(u))^2.$$

Differentiating this and on using (2.44), we obtain

$$\begin{aligned}(1-u)V'(u) - V(u) &= Q^2(u) - 2(1-u)(M(u)+Q(u))(M'(u)+Q'(u)) \\ &\quad +(M(u)+Q(u))^2 \\ &= -M^2(u),\end{aligned}$$

where $V'(u)$ is the derivative of $V(u)$ with respect to u. Thus, the mean residual quantile function is determined from the residual variance quantile function as

$$M^2(u) = V(u) - (1-u)V'(u). \tag{2.45}$$

Integrating this over $(u,1)$, we get

$$V(u) = (1-u)^{-1}\int_u^1 M^2(p)\,dp. \tag{2.46}$$

Equation (2.46) expresses the fact that the residual variance quantile function is determined from the mean residual quantile function. In view of the characterization of the distribution by $M(u)$, it follows from (2.45) and (2.46) that $V(u)$ characterizes the life distribution. This result is stronger than the one currently available in the literature and mentioned earlier in Sect. 2.1.3. Notice also that both $h(x)$ and $m(x)$ are present in the identity (2.12), but (2.46) needs only the knowledge of $M(u)$ to derive $V(u)$. With simple forms of $M(u)$ being not available for many common distributions, and even where they are available, the integral in (2.46) leads to no closed-form solutions, characterizations are rare to find.

The coefficient of variation $C^*(u)$ is defined in the quantile formulation as

$$C^{*2}(u) = \frac{V(u)}{M^2(u)}. \tag{2.47}$$

Then, we have

$$\begin{aligned}\frac{1}{C^{*2}(u)} &= \frac{V(u)-(1-u)V'(u)}{V(u)} \\ &= 1-(1-u)\frac{V'(u)}{V(u)}\end{aligned}$$

or

$$\frac{d \log V(u)}{du} = (1-u)^{-1}(1-(C^*(u))^{-2}). \tag{2.48}$$

Example 2.14. For the generalized Pareto distribution, we have

$$Q(u) = \frac{b}{a}[(1-u)^{-\frac{a}{a+1}} - 1],$$
$$M(u) = b(1-u)^{-\frac{a}{a+1}}.$$

Hence, from (2.46), we find

$$\begin{aligned} V(u) &= (1-u)^{-1} \int_u^1 b^2 (1-p)^{-\frac{2a}{a+1}} dp \\ &= \frac{1+a}{1-a} b^2 (1-u)^{-\frac{2a}{a+1}}; \end{aligned}$$

also, we have

$$V(u) = KM^2(u), \quad K = \frac{1+a}{1-a}.$$

Compare these with the results in Example 2.5 when the definitions based on the distribution function are applied to the corresponding functions. The coefficient of variation in this case is $C(u) = K^{\frac{1}{2}}$, a constant.

2.6 Other Quantile Functions

We briefly mention some other quantile functions required for the discussions in the sequel. The αth percentile residual quantile function is obtained from (2.19) as

$$P_\alpha(u) = p_\alpha(Q(u)) = Q[1-(1-\alpha)(1-u)] - Q(u). \tag{2.49}$$

Lillo [399] has pointed out that $Q(u)$ is uniquely determined from the knowledge of $P_\alpha(u)$ and the quantile function in an interval, viz.,

$$G(u) = Q(u), \quad 0 \le u \le B_1 = W_\alpha(0),$$

where $G(u)$ is a continuous increasing function defined on $[0, W_\alpha(0))$, satisfying $G(0) = 0$ and $W_\alpha(u) = Q[1-(1-\alpha)(1-u)]$.

Table 2.6 Reversed hazard quantile functions of distributions in Table 2.3

Distribution	$Q(u)$	$\Lambda(u)$
Power	$\alpha u^{\frac{1}{\beta}}$	$\beta(\alpha u^{\frac{1}{\beta}})^{-1}$
Reciprocal exponential	$-\frac{1}{\lambda}(\log u)^{-1}$	$\lambda^{-1}(\log u)^2$
Reciprocal beta	$[R(1-u^{\frac{1}{c}})]^{-1}$	$Rc(1-u^{\frac{1}{c}})^2 u^{-\frac{1}{c}}$
Reciprocal Pareto II (Lomax)	$[\alpha(u^{-\frac{1}{c}}-1)]^{-1}$	$c\alpha(1-u^{\frac{1}{c}})^2 u^{-\frac{1}{c}}$
Reciprocal Weibull	$\sigma(-\log u)^{\frac{1}{\lambda}}$	$\sigma\lambda(-\log u)^{1+\frac{1}{\lambda}}$
Generalized exponential	$-\lambda^{-1}\log(1-u^{\frac{1}{\theta}})$	$\lambda(1-u^{\frac{1}{\theta}})u^{-\frac{1}{\theta}}$
Burr	$(u^{-\frac{1}{k}}-1)^{-\frac{1}{c}}$	$ck(u^{\frac{1}{k}}-1)^{1+\frac{1}{c}}u^{\frac{1}{k}}$
Generalized power	$(1-u^{\frac{1}{\theta}})^{-\frac{1}{\beta}}$	$\theta\beta(1-u^{\frac{1}{\theta}})^{1+\frac{1}{\beta}}u^{-\frac{1}{\theta}}$
Negative Weibull	$(1-\log u^{\frac{1}{\theta}})^{-\frac{1}{\beta}}$	$\beta\theta(1-\log u^{\frac{1}{\theta}})^{1+\frac{1}{\beta}}$

Various reliability functions in reversed time can also be defined in a manner similar to those in Sect. 2.2. Since the algebra is almost parallel, we give only the relevant results.

The reversed hazard quantile function is

$$\Lambda(u) = \lambda(Q(u)) = [uq(u)]^{-1},$$

and it determines the distribution through the formula

$$Q(u) = \int_0^u [p\Lambda(p)]^{-1} dp. \tag{2.50}$$

The reversed hazard quantile functions of some distributions are presented in Table 2.6.

Similarly, the reversed mean residual quantile function is given by

$$\begin{aligned} R(u) = r(Q(u)) &= u^{-1} \int_0^u [Q(u) - Q(p)] dp \\ &= u^{-1} \int_0^u p q(p) dp. \end{aligned} \tag{2.51}$$

Furthermore, we have

$$Q(u) = R(u) + \int_0^u p^{-1} R(p) dp, \tag{2.52}$$

$$\begin{aligned} [\Lambda(u)]^{-1} &= R(u) + uR'(u), \\ R(u) &= u^{-1} \int_0^u [\Lambda(p)]^{-1} dp, \\ H(u) &= (1-u)^{-1} u \Lambda(u), \\ (1-u)M(u) &= \mu + uR(u) - Q(u). \end{aligned}$$

The reversed variance residual quantile function given by

$$D(u)=u^{-1}\int_0^u Q^2(p)dp-(Q(u)-R(u))^2 \tag{2.53}$$

satisfies the relation

$$R^2(u)=D(u)+uD'(u),$$

and so

$$D(u)=u^{-1}\int_0^u R^2(p)dp.$$

where $D'(u)$ is the derivative of $D(u)$ with respect to u.

Example 2.15. The one-parameter family with

$$Q(u)=u^{1+\theta}(\theta u+1-\theta),\quad 0\le u\le 1;\, 0<\theta\le 1,$$

has its quantile density function as

$$q(u)=u^{\theta}(\theta u+(\theta u+(\theta u+1-\theta)(1+\theta)).$$

Hence,

$$\Lambda(u)=[u^{\theta+1}(1+\theta^2+\theta(\theta+2)u)]$$

and so

$$\begin{aligned}R(u)&=\int_0^u p\,q(p)dp\\&=\frac{u^{\theta+1}}{(\theta+2)(\theta+3)}[(1-\theta^2)(3+\theta)+(\theta+2)^2u].\end{aligned}$$

Two other important concepts of interest in quantile-based reliability theory are the total time on test transforms and the *L*-moments of residual life. These will be discussed separately in Chaps. 5 and 6, respectively.

Chapter 3
Quantile Function Models

Abstract One of the objectives of quantile-based reliability analysis is to make use of quantile functions as models in lifetime data analysis. Accordingly, in this chapter, we discuss the characteristics of certain quantile functions known in the literature. The models considered are the generalized lambda distribution of Ramberg and Schmeiser, the generalized Tukey lambda family of Freimer, Kollia, Mudholkar and Lin, the four-parameter distribution of van Staden and Loots, the five-parameter lambda family and the power-Pareto model of Gilchrist, the Govindarajulu distribution and the generalized Weibull family of Mudholkar and Kollia.

The shapes of the different systems and their descriptive measures of location, dispersion, skewness and kurtosis in terms of conventional moments, L-moments and percentiles are provided. Various methods of estimation based on moments, percentiles, L-moments, least squares and maximum likelihood are reviewed. Also included are the starship method, the discretized approach specifically introduced for the estimation of parameters in the quantile functions and details of the packages and tables that facilitate the estimation process.

In analysing the reliability aspects, one also needs various functions that describe the ageing phenomenon. The expressions for the hazard quantile function, mean residual quantile function, variance residual quantile function, percentile residual life function and their counter parts in reversed time given in the preceding chapters provide the necessary tools in this direction. Some characterization theorems show the relationships between reliability functions unique to various distributions. Applications of selected models and the estimation procedures are also demonstrated by fitting them to some data on failure times.

3.1 Introduction

Probability distributions facilitate characterization of the uncertainty prevailing in a data set by identifying the patterns of variation. By summarizing the observations into a mathematical form that contains a few parameters, distributions also provide

N.U. Nair et al., *Quantile-Based Reliability Analysis*, Statistics for Industry and Technology, DOI 10.1007/978-0-8176-8361-0_3,

means to analyse the basic structure that generates the observations. In finding appropriate distributions that adequately describe a data set, there are in general two approaches. One is to make assumptions about the physical characteristics that govern the data generating mechanism and then to find a model that satisfies such assumptions. This can be done either by deriving the model from the basic assumptions and relations or by adapting one of the conventional models from other disciplines, such as physical, biological or social sciences with appropriate conceptual interpretations. These theoretical models are later tested against the observations by the use of a goodness-of-fit test, for example. A second approach to modeling is entirely data dependent. Models derived in this manner are called empirical or black box models. In situations wherein there is a lack of understanding of the data generating process, the objective is limited to finding the best approximation to the data or because of the complexity of the model involved, a distribution is selected to fit the data. The usual procedure in such cases is to first make a preliminary assessment of the features of the available observations and then decide upon a mathematical formulation of the distribution that can approximate it. Empirical modelling problems usually focus attention on flexible families of distributions with enough parameters capable of producing different shapes and characteristics. The Pearson family, Johnson system, Burr family of distributions, and some others, which include several commonly occurring distributions, provide important aids in this regard. In this chapter, we discuss some families of distributions specified by their quantile functions that can be utilized for modelling lifetime data. Various quantile-based properties of distributions and concepts in reliability presented in the last two chapters form the background material for the ensuing discussion. The main distributions discussed here are the lambda distributions, power-Pareto model, Govindarajulu distribution and the generalized Weibull family. We also demonstrate that these models can be used as lifetime distributions while modelling real lifetime data.

3.2 Lambda Distributions

A brief historical account of developments on the lambda distributions was provided in Sect. 1.1. During the past 60 years, considerable efforts were made to generalize the basic model of Hastings et al. [264] and Tukey [567] and also to find new applications and inferential procedures. In general, the applications of different versions span a variety of fields such as inventory control (Silver [540]), logistic regression (Pregibon [497]), meteorology (Osturk and Dale [476]), survival analysis (Lefante Jr. [380]), queueing theory (Robinson and Chan [508]), random variate generation and goodness-of-fit tests (Cao and Lugosi [128]), fatigue studies (Bigerelle et al. [100]) process control (Fournier et al. [200]), biochemistry (Ramos-Fernandez et al. [505]), economics (Haritha et al. [260]), corrosion (Najjar et al. [456]) and reliability analysis (Nair and Vineshkumar [452]).

The basic model from which all other generalizations originate is the Tukey lambda distribution with quantile function

$$Q(u) = \frac{u^\lambda - (1-u)^\lambda}{\lambda}, \quad 0 \le u \le 1, \tag{3.1}$$

defined for all non-zero lambda values. As $\lambda \to 0$, we have

$$Q(u) = \log\left(\frac{u}{1-u}\right)$$

corresponding to the logistic distribution. van Dyke [571] compared a normalized version of (3.1) with the t-distribution. Model (3.1) was studied by Filliben [197] who used it to approximate symmetric distributions with varying tail weights. Joiner and Rosenblatt [304] studied the sample range and Ramberg and Schmeiser [503] discussed the application of the distribution in generating symmetric random variables. For $\lambda = 1$ and $\lambda = 2$, it is easy to verify that (3.1) becomes uniform over $(-1,1)$ and $(-\frac{1}{2}, \frac{1}{2})$, respectively. The density functions are U shaped for $1 < \lambda < 2$ and unimodal for $\lambda < 1$ or $\lambda > 2$. With (3.1) being symmetric and having range for negative values of X, it has limited use as a lifetime model.

Remark 3.1. The Tukey lambda distribution defined in (3.1) is an extremal distribution that gets characterized by means of largest order statistics. To see this, let $X_{1:n} < \cdots < X_{n:n}$ be the order statistics from a random sample of size n from a symmetric distribution F with mean 0 and variance σ^2. Then, due to the symmetry of the distribution, we have $E(X_{n:n}) = -E(X_{1:n})$, and so we can write from (1.23) and (1.24) that

$$E(X_{n:n}) = \frac{1}{2}\int_0^1 Q(u)n(u^{n-1} - (1-u)^{n-1})du. \tag{3.2}$$

By applying Cauchy–Schwarz inequality to (3.2), we readily find

$$\begin{aligned} E(X_{n:n}) &\le \frac{\sigma}{2}\left\{\int_0^1 n^2\left(u^{2n-2} + (1-u)^{2n-2} - 2u^{n-1}(1-u)^{n-1}\right)du\right\}^{1/2} \\ &= \frac{\sigma n}{\sqrt{2}}\left\{\frac{1}{2n-1} - B(n,n)\right\}^{1/2}, \end{aligned} \tag{3.3}$$

where $B(a,b) = \Gamma(a)\Gamma(b)/\Gamma(a+b)$, $a,b > 0$, is the complete beta function. Note that, from (3.3), by setting $n = 2$ and $n = 3$, we obtain the bounds

$$E(X_{2:2}) \le \frac{\sigma}{\sqrt{3}} \quad \text{and} \quad E(X_{3:3}) \le \frac{\sigma\sqrt{3}}{2}.$$

The bound in (3.3) was established originally by Hartley and David [263] and Gumbel [229]. It is useful to note that the bound in (3.3), derived from (3.2), is attained if and only if

$$Q(u) \propto u^{n-1} - (1-u)^{n-1}, \; u \in (0,1).$$

When $n = 2$ and 3, we thus find $Q(u) \propto 2u - 1$, which corresponds to the uniform distribution; see Balakrishnan and Balasubramanian [50] for some additional insight into this characterization result. Thus, we observe from (3.2) that the Tukey lambda distribution with integral values of λ is an extremal distribution and is characterized by the mean of the largest order statistic in (3.3). The same goes for the Tukey lambda distribution in (3.1) for positive real values in terms of fractional order statistics, in view of Remark 1.1.

3.2.1 Generalized Lambda Distribution

Asymmetric versions of (3.1) in various forms such as

$$Q(u) = Au^{\lambda} + B(1-u)^{\theta} + C$$

and

$$Q(u) = au^{\lambda} - (1-u)^{\lambda}$$

were studied subsequently (Joiner and Rosenblatt [304], Shapiro and Wilk [536]). All such versions are subsumed in the more general form

$$Q(u) = \lambda_1 + \frac{1}{\lambda_2}(u^{\lambda_3} - (1-u)^{\lambda_4}) \tag{3.4}$$

introduced by Ramberg and Schmeiser [503], which is called the generalized lambda distribution. This is the most discussed member of the various lambda distributions, because of its versatility and special properties. In (3.4), λ_1 is a location parameter, λ_2 is a scale parameter, while λ_3 and λ_4 determine the shape. The distribution takes on different supports depending on the parameters λ_2, λ_3 and λ_4, while λ_1, being the location parameter, can take values on the real line in all cases (Table 3.1).

As a life distribution, the required constraint on the parameters is

$$Q(0) = \lambda_1 - \frac{1}{\lambda_2} \geq 0.$$

The quantile density function is

$$q(u) = \lambda_2^{-1}[\lambda_3 u^{\lambda_3 - 1} + \lambda_4(1-u)^{\lambda_4 - 1}] \tag{3.5}$$

and accordingly the density quantile function is

$$f(Q(u)) = \lambda_2[\lambda_3 u^{\lambda_3 - 1} + \lambda_4(1-u)^{\lambda_4 - 1}]^{-1} \tag{3.6}$$

Table 3.1 Supports for the generalized lambda distribution

Region	λ_2	λ_3	λ_4	Support
1	<0	<-1	>1	$(-\infty, \lambda_1 + \frac{1}{\lambda_2})$
2	<0	>1	<-1	$(\lambda_1 - \frac{1}{\lambda_2}, \infty)$
	>0	>0	>0	$(\lambda_1 - \frac{1}{\lambda_2}, \lambda_1 + \frac{1}{\lambda_2})$
3	>0	$=0$	>0	$(\lambda_1, \lambda_1 + \frac{1}{\lambda_2})$
	>0	>0	$=0$	$(\lambda_1 - \frac{1}{\lambda_2}, \lambda_2)$
	<0	<0	<0	$(-\infty, \infty)$
4	<0	$=0$	<0	(λ_1, ∞)
	<0	<0	$=0$	$(-\infty, \lambda_1)$

which has to remain non-negative for (3.4) to represent a proper distribution. This places constraints on the parameter space. A special feature of (3.4) is that it is a valid distribution only in the regions ($\lambda_3 \leq -1$, $\lambda_4 \geq 1$), ($\lambda_3 \geq 1$, $\lambda_4 \leq -1$), ($\lambda_3 \geq 0$, $\lambda_4 \geq 0$), ($\lambda_3 \leq 0$, $\lambda_4 \leq 1$), and for values in ($-1 < \lambda_3 < 0$, $\lambda_4 > 0$) for which

$$\frac{(1-\lambda_3)^{1-\lambda_3}}{(\lambda_4-\lambda_3)^{\lambda_4-\lambda_3}}(\lambda_4-1)^{\lambda_4-1} < -\frac{\lambda_3}{\lambda_4},$$

and values in $(\lambda_3 > 1, -1 < \lambda_4 < 0)$ for which

$$\frac{(1-\lambda_4)^{1-\lambda_4}}{(\lambda_3-\lambda_4)^{\lambda_3-\lambda_4}}(\lambda_3-1)^{\lambda_3-1} < -\frac{\lambda_4}{\lambda_3};$$

see Karian and Dudewicz [314] for a detailed study in this respect. Since

$$E(X^r) = \int_0^1 \left[\lambda_1 + \frac{p^{\lambda_3} - (1-p)^{\lambda_4}}{\lambda_2}\right]^r dp$$

from (1.30), the mean is simply

$$E(X) = \mu = \lambda_1 + \frac{1}{\lambda_2}\left(\frac{1}{\lambda_3+1} - \frac{1}{\lambda_4+1}\right). \tag{3.7}$$

Since λ_1 is not present in the central moments, we set $\lambda_1 = 0$. Ramberg et al. [502] find that

$$E(X^r) = \lambda_2^{-r}\sum_{i=0}^{r}\binom{r}{i}(-1)^i B(\lambda_3(r-i)+1, \lambda_4 i+1)$$

from which we obtain the following central moments:

$$\sigma^2 = \frac{B - A^2}{\lambda_2^2}, \tag{3.8}$$

$$\mu_3 = \frac{C - 3AB + A^3}{\lambda_2^3}, \tag{3.9}$$

$$\mu_4 = \frac{D - 4AC + 6A^2B - 3A^4}{\lambda_2^4}, \tag{3.10}$$

where

$$A = \frac{1}{\lambda_3 + 1} - \frac{1}{\lambda_4 + 1},$$

$$B = \frac{1}{2\lambda_3 + 1} + \frac{1}{2\lambda_4 + 1} - 2B(\lambda_3 + 1, \lambda_4 + 1),$$

$$C = \frac{1}{3\lambda_3 + 1} - 3B(2\lambda_3 + 1, \lambda_4 + 1) + 3B(\lambda_3 + 1, 2\lambda_4 + 1) - \frac{1}{3\lambda_4 + 1},$$

and

$$D = \frac{1}{4\lambda_3 + 1} - 4B(2\lambda_3 + 1, \lambda_4 + 1) + 6B(2\lambda_3 + 1, 2\lambda_4 + 1) - 4B(\lambda_3 + 1, 3\lambda_4 + 1) + \frac{1}{4\lambda_4 + 1}.$$

The rth moment exists only if $-\frac{1}{r} < \min(\lambda_3, \lambda_4)$. When $\lambda_3 = \lambda_4$, it is verified that $\mu_3 = 0$ and the generalized lambda distribution is symmetric in this case. A detailed study of the skewness and kurtosis for different values of λ_3 and λ_4 is given in Karian and Dudewicz [315]. The (β_1, β_2) diagram includes the skewness values corresponding to the uniform, t, F, normal, Weibull, lognormal and some beta distributions. One limitation that needs to be mentioned regarding skewness is that the generalized lambda family does not cover the entire area as some other systems (like the Pearson system) do; but, it also covers some new areas that are not covered by others. This four-parameter distribution includes a wide range of shapes for its density function; see Fig. 3.1 for some selection shapes.

The basic characteristics of the distribution can also be expressed in terms of the percentiles. Using (1.6)–(1.9), we have the following:

the median

$$M = \lambda_1 + \frac{1}{\lambda_2}\left[\left(\frac{1}{2}\right)^{\lambda_3} - \left(\frac{1}{2}\right)^{\lambda_4}\right], \tag{3.11}$$

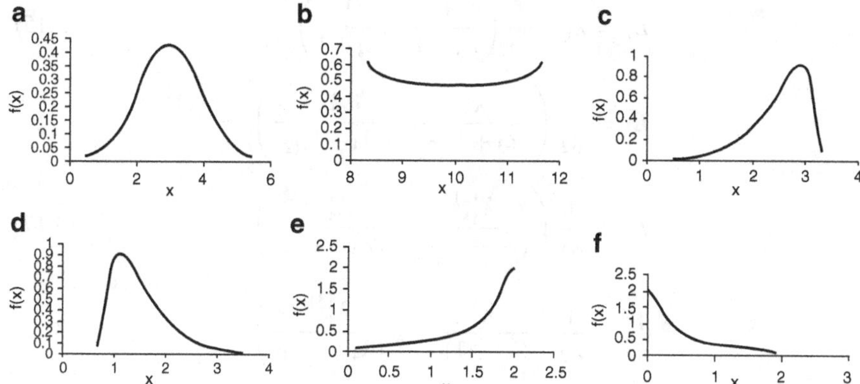

Fig. 3.1 Density plots of the generalized lambda distribution (Ramberg and Schmeiser [503] model) for different choices of $(\lambda_1, \lambda_2, \lambda_3, \lambda_4)$. **(a)** (1,0.2,0.13,0.13); **(b)** (1,0.6,1.5,-1.5); **(c)** (1,0.6,1.75,1.2); **(d)** (1,0.2,0.13,0.013); **(e)** (1,0.2,0.0013,0.13); **(f)** (1,1,0.5,4)

the interquantile range

$$\mathrm{IQR} = \frac{1}{\lambda_2}\left[\frac{3^{\lambda_3}-1}{4^{\lambda_3}} + \frac{3^{\lambda_4}-1}{4^{\lambda_4}}\right], \tag{3.12}$$

Galton's measure of skewness

$$S = \frac{4^{-\lambda_3}(3^{\lambda_3} - 2^{\lambda_3+1} - 1) - 4^{\lambda_4}(1 + 3^{\lambda_4} - 2^{\lambda_4+1})}{\frac{3^{\lambda_3}-1}{4^{\lambda_3}} + \frac{3^{\lambda_4}-1}{4^{\lambda_4}}}, \tag{3.13}$$

and Moors' measure of kurtosis

$$T = \frac{8^{-\lambda_3}(1 + 3^{\lambda_3} + 5^{\lambda_3} + 7^{\lambda_3}) - 8^{-\lambda_4}(1 + 3^{\lambda_4} + 5^{\lambda_4} + 7^{\lambda_4})}{4^{-\lambda_3}(3^{\lambda_3} - 1) + 4^{-\lambda_4}(3^{\lambda_4} - 1)}. \tag{3.14}$$

For this distribution, the L-moments have comparatively simpler expressions than the conventional moments. One can use (1.34)–(1.37) to calculate these. To simplify their expressions, we employ the notation

$$(n)_{(r)} = n(n+1)\cdots(n+r-1)$$

and

$$(n)^{(r)} = n(n-1)\cdots(n-r+1)$$

to denote the ascending and descending factorials, respectively. Then, the first four L-moments are as follows (Asquith [40]):

$$L_1 = \lambda_1 + \frac{1}{\lambda_2}\left(\frac{1}{\lambda_3+1} - \frac{1}{\lambda_4+1}\right), \tag{3.15}$$

$$L_2 = \frac{1}{\lambda_2}\left(\frac{\lambda_3}{(\lambda_3+1)_{(2)}} + \frac{\lambda_4}{(\lambda_4+1)_{(2)}}\right), \tag{3.16}$$

$$L_3 = \frac{1}{\lambda_2}\left(\frac{\lambda_3^{(2)}}{(\lambda_3+1)_{(3)}} - \frac{\lambda_4^{(2)}}{(\lambda_4+1)_{(3)}}\right), \tag{3.17}$$

$$L_4 = \frac{1}{\lambda_2}\left(\frac{\lambda_3^{(3)}}{(\lambda_3+1)_{(4)}} - \frac{\lambda_4^{(3)}}{(\lambda_4+1)_{(4)}}\right). \tag{3.18}$$

Thus, the L-skewness and L-kurtosis become

$$\tau_3 = \frac{\lambda_3^{(2)}(\lambda_4+1)_{(3)} - \lambda_4^{(2)}(\lambda_3+1)_{(3)}}{\lambda_3(\lambda_3+3)(\lambda_4+1)_{(3)} + \lambda_4(\lambda_4+3)(\lambda_3+1)_{(3)}} \tag{3.19}$$

and

$$\tau_4 = \frac{(\lambda_3)^{(3)}(\lambda_4+1)_{(4)} + (\lambda_4)^{(3)}(\lambda_3+1)_{(4)}}{\lambda_3(\lambda_3+3)(\lambda_3+4)(\lambda_4+1)_{(4)} - \lambda_4(\lambda_4+3)(\lambda_3+1)_{(4)}}. \tag{3.20}$$

All the L-moments exist for every $\lambda_3, \lambda_4 > -1$. On the other hand, the conventional moments require $\lambda_3, \lambda_4 > -\frac{1}{4}$ for the evaluation of Pearson's skewness β_1 and kurtosis β_2. Thus, L-skewness and kurtosis permit a larger range of values in the parameter space. The problem of characterizing the generalized lambda distribution has been considered in Karvanen and Nuutinen [313]. For the symmetric case, they have derived the boundaries analytically and in the general case, numerical methods have been used. They found that with an exception of the smallest values of τ_4, the family (3.4) covers all possible (τ_3, τ_4) pairs and often there are two or more distributions sharing the same τ_3 and τ_4. A wider set of generalized lambda distributions can be characterized when L-moments are used than by the conventional moments. This is an important advantage in the context of data analysis while seeking appropriate models.

The moments of order statistics have closed forms as well. For example, the expectation of order statistics from a random sample of size n is obtained from (1.28) as

$$E(X_{r:n}) = \lambda_1 + \frac{1}{\lambda_2}\frac{\Gamma(\lambda_3+r)}{\Gamma(r)}\frac{\Gamma(n+1)}{\Gamma(\lambda_3+n+1)} + \frac{1}{\lambda_2}\frac{\Gamma(n+\lambda_4-r+1)\Gamma(n+1)}{\Gamma(n+\lambda_4+1)\Gamma(n-r)}. \tag{3.21}$$

In particular, from (3.21), we obtain

$$E(X_{n:n}) = \lambda_1 + \frac{n}{\lambda_2(\lambda_3+n)} - \frac{n!}{\lambda_2(\lambda_4+1)_{(n)}},$$

$$E(X_{1:n}) = \lambda_1 + \frac{n!}{\lambda_2(\lambda_3+1)_{(n)}} - \frac{n}{\lambda_2(n+\lambda_4)}.$$

Also, the distributions of $X_{1:n}$ and $X_{n:n}$ are given by

$$Q_1(u) = \lambda_1 + \frac{1}{\lambda_2}\left[(1-(1-u)^{\frac{1}{n}})^{\lambda_3} - (1-u)^{\frac{\lambda_4}{n}}\right],$$

$$Q_n(u) = \lambda_1 + \frac{1}{\lambda_2}\left[u^{\frac{\lambda_3}{n}} - (1-u^{\frac{1}{n}})^{\lambda_4}\right].$$

Since there exist members of generalized lambda family with support on the positive real line, its scope as a lifetime model is apparent. However, this fact has not been exploited much. The hazard quantile function (2.30) has the simple form

$$H(u) = \frac{\lambda_2}{(1-u)[\lambda_3 u^{\lambda_3-1} + \lambda_4(1-u)^{\lambda_4-1}]}. \tag{3.22}$$

Similarly, the mean residual quantile function is obtained from (2.43) as

$$\begin{aligned} M(u) &= \frac{1}{1-u}\int_u^1 (1-p)q(p)dp \\ &= \frac{1}{\lambda_2(1-u)}\left[\frac{\lambda_4}{\lambda_4+1}(1-u)^{\lambda_4+1} + \frac{1-u^{\lambda_3+1}}{\lambda_3+1} - (1-u)u^{\lambda_3}\right]. \end{aligned}$$

Note that, in this case,

$$M(0) = \int_0^1 Q(p)dp - Q(0)$$

or

$$\mu = Q(0) + M(0).$$

The above expression is a general condition to be used whenever the left end of the support is greater than zero. The variance residual quantile function is calculated as

$$V(u) = \frac{1}{1-u}\int_u^1 Q^2(p)dp - \left[\frac{1}{1-u}\int_u^1 Q(p)dp\right]^2$$
$$= A_1(u) - A_2^2(u),$$

where

$$A_1(u) = \frac{1}{\lambda_2^2(1-u)}\left[\frac{1-u^{2\lambda_3+1}}{2\lambda_3+1} + \frac{(1-u)^{2\lambda_4+1}}{2\lambda_4+1} - 2B_{1-u}(\lambda_4+1, \lambda_3+1)\right],$$

$$A_2(u) = \frac{1}{\lambda_2(1-u)}\left[\frac{1-u^{\lambda_3+1}}{\lambda_3+1} - \frac{(1-u)^{\lambda_4+1}}{\lambda_4+1}\right]$$

and $B_x(m,n) = \int_0^x t^{m-1}(1-t)^{n-1}dt$ is the incomplete beta function.

The term

$$\mu(u) = \frac{1}{1-u}\int_u^1 Q(p)dp \tag{3.23}$$

is of interest in reliability analysis, being the quantile version of $E(X|X > x)$. It is called the conditional mean life or the vitality function. One may refer to Kupka and Loo [363] for a detailed exposition of the properties of the vitality function and its role in explaining the ageing process. We see that from (3.23), $Q(u)$ can be recovered up to an additive constant as

$$Q(u) = -\frac{d}{du}(1-u)\mu(u),$$

and therefore functional forms of $\mu(u)$ will enable us to identify the life distribution. Thus, a generalized lambda distribution is determined as

$$a - \frac{d}{du}(1-u)\mu(u)$$

if the conditional mean quantile function $\mu(u)$ satisfies

$$\mu(u) = a + b\left[\frac{1-u^c}{c} - \frac{(1-u)^d}{d}\right]$$

for real a, b, c and d for which $Q(0) \geq 0$.

The αth percentile residual life is calculated from (2.50) as

$$P_\alpha(u) = \frac{1}{\lambda_2}[(\alpha + u - \alpha u)^{\lambda_3} - u^{\lambda_3} - (1-u)^{\lambda_4}(1-(1-\alpha)^{\lambda_4})].$$

Various functions in reversed time presented in (2.50), (2.51) and (2.53) yield

$$\Lambda(u) = \lambda_2[u(\lambda_3 u^{\lambda_3 - 1} + \lambda_4(1-u)^{\lambda_4 - 1})]^{-1},$$

$$R(u) = \frac{1}{\lambda_2}\left[\frac{\lambda_3}{\lambda_3+1}u^{\lambda_3} - (1-u)^{\lambda_4} + \frac{1-(1-u)^{\lambda_4+1}}{(\lambda_4+1)^u}\right],$$

$$D^*(u) = B_1(u) - B_2^2(u),$$

where

$$B_1(u) = \frac{1}{\lambda^2 u}\left[\frac{u^{2\lambda_3+1}}{2\lambda_3+1} - \frac{(1-u)^{2\lambda_4+1}-1}{2\lambda_4+1} - 2B_u(\lambda_3+1,\lambda_4+1)\right]$$

and

$$B_2(u) = \frac{1}{\lambda_2 u}\left[\frac{u^{\lambda_3+1}}{\lambda_3+1} - \frac{(1-u)^{\lambda_4+1}-1}{\lambda_4+1}\right].$$

Like the function $\mu(u)$, one can also consider

$$\theta(u) = \frac{1}{u}\int_0^u Q(p)dp \tag{3.24}$$

which is the quantile formulation of $E(X|X \le x)$. This latter function's relationship with reversed hazard function has been used in Nair and Sudheesh [451] to characterize distributions. It has applications in several other fields like economics and risk analysis. For example, when X is interpreted as the income and x is the poverty level, the above expectation denotes the average income of the poor people and is an essential component for the evaluation of poverty index and income inequality. The form of (3.24) is convenient in identifying models, like

$$\theta(u) = a + b\left[\frac{u^{c-1}}{c} + \frac{(1-u)^d - 1}{du}\right]$$

determining the generalized lambda distribution. The formula for calculating $Q(u)$ from $\theta(u)$ is

$$Q(u) = a + \frac{d}{du}u\theta(u). \tag{3.25}$$

Finally, the reversed percentile residual life function is (2.50)

$$q_\alpha(u) = \frac{1}{\lambda_2}\left[u^{\lambda_3} - ((1-\alpha)u)^{\lambda_3} - \{(1-u)^{\lambda_4} - (1-(1-\alpha)u)^{\lambda_4}\}\right]$$
$$= \frac{1}{\lambda_2}\left[u^{\lambda_3} - (1-(1-\alpha)^{\lambda_3}) - (1-u)^{\lambda_4} + (1-u+\alpha u)^{\lambda_4}\right].$$

There is no conflict of opinion regarding the potential of the generalized lambda family in empirical data modelling because of its flexibility to represent different kinds of data situations. However, the difficulties experienced in the estimation problem, especially on the computational front, have stimulated extensive research on various methods, conventional as well as new. A popular approach for estimation of parameters of quantile functions is the method of moments, in which the first four moments of the generalized lambda distribution are matched with the corresponding moments of the sample. Instead of choosing the first four moments directly, Ramberg and Schmeiser [504] opted for the equations

$$\mu = \frac{1}{n}\sum_{i=1}^{n} x_i\,, \tag{3.26}$$

$$\sigma^2 = \frac{1}{n}\sum_{i=1}^{n}(x_i - \bar{x})^2\,, \tag{3.27}$$

$$r_1 = \frac{n^{1/2}\sum(x_i - \bar{x})^3}{[\sum(x_i - \bar{x})^2]^{3/2}}\,, \tag{3.28}$$

$$r_2 = \frac{n\sum(x_i - \bar{x})^4}{[\sum(x_i - \bar{x})^2]^2}\,, \tag{3.29}$$

where μ and σ^2 are as given in (3.7) and (3.8), $\gamma_1 = \frac{\mu_3}{\sigma^3}$ and $\gamma_2 = \frac{\mu_4}{\sigma^4}$ with values for μ_3 and μ_4 as in (3.9) and (3.10). Since γ_1 and γ_2 contain only λ_3 and λ_4, the solutions of (3.28) and (3.29) give λ_3 and λ_4. From the remaining two equations, λ_1 and λ_2 can be readily found. Even though theoretically the method looks simple, in practice, one has to apply numerical methods to solve the equations as they are nonlinear. Dudewicz and Karian [181] have provided extensive tables from which the parameters can be determined for a given choice of skewness and kurtosis of the data. They also describe an algorithm that summarizes the steps in the calculation. A second method to obtain a best solution is to use computer programs that ensure the solutions of (3.26)–(3.29) to satisfy

$$\max(|\mu - \hat{\mu}|, |\sigma^2 - \hat{\sigma}^2|, |\gamma_1 - \hat{\gamma}_1|, |\gamma_2 - \hat{\gamma}_2|) < \varepsilon \tag{3.30}$$

for some prefixed tolerance $\varepsilon > 0$. This is accomplished by starting with a good set of initial values for the parameters. Then search is made through algorithms that satisfy (3.30). However, there is no guarantee that a given set of initial values necessarily end up resulting in a solution nor that it improves upon the value of ε in each iteration. See Karian and Dudewicz [315] for such a computational program. In both the methods described above, the region specified by $1+\gamma_1^2 < \gamma_2 < 1.8+1.7\gamma_1^2$ is not attained and one may not arrive at a set of lambda values that satisfy a goodness-of-fit test. These and other problems are explained in Karian and Dudewicz [315, 317].

A similar logic applies to the method of L-moments prescribed in Asquith [40] and Karian and Dudewicz [315]. The equations to be solved in the latter work are

$$L_i = l_i, \quad i = 1,2, \tag{3.31}$$

$$\tau_3 = t_3, \tag{3.32}$$

$$\tau_4 = t_4, \tag{3.33}$$

where L_1, L_2, τ_3 and τ_4 have the expressions in (3.15), (3.16), (3.19) and (3.20), where

$$t_3 = \frac{l_3}{l_2}, \quad t_4 = \frac{l_4}{l_2},$$

$$l_r = \sum_{j=0}^{r-1} p_{rj} b_j, \quad r = 1,2,\ldots,n$$

where

$$b_j = \frac{1}{n} \sum_{i=j+1}^{n} \frac{(i-1)^{(j)}}{(n-1)^{(r)}} x_{j:n}.$$

Clearly, (3.32) and (3.33) do not contain λ_1 and λ_2 and are therefore solvable for λ_3 and λ_4. The other two parameters are then found from (3.31) by using the estimates of λ_3 and λ_4.

In the work of Asquith [40], estimates of λ_3 and λ_4 are values that minimize

$$\varepsilon = (t_3 - \hat{\tau}_3)^2 + (t_4 - \hat{\tau}_4)^2, \tag{3.34}$$

where $\hat{\tau}_i$ $(i = 3,4)$ is the estimated value of τ_i. After choosing initial values of λ_3 and λ_4, we arrive at the optimal value according to (3.34) and then check whether the solutions obtained meet the requirements $-1 < \tau_3 < 1$ and $\frac{1}{4}(5\tau_3^2 - 1) \le \tau_4 < 1$. If not, we need to choose another set of initial values and repeat the above steps. After solving for λ_2 from (3.31), compute $\hat{\tau}_5$ using the expression

$$\tau_5 = \frac{(\lambda_3)^{(4)}(\lambda_4+1)_{(4)} - (\lambda_4)^{(4)}(\lambda_3+1)_{(4)}}{(\lambda_3+3)_{(3)}(\lambda_4+3)_{(3)}[\lambda_3(\lambda_4+1)_{(2)}\lambda_4(\lambda_3+1)_{(2)}]}$$

and seek the values that minimize $(t_5 - \hat{\tau}_5)$. Finally, we need to substitute it into (3.31) to find $\hat{\lambda}_1$.

A third method is to match the percentiles of the distribution with those of the data. As a first step, the sample percentiles are computed as

$$\xi_p = X_{r:n} + \frac{a}{b}(X_{r+1:n} - X_{r:n}),$$

where $(n+1)p = r + \frac{a}{b}$ in which r is a positive integer and $0 < \frac{a}{b} < 1$. Karian and Dudewicz [315] considered the following four equations:

$$\xi_{0.5} = Q(0.5) = \lambda_1 + \frac{(0.5)^{\lambda_3} - (0.5)^{\lambda_4}}{\lambda_2},$$

$$\xi_{0.9} - \xi_{0.1} = Q(0.9) - Q(0.1) = \frac{(0.9)^{\lambda_3} - (0.1)^{\lambda_4} + (0.9)^{\lambda_4} - (0.1)^{\lambda_3}}{\lambda_2},$$

$$\frac{\xi_{0.5} - \xi_{0.1}}{\xi_{0.9} - \xi_{0.5}} = \frac{Q(0.5) - Q(0.1)}{Q(0.9) - Q(0.5)} = \frac{(0.9)^{\lambda_4} - (0.1)^{\lambda_3} + (0.5)^{\lambda_3} - (0.5)^{\lambda_4}}{(0.9)^{\lambda_3} - (0.1)^{\lambda_4} + (0.5)^{\lambda_4} - (0.5)^{\lambda_3}},$$

$$\frac{\xi_{0.75} - \xi_{0.25}}{\xi_{0.9} - \xi_{0.1}} = \frac{Q(0.75) - Q(0.25)}{Q(0.9) - Q(0.5)} = \frac{(0.75)^{\lambda_3} - (0.25)^{\lambda_4} + (0.75)^{\lambda_4} - (0.25)^{\lambda_3}}{(0.9)^{\lambda_3} - (0.1)^{\lambda_4} + (0.9)^{\lambda_4} - (0.1)^{\lambda_3}}.$$

Solving the above system of equations, we obtain the percentile-based estimates. For this purpose, either numerical methods have to be resorted to or refer to the tables in Appendix D of Karian and Dudewicz [315] which gives the values of λ_1, λ_2, λ_3 and λ_4 based on the sample values for the LHS of the above four equations.

In all the three methods discussed so far, the question of more than one set of lambda values in the admissible regions may be possible. The choice of the appropriate set depends on the data and some goodness-of-fit procedure. Karian and Dudewicz [314] compared the relative merits of the two-moment approaches and the percentile method. Using the p-values of the chi-square goodness-of-fit test, the quality of fit was ascertained. They noted that, in general, percentile and L-moment methods gave better fits more frequently. Further, in terms of the L^2-norm, which measures the discrepancy between two functions $f(x)$ and $g(x)$ by

$$\int |g(x) - f(x)|^2 dx,$$

the method of percentiles was found to be better than the method of moments over a broad range of values in the (r_1, r_2) space in samples of size 1,000.

Another useful estimation procedure based on the least-square approach was proposed by Osturk and Dale [477]. Let $X_{r:n}$ $(r = 1, \ldots, n)$ denote the order statistics of the data and $U_{r:n}$ the order statistics of the corresponding uniformly distributed random variable $F(X)$ for $r = 1, 2, \ldots, n$. The least-square method is to find λ_i such that the sum of squared differences between the observed and expected order statistics is minimum. This is achieved by minimizing

$$A(\lambda_1,\lambda_2,\lambda_3,\lambda_4)=\sum_{r=1}^{n}\left\{x_{r:n}-\lambda_1-\frac{1}{\lambda_2}(E(U_{r:n}^{\lambda_3}-(1-U_{r:n})^{\lambda_4}))\right\}^2. \tag{3.35}$$

From the density function of uniform order statistics given by (Arnold et al. [37])

$$f_r(x_r)=\frac{1}{B(r,n-r+1)}x_r^{r-1}(1-x_r)^{n-r},\ 0<x_r<1, \tag{3.36}$$

we have

$$M_r=E(U_{r:n}^{\lambda_3})=\frac{\Gamma(n+1)\Gamma(\lambda_3+r)}{\Gamma(r)\Gamma(n+\lambda_3+1)}=\frac{n!}{(r-1)(\lambda_3+r)_{(n+1)}}$$

and similarly

$$N_r=E(1-U_{r:n})^{\lambda_4}=\frac{n!}{(n-r)!(\lambda_4+n)^{(r)}}.$$

Owing to the difficulties in simultaneously minimizing (3.35) with respect to the four parameters, first minimize (3.35) with respect to λ_1 and λ_2 by treating λ_3 and λ_4 as constants. As in the case of simple linear regression, setting the derivatives of (3.35) to zero, we can solve for λ_1 and λ_2 as

$$\hat{\lambda}_2=\frac{\sum_{r=1}^{n}(x_{r:n}-\bar{x})(v_r-\bar{v})}{\sum_{r=1}^{n}(v_r-\bar{v})^2} \tag{3.37}$$

and

$$\hat{\lambda}_1=\bar{x}=\bar{v}\bar{\lambda}_2, \tag{3.38}$$

where $v_r=M_r-N_r$ and $\bar{v}=\frac{1}{n}\sum v_r$. Then, upon substituting (3.37) and (3.38) in (3.35), we get

$$A(\lambda_3,\lambda_4)=\sum_{r=1}^{n}(x_{r:n-\bar{x}})^2\left[1-\frac{(\sum_{r=1}^{n}(x_{r:n}-\bar{x})(v_r-\bar{v}))^2}{\sum_{r=1}^{n}(v_r-\bar{v})^2\sum_{r=1}^{n}(x_{r:n}-\bar{x})^2}\right].$$

Thus, λ_3 and λ_4 are found by minimizing

$$-\frac{[\sum_{r=1}^{n}(x_{r:n}-\bar{x})(v_r-\bar{v})]^2}{\sum_{r=1}^{n}(v_r-\bar{v})^2\sum_{r=1}^{n}(x_{r:n}-\bar{x})^2}. \tag{3.39}$$

Finally, the solutions from (3.39), when substituted into (3.37) and (3.38), give $\hat{\lambda}_1$ and $\hat{\lambda}_2$.

A second version of percentile method in Karian and Dudewicz [314] proposes equating the population median M, the interdecile range

$$\text{IDR} = Q(1-u) - Q(u),$$

the tail weight ratio

$$\text{TWR} = \frac{Q(\frac{1}{2}) - Q(u)}{Q(1-u) - Q(\frac{1}{2})},$$

and the tail weight factor

$$\text{TWF} = \frac{\text{IQR}}{\text{IDR}}$$

with the corresponding sample quantities. These give rise to the equations

$$\lambda_1 + \frac{(0.5)^{\lambda_3} - (0.5)^{\lambda_4}}{\lambda_2} = m, \tag{3.40}$$

$$\frac{1}{\lambda_2}[(1-u)^{\lambda_3} - u^{\lambda_4} + (1-u)^{\lambda_4} - u^{\lambda_3}] = \xi_{1-u} - \xi_u, \tag{3.41}$$

$$\frac{(1-u)^{\lambda_4} - u^{\lambda_3} + (0.5)^{\lambda_3} - (0.5)^{\lambda_4}}{(1-u)^{\lambda_3} - u^{\lambda_4} + (0.5)^{\lambda_4} - (0.5)^{\lambda_3}} = \frac{\xi_{0.5} - \xi_u}{\xi_{1-u} - \xi_u}, \tag{3.42}$$

$$\frac{(0.75)^{\lambda_3} - (0.25)^{\lambda_4} + (0.75)^{\lambda_4} - (0.25)^{\lambda_3}}{(1-u)^{\lambda_3} - u^{\lambda_4} + (1-u)^{\lambda_4} - u^{\lambda_3}} = \frac{\xi_{0.75} - \xi_{0.25}}{\xi_{1-u} - \xi_u}. \tag{3.43}$$

Since (3.42) and (3.43) involve only λ_3 and λ_4, they are solved first and then insert these values in (3.40) and (3.41) to estimate λ_1 and λ_2. All the equations involve u and therefore a choice of u lying between 0 and $\frac{1}{4}$ is suggested by Karian and Dudewicz [314]. They also provide a table of

$$\left[\frac{\xi_{0.5} - \xi_u}{\xi_{1-u} - \xi_{0.5}}, \frac{\xi_{0.75} - \xi_{0.25}}{\xi_{1-u} - \xi_u}\right]$$

as pairs of values and the corresponding solutions, the algorithm and illustrations of how to use the tables.

King and MacGillivray [326] have introduced a new procedure called the starship method, which involves estimation of the parameters along with a goodness-of-fit test. Laying a four-dimensional grid over a region in the four-dimensional space that covers the range of the parameter values, a goodness of fit is performed over the points in the grid. If the fit is not satisfied with one point, another is selected and so on, with the procedure terminating with parameter values that have the best measure of fit. Lakhany and Mausser [371] and Fournier et al. [201] have pointed out that the starship method is quite time consuming especially for large samples.

In practice, in most of the methods described above, the parameters obtained need not produce an adequate model. There can also be cases where multiple solutions exist and the solutions do not span the entire data set. So, goodness-of-fit tests have to be carried out separately after estimation or such a test must be embedded in the procedure as with the starship method. There have been several attempts to device procedures that automate the restart of the algorithms and also do the necessary tests. Lakhany and Mausser [371] devised a modification to the starship method. Instead of using a full four-dimensional grid, they used successive simplex from random starting points until the goodness of fit does not reject the distribution. It cannot, however, be said that always the best fit is realized. The GLIDEX package provides fitting methods using discretized and numerical maximum likelihood approach (Su [549]) and the starship methods. King and MacGillvray [327] have suggested a method of estimation with the aid of location and scale free shape functionals

$$S(u) = \frac{Q(u)+Q(1-u)-2M}{Q(u)-Q(1-u)}$$

and

$$d(u,v) = Q(u)+Q(1-u) - \frac{Q(v)+Q(1-v)}{Q(v)-Q(1-v)}$$

by minimizing the distance between the sample and population values of the functionals. Fournier et al. [201] proposed another method that minimizes the $D = \max|S_n(x) - F(x)|$, where $S_n(x)$ is the empirical distribution function in a two-dimensional grid representing the (λ_3, λ_4) space. Two other works in this context are the estimation of parameters for grouped data (Tarsitano [564]) and for censored data (Mercy and Kumaran [416]). Karian and Dudewicz [316] discuss the computational difficulties encountered in the estimation procedure of the generalized lambda distribution.

3.2.2 *Generalized Tukey Lambda Family*

A major limitation of the generalized lambda family discussed above is that the distribution is valid only for certain regions in the parameter space. Freimer et al. [203] introduced a modified generalized lambda distribution defined by

$$Q(u) = \lambda_1 + \frac{1}{\lambda_2}\left[\frac{u^{\lambda_3}-1}{\lambda_3} - \frac{(1-u)^{\lambda_4}-1}{\lambda_4}\right] \tag{3.44}$$

which is well defined for the values of the shape parameters λ_3 and λ_4 over the entire two-dimensional space. The quantile density function has the simple form

$$q(u) = \frac{1}{\lambda_2}[u^{\lambda_3-1} + (1-u)^{\lambda_4-1}].$$

Since our interest in (3.44) is as a life distribution, we should have

$$Q(0) = \lambda_1 - \frac{1}{\lambda_2\lambda_3} \geq 0$$

in which case the support becomes $(\lambda_1 - \frac{1}{\lambda_2\lambda_3}, \lambda_1 + \frac{1}{\lambda_2\lambda_4})$ whenever $\lambda_3 > \lambda_4 > 0$ and $(\lambda_1 - \frac{1}{\lambda_2\lambda_3}, \infty)$ if $\lambda_3 > 0$ and $\lambda_4 \leq 0$. This is a crucial point to be verified when the distribution is used to model data pertaining to non-negative random variables. The exponential distribution is a particular case of the family as $\lambda_3 \to \infty$ and $\lambda_4 \to 0$. All the approximation that are valid for the modified generalized lambda family are valid in (3.44) as well.

The first four raw moments of this distribution are as follows:

$$\mu = \lambda_1 - \frac{1}{\lambda_2}\left[\frac{1}{\lambda_3+1} - \frac{1}{\lambda_4+1}\right],$$

$$\mu_2' = \frac{1}{\lambda_2^2}\left[\frac{1}{\lambda_3^2(2\lambda_3+1)} - \frac{1}{\lambda_4^2(2\lambda_4+1)} - \frac{2}{\lambda_3\lambda_4}B(\lambda_3+1,\lambda_4+1)\right],$$

$$\mu_3' = \frac{1}{\lambda_2^3}\left[\frac{1}{\lambda_3^3(3\lambda_3+1)} - \frac{1}{\lambda_4^3(3\lambda_4+1)} - \frac{3}{\lambda_3^2\lambda_4}B(2\lambda_3+1,\lambda_4+1)\right.$$
$$\left.+\frac{3}{\lambda_3\lambda_4^2}B(\lambda_3+1,2\lambda_4+1)\right],$$

$$\mu_4' = \frac{1}{\lambda_2^4}\left[\frac{1}{\lambda_3^4(4\lambda_3+1)} + \frac{1}{\lambda_4^4(4\lambda_4+1)} + \frac{6}{\lambda_3^2\lambda_4^2}B(2\lambda_3+1,2\lambda_4+1)\right.$$
$$\left.-\frac{4}{\lambda_3^3\lambda_4}B(3\lambda_3+1,\lambda_4+1) + \frac{4}{\lambda_3\lambda_4^3}B(\lambda_3+1,3\lambda_4+1)\right].$$

In order to have a finite moment of order k, it is necessary that $\min(\lambda_3,\lambda_4) > -\frac{1}{k}$. An elaborate discussion on the skewness and kurtosis has been carried out in Freimer et al. [203]. The family completely covers the β_1 values with two disjoint curves corresponding to any $\sqrt{\beta_1}$ except zero. As one of the parameters is held fixed, the behaviour of skewness is as follows. At $\lambda_3 = -\frac{1}{3}$, $\sqrt{\beta_1} = -\infty$ then increases monotonically to zero for λ_3 in $(-\frac{1}{3}, 1)$ and then tends to ∞ as $\lambda_3 \to \infty$. Similarly, as λ_2 increases from $-\frac{1}{3}$ to 1 to ∞, $\sqrt{\beta_1}$ decreases from ∞ to 0 and to $-\infty$. The family attains symmetry at $\lambda_3 = \lambda_4$, but $\sqrt{\beta_1}$ may be zero even if $\lambda_3 \neq \lambda_4$. Considerable richness is seen in density shapes, there being members that are unimodal, U-shaped, J-shaped and monotone, which are symmetric or skew with short, medium and long tails; see, e.g., Fig. 3.2. Also, there are members with arbitrarily large values for kurtosis, though it does not contain the lowest possible β_2 for a given β_1. There can be more than one set of (λ_3,λ_4) corresponding to a given (β_1,β_2).

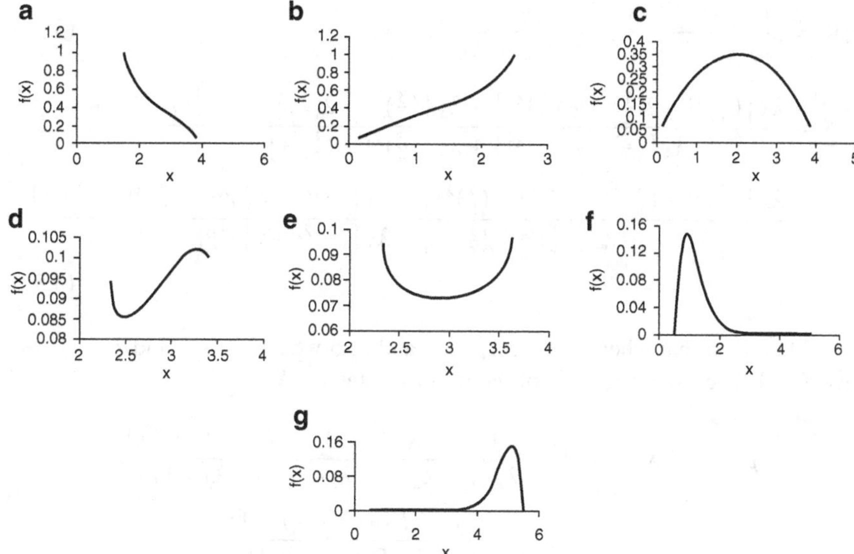

Fig. 3.2 Density plots of the GLD (Freimer et al. model) for different choices of $(\lambda_1, \lambda_2, \lambda_3, \lambda_4)$. (**a**) (2,1,2,0.5); (**b**) (2,1,0.5,2); (**c**) (2,1,0.5,0.5); (**d**) (3,1,1.5,2.5); (**e**) (3,1,1.5,1.6,); (**f**) (1,1,2,0.1); (**g**) (5,1,0.1,2)

Compared to the conventional central moments, the L-moments have much simpler expressions:

$$L_1 = \mu = \lambda_1 - \frac{1}{\lambda_2}\left[\frac{1}{\lambda_3+1} - \frac{1}{\lambda_4+1}\right], \tag{3.45}$$

$$L_2 = \frac{1}{\lambda_2}\left[\frac{1}{(\lambda_3+1)_{(2)}} - \frac{1}{(\lambda_4+1)_{(2)}}\right], \tag{3.46}$$

$$L_3 = \frac{1}{\lambda_2}\left[\frac{\lambda_3-1}{(\lambda_3+1)_{(3)}} - \frac{\lambda_4-1}{(\lambda_4+1)_{(3)}}\right], \tag{3.47}$$

$$L_4 = \frac{1}{\lambda_2}\left[\frac{(\lambda_3-1)^{(2)}}{(\lambda_3+1)_{(4)}} - \frac{(\lambda_4-1)^{(2)}}{(\lambda_4+1)_{(4)}}\right]. \tag{3.48}$$

The measures of location, spread, skewness and kurtosis based on percentiles are as follows:

$$M = \lambda_1 + \frac{1}{\lambda_2}\left[\frac{(\frac{1}{2})^{\lambda_3}-1}{\lambda_3} - \frac{(\frac{1}{2})^{\lambda_4}-1}{\lambda_4}\right], \tag{3.49}$$

$$\text{IQR} = \frac{1}{2\lambda_2}\left(\frac{(\frac{3}{4})^{\lambda_3} - (\frac{1}{4})^{\lambda_3}}{\lambda_3} - \frac{(\frac{3}{4})^{\lambda_4} - (\frac{1}{4})^{\lambda_4}}{\lambda_2}\right), \tag{3.50}$$

$$S = \frac{\lambda_4\{(\frac{3}{4})^{\lambda_3} - 2(\frac{1}{2})^{\lambda_3} + (\frac{1}{4})^{\lambda_3}\} - \lambda_3\{(\frac{3}{4})^{\lambda_4} - 2(\frac{1}{2})^{\lambda_4} + (\frac{1}{4})^{\lambda_4}\}}{\lambda_4\{(\frac{3}{4})^{\lambda_3} - (\frac{1}{4})^{\lambda_3}\} + \lambda_3\{(\frac{3}{4})^{\lambda_4} - (\frac{1}{4})^{\lambda_4}\}}, \tag{3.51}$$

$$T = \frac{\lambda_4\{(\frac{7}{8})^{\lambda_3} - (\frac{5}{8})^{\lambda_3} + (\frac{3}{8})^{\lambda_3} - (\frac{1}{8})^{\lambda_3}\} - \lambda_3\{(\frac{7}{8})^{\lambda_4} - (\frac{5}{8})^{\lambda_4} + (\frac{3}{8})^{\lambda_4} - (\frac{1}{8})^{\lambda_4}\}}{\lambda_4\{(\frac{3}{4})^{\lambda_3} - (\frac{1}{4})^{\lambda_3}\} + \lambda_3\{(\frac{3}{4})^{\lambda_4} - (\frac{1}{4})^{\lambda_4}\}}. \tag{3.52}$$

It could be seen that when $\lambda_3 = 1$, $\lambda_4 \to \infty$ and also when $\lambda_3 \to \infty$ and $\lambda_4 = 1$, we have $S = 0$. The expected value of the rth order statistic $X_{r:n}$ is

$$\mu_{r:n} = E(X_{r:n}) = \lambda_1 - \frac{1}{\lambda_2\lambda_3} + \frac{1}{\lambda_2\lambda_4} + \frac{1}{\lambda_2\lambda_3}\frac{\Gamma(\lambda_3 + r)}{\Gamma(n + \lambda_3 + 1)}\frac{n!}{r!}$$
$$- \frac{1}{\lambda_2\lambda_4}\frac{n!}{(n-r)!}\frac{\Gamma(n + \lambda_3 - r + 1)}{\Gamma(n + \lambda_4 + 1)}.$$

Setting $r = 1$ and n, we get

$$E(X_{1:n}) = \lambda_1 - \frac{1}{\lambda_2\lambda_3} + \frac{1}{\lambda_2\lambda_4} + \frac{n!}{\lambda_2(\lambda_3)_{(n+1)}} - \frac{n}{\lambda_2\lambda_4(\lambda_4 + n)}$$

and

$$E(X_{n:n}) = \lambda_1 - \frac{1}{\lambda_2\lambda_3} + \frac{1}{\lambda_2\lambda_4} + \frac{n}{\lambda_2\lambda_3(\lambda_3 + n)} - \frac{n!}{\lambda_2(\lambda_4)_{(n+1)}}.$$

The distributions of $X_{1:n}$ and $X_{n:n}$ are given by

$$Q_1(u) = \lambda_1 + \frac{1}{\lambda_2}\left[\frac{[1 - (1 - u^{1/n})]^{\lambda_3} - 1}{\lambda_3} - \frac{(1-u)^{\frac{\lambda_4}{n}} - 1}{\lambda_4}\right]$$

and

$$Q_n(u) = \lambda_1 + \frac{1}{\lambda_2}\left(\frac{u^{\frac{\lambda_3}{n}} - 1}{\lambda_3} - \frac{(1 - u^{\frac{1}{n}})^{\lambda_4} - 1}{\lambda_4}\right).$$

Various reliability functions of the model have closed-form algebraic expressions, except for the variances which contain beta functions. The hazard quantile function is

$$H(u) = \lambda_2[(1-u)^{\lambda_4} + (1-u)u^{\lambda_3-1}]. \tag{3.53}$$

Mean residual quantile function simplifies to

$$M(u) = \frac{(1-u)^{\lambda_4}}{\lambda_2(\lambda_4+1)} + \frac{1-u^{\lambda_3+1}}{\lambda_2(1+\lambda_3)(1-u)} - \frac{u^{\lambda_3}}{\lambda_2\lambda_3}. \tag{3.54}$$

The variance residual quantile function is

$$V(u) = A_1(u) - A_2^2(u),$$

where

$$A_1(u) = \frac{1-u^{2\lambda_3+1}}{\lambda_2^2(2\lambda_3+1)(1-u)} + \frac{(1-u)^{2\lambda_4}}{\lambda_2\lambda_4(2\lambda_4+1)} - \frac{2B_{1-u}(\lambda_4+1,\lambda_3+1)}{\lambda_2^2\lambda_3\lambda_4(1-u)}$$

and

$$A_2(u) = \frac{1-u^{\lambda_3+1}}{\lambda_2\lambda_3(1+\lambda_3)(1-u)} - \frac{(1-u)^{\lambda_4+1}}{\lambda_2\lambda_4(\lambda_4+1)}.$$

Percentile residual life function becomes

$$P_\alpha(u) = \frac{1}{\lambda_2}[(1-(1-\alpha)(1-u))^{\lambda_3} + (1-u)^{\lambda_4}(1-(1-\alpha)^{\lambda_4}) - u^{\lambda_3}].$$

Expression for the reversed hazard quantile function is

$$\Lambda(u) = \left[\frac{u}{\lambda_2}(u^{\lambda_3-1} + (1-u)^{\lambda_4-1})\right]^{-1}.$$

The reversed mean residual quantile function is

$$R(u) = \frac{1}{\lambda_2}\left[\frac{u^{\lambda_3}}{\lambda_3+1} - \frac{(1-u)^{\lambda_4}}{\lambda_4+1} - \frac{(1-u)^{\lambda_4+1}}{\lambda_4(\lambda_4+1)u} + \frac{1}{\lambda_4(\lambda_4+1)u}\right],$$

the reversed percentile residual life function is

$$q_\alpha(u) = \frac{u^{\lambda_3}}{\lambda_2\lambda_3}(1-(1-\alpha)^{\lambda_3}) - \frac{1}{\lambda_2\lambda_4}[(1-u(1-\alpha))^{\lambda_4} - (1-u)^{\lambda_4}],$$

and the reversed variance residual quantile function is

$$D^*(u) = B_1(u) - B_2^2(u),$$

where

$$B_1(u) = \frac{u^{2\lambda_3}}{\lambda_2^2\lambda_3^2(2\lambda_3+1)} + \frac{(1-u)^{2\lambda_4+1}-1}{\lambda_2^2\lambda_4^2(2\lambda_4+1)u} - \frac{2B_u(\lambda_3+1,\lambda_4+1)}{u\lambda_2^2\lambda_3\lambda_4}$$

and

$$B_2(u) = \frac{u^{\lambda_3}}{\lambda_2\lambda_3(\lambda_3+1)} - \frac{(1-u)^{\lambda_4+1}-1}{\lambda_2\lambda_3(\lambda_4+1)u}.$$

Although the problem of estimating the parameters of (3.44) is quite similar and all the methods described earlier for the generalized lambda distribution are applicable in this case also, there is comparatively less literature available on this subject. The moment matching method and the least-square approach were discussed by Lakhany and Massuer [371]. Since these methods involved only replacement of the corresponding expressions for (3.44) in the previous section, the details are not presented here for the sake of brevity. Su [550] discussed two new approaches—the discretized approach and the method of maximum likelihood for the estimation problem, by tackling it on two fronts: (a) finding suitable initial values and (b) selecting the best fit through an optimization scheme. For the distribution in (3.44), the initial values of λ_3 and λ_4 consist of low discrepancy quasi-random numbers ranging from -0.25 to 1.5. After generating these random values, they were used to derive λ_1 and λ_2 by the method of moments as in Lakhany and Massuer [371]. From these initial values, the GLDEX package (Su [551]) is employed to find the best set of initial values for the optimization process. In the discretized approach, the range of the data is divided into equally spaced classes, and after arranging the observations in ascending order of magnitude, the proportion falling in each class is ascertained. Then, the differences between the observed (d_i) and theoretical (t_i) proportions are minimized through either

$$\sum_{i=1}^{k}(d_i - t_i)^2 \quad \text{or} \quad \sum_{i=1}^{k} d_i(d_i - t_i)^2,$$

where k is the number of classes.

In the maximum likelihood method, the u_i values corresponding to each x_i in the data are to be computed first using $Q(u)$. A numerical method such as the Newton-Raphson can be employed for this purpose. Then, with the help of Nelder–Simplex algorithm, the log likelihood function

$$\log L = \sum_{i=1}^{n} \log\left(\frac{\lambda_2}{u_i^{\lambda_3-1} + (1-u_i)^{\lambda_4-1}}\right)$$

is maximized to get the final estimates. The GLDEX package provides diagnostic tests that assess the quality of the fit through the Kolmogorov–Smirnov test, quantile plots and agreement between the measures of location, spread, skewness and kurtosis of the data with those of the model fitted to the observations.

Haritha et al. [260] adopted a percentile method in which they matched the measures of location (median, M), spread (interquartile range), skewness (Galton's coefficient, S) and kurtosis (Moors' measure, T) of the population in (3.49)–(3.52) and the data. Among the solutions of the resulting equations, they chose the parameter values that gave

$$e = \max(|\hat{M} - m|, |\hat{\text{IQR}} - iqr|, |\hat{S} - \Delta|, |\hat{T} - t|) < \varepsilon$$

for the smallest ε.

3.2.3 van Staden–Loots Model

A four-parameter distribution that belongs to lambda family proposed by van Staden and Loots [572], but different from the two versions discussed in Sects. 3.2.1 and 3.2.2, will be studied in this section. The distribution is generated by considering the generalized Pareto model in the form

$$Q_1(u) = \begin{cases} \frac{1}{\lambda_4}(1 - (1-u)^{\lambda_4}) & \lambda_4 \neq 0 \\ -\log(1-u) & \lambda_4 = 0 \end{cases}$$

and its reflection

$$Q_2(u) = \begin{cases} \frac{1}{\lambda_4}(u^{\lambda_4} - 1) & \lambda_4 \neq 0 \\ \log u & \lambda_4 = 0. \end{cases}$$

A weighted sum of these two quantile functions with respective weights λ_3 and $1-\lambda_3$, $0 \le \lambda_3 \le 1$, along with the introduction of a location parameter λ_1 and a scale parameter λ_2, provide the new form. Thus, the quantile function of this model is

$$Q(u) = \lambda_1 + \lambda_2 \left[(1-\lambda_3)\frac{u^{\lambda_4} - 1}{\lambda_4} - \lambda_3 \frac{(1-u)^{\lambda_4} - 1}{\lambda_4} \right], \quad \lambda_2 > 0. \tag{3.55}$$

Equation (3.55) includes the exponential, logistic and uniform distributions as special cases. The support of this distribution is as follows for different choices of λ_3 and λ_4:

Region	λ_3	λ_4	Support
1	0	≤ 0	$(-\infty, \lambda_1)$
		> 0	$\left(\lambda_1 - \frac{\lambda_2}{\lambda_4}, \lambda_1\right)$
2	$(0,1)$	≤ 0	$(-\infty, \infty)$
		> 0	$\left(\lambda_1 - \frac{\lambda_2(1-\lambda_3)}{\lambda_4}, \lambda_1 + \frac{\lambda_3\lambda_2}{\lambda_4}\right)$
3	1	≤ 0	(λ_1, ∞)
		> 0	$\left(\lambda_1, \lambda_1 + \frac{\lambda_2}{\lambda_4}\right)$

For (3.55) to be a life distribution, one must have $\lambda_1 - \lambda_2(1-\lambda_3)\lambda_4^{-1} \geq 0$. This gives members with both finite and infinite support, depending on whether λ_4 is positive or negative.

As for descriptive measures, the mean and variance are given by

$$\mu = \lambda_1 - \frac{\lambda_2}{(1+\lambda_4)}(1-2\lambda_3)$$

and

$$\sigma^2 = \frac{\lambda_2^2}{(1+\lambda_4)^2}\left[\frac{\lambda_3^2+(1-\lambda_3)^2}{1+2\lambda_4} - \frac{2\lambda_3(1-\lambda_3)}{\lambda_3}((1+\lambda_4)^2 B(1+\lambda_4, 1+\lambda_4)-1)\right].$$

One attractive feature of this family is that its L-moments have very simple forms, and they exist for all $\lambda_4 > -1$, and are as follows:

$$L_1 = \mu,$$

$$L_2 = \frac{\lambda_2}{(\lambda_4+1)(\lambda_4+2)},$$

$$L_r = \lambda_2(1-2\lambda_3)^S \frac{(\lambda_4-1)^{(r-2)}}{(\lambda_4+1)_{(r)}}, \quad r = 3, 4, \ldots,$$

where $S = 1$ when r is odd and $S = 0$ when r is even. These values give L-skewness and L-kurtosis to be

$$\tau_3 = \frac{(\lambda_4-1)(1-2\lambda_3)}{\lambda_4+3}$$

and

$$\tau_4 = \frac{(\lambda_4-1)(\lambda_4-2)}{(\lambda_4+3)(\lambda_4+4)},$$

respectively. van Staden and Loots [572] note that, as in the case of the two four-parameter lambda families discussed in the last two sections, there is no unique (λ_3, λ_4) pair for a given value of (τ_3, τ_4). When $\lambda_3 = -\frac{1}{2}$, the distribution is symmetric.

L-skewness covers the entire permissible span $(-1,1)$ and the kurtosis is independent of λ_3 with a minimum attained at $\lambda_4 = \sqrt{6}-1$. The percentile-based measures also have simple explicit forms and are given by

$$M = \lambda_1 + \frac{\lambda_2(1-2\lambda_3)}{\lambda_4}\left(\left(\frac{1}{2}\right)^{\lambda_4} - 1\right), \tag{3.56}$$

In the symmetric case, $$M = \mu = \lambda_1, \tag{3.57}$$

$$\text{IQR} = \frac{\lambda_2(3^{\lambda_4}-1)}{\lambda_4 4^{\lambda_4}}, \tag{3.58}$$

$$S = \frac{(1-2\lambda_3)(1+3^{\lambda_4}-2^{\lambda_4+1})}{3^{\lambda_4}-1}, \tag{3.59}$$

$$T = \frac{(1-2\lambda_3)2^{\lambda_4}(7^{\lambda_4}+5^{\lambda_4}+3^{\lambda_4}+1)}{3^{\lambda_4}-1}. \tag{3.60}$$

The quantile density function is

$$q(u) = \lambda_2[(1-\lambda_3)u^{\lambda_4-1} + \lambda_3(1-u)^{\lambda_4-1}]$$

and so the density quantile function is

$$f(Q(u)) = \lambda_2^{-1}[(1-\lambda_3)u^{\lambda_4-1} + \lambda_3(1-u)^{\lambda_4-1}]^{-1}.$$

Figure 3.3 displays some shapes of the density function.

The expectations of the order statistics from (3.55) are as follows:

$$E(X_{r:n}) = \lambda_1 + \frac{\lambda_2}{\lambda_4}\left[\frac{(1-\lambda_3)\Gamma(\lambda_4+1)}{\Gamma(r)} - \frac{\lambda_3\Gamma(n+\lambda_4-r+1)}{\Gamma(n-r+1)}\right]\frac{n!}{\Gamma(\lambda_4+n+1)}$$
$$+\frac{\lambda_2}{\lambda_4}(2\lambda_3-1), \quad r=1,2,\ldots,n,$$

$$E(X_{1:n}) = \lambda_1 + \frac{\lambda_2}{\lambda_4}(2\lambda_3-1) + \frac{\lambda_2}{\lambda_4}\left[\frac{n!(1-\lambda_3)}{(\lambda_4+1)_{(n)}} - \frac{n\lambda_3}{\lambda_4+n}\right],$$

$$E(X_{n:n}) = \lambda_1 + \frac{\lambda_2}{\lambda_4}(2\lambda_3-1) + \frac{\lambda_2}{\lambda_4}\left[\frac{n(1-\lambda_3)}{\lambda_4+n} - \frac{\lambda_3 n!}{(\lambda_4+1)_{(n)}}\right].$$

Since there are members of the family with support on the positive real line, the model will be useful for describing lifetime data. In this context, the hazard quantile function is

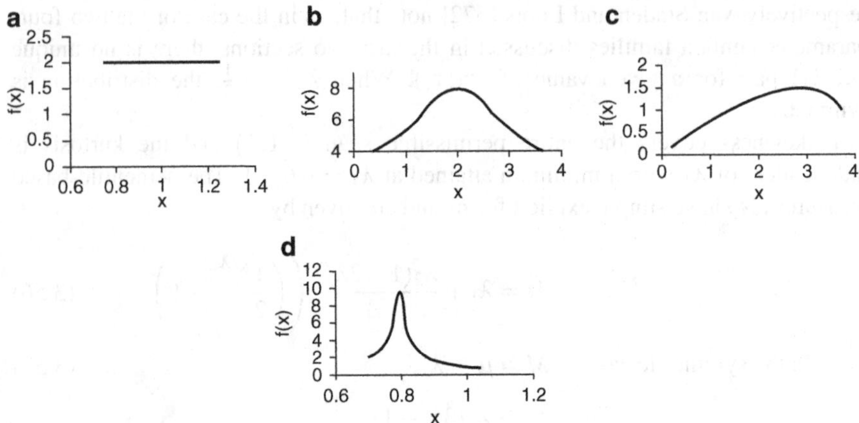

Fig. 3.3 Density plots of the GLD proposed by van Staden and Loots [572] for varying $(\lambda_1, \lambda_2, \lambda_3, \lambda_4)$. **(a)** (1,1,0.5,2); **(b)** (2,1,0.5,3); **(c)** (3,2,0.25,0.5); **(d)** (1,2,0.1,−1)

$$H(u) = \{\lambda_2(1-u)((1-\lambda_3)u^{\lambda_4-1} + \lambda_3(1-u)^{\lambda_4-1})\}^{-1}. \tag{3.61}$$

Similarly, the mean residual quantile function is

$$M(u) = \lambda_2 \left[\frac{1-\lambda_3}{\lambda_4}\left(\frac{1-u^{\lambda_4+1}}{(1-u)(\lambda_4+1)} - u^{\lambda_4}\right) - \frac{\lambda_3}{\lambda_4+1}(1-u)^{\lambda_4}\right], \tag{3.62}$$

the reversed hazard quantile function is

$$\Lambda(u) = \left[\lambda_2 u\left\{(1-\lambda_3)u^{\lambda_4-1} + \lambda_3(1-u)^{\lambda_4-1}\right\}\right]^{-1},$$

and the reversed mean residual quantile function is

$$R(u) = \lambda_2\left[\frac{(1-\lambda_3)}{\lambda_4+1}u^{\lambda_4} - \frac{(1-u)^{\lambda_4}}{\lambda_4} + \frac{(1-u)^{\lambda_4+1} - 1}{u\lambda_4(\lambda_4+1)}\right].$$

Further, the form

$$u\,\theta(u) = A + B((1-\alpha)u^C + \alpha(1-u)^c)$$

determines the quantile function in (3.55) as

$$Q(u) = A + \frac{d}{du}u\,\theta(u),$$

where $\theta(u)$ is as defined in (3.24).

van Staden and Loots [572] prescribed the method of L-moments for the estimation of the parameters. With the aid of

$$\hat{\lambda}_4 = \frac{3 + 7t_4 \pm (t_4^2 + 98t_4 + 1)^{\frac{1}{2}}}{2(1 - t_4)},$$

where t_4 is the sample L-kurtosis, λ_4 can be estimated. Using $\hat{\lambda}_4$, the estimate $\hat{\lambda}_3$ of λ_3 can be determined from

$$\hat{\lambda}_3 = \begin{cases} \frac{1}{2}[1 - \frac{t_3(\hat{\lambda}_3+3)}{\hat{\lambda}_4-1}] & , \hat{\lambda}_4 \neq 1 \\ \frac{1}{2} & , \hat{\lambda}_4 = 1 \end{cases},$$

where t_3 is the sample L-skewness. The other two-parameter estimates are computed as

$$\hat{\lambda}_2 = l_2(\hat{\lambda}_4 + 1)(\hat{\lambda}_4 + 2),$$

$$\hat{\lambda}_1 = l_1 + \frac{\hat{\lambda}_2(1 - 2\hat{\lambda}_3)}{\hat{\lambda}_4 + 1},$$

with l_1 and l_2 being the usual first two sample L-moments.

The method of percentiles can also be applied for parameter estimation. In fact,

$$\frac{t}{s} = \frac{2^{\lambda_4}(7^{\lambda_4} + 5^{\lambda_4} + 3^{\lambda_4} + 1)}{3^{\lambda_4} + 1 - 2^{\lambda_4+1}}$$

provides λ_4, t and s being the Moors and Galton measures, respectively, evaluated from the data. This is used in (3.59) to find $\hat{\lambda}_3$, and then $\hat{\lambda}_2$ and $\hat{\lambda}_1$ are determined from (3.56) and (3.58) by equating IQR and M with iqr and m, respectively.

3.2.4 Five-Parameter Lambda Family

Gilchrist [215] proposed a five-parameter family of distributions with quantile function

$$Q(u) = \lambda_1 + \frac{\lambda_2}{2}\left[(1 - \lambda_3)\left(\frac{u^{\lambda_4} - 1}{\lambda_4}\right) - (1 + \lambda_3)\left(\frac{(1-u)^{\lambda_5} - 1}{\lambda_5}\right)\right] \tag{3.63}$$

as an extension to the Freimer et al. [203] model in (3.44). Tarsitano [564] studied this model and evaluated various estimation methods for this family. The family has its quantile density function as

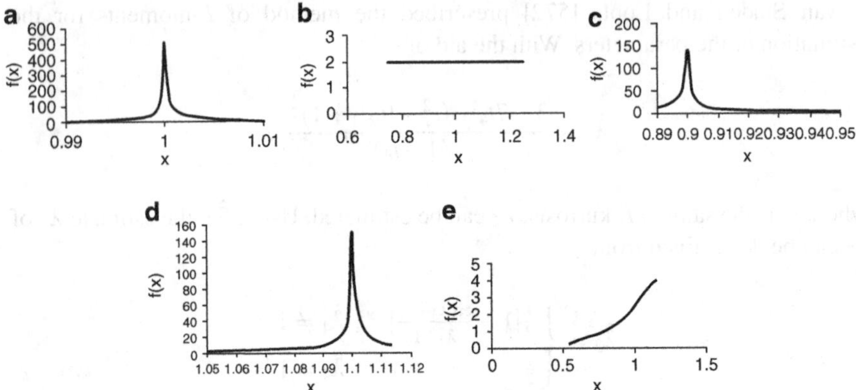

Fig. 3.4 Density plots of the five-parameter GLD for different choices of $(\lambda_1, \lambda_2, \lambda_3, \lambda_4, \lambda_5)$. (**q**) (1,1,0,10,10); (**b**) (1,1,0,2,2); (**c**) (1,1,0.5,-0.6,-0.5); (**d**) (1,1,-0.5,-0.6,-0.5); (**e**) (1,1,0.5,0.5,5)

$$q(u) = \lambda_2 \left[\frac{1-\lambda_3}{2} u^{\lambda_4 - 1} + \frac{1-\lambda_3}{2} (1-u)^{\lambda_5 - 1} \right].$$

In (3.63), λ_1 controls the location parameter though not exclusively, $\lambda_2 \geq 0$ is a scale parameter and λ_3, λ_4 and λ_5 are shape parameters. It is evident that the generalized Tukey lambda family in (3.44) is a special case when $\lambda_3 = 0$. The support of the distribution is given by

$$\left(\lambda_1 - \lambda_2 \frac{(1-\lambda_3)}{2}, \lambda_1 + \lambda_2 \frac{(1+\lambda_3)}{2}\right) \text{ when } \lambda_4 > 0,\ \lambda_5 > 0,$$

$$\left(\lambda_1 - \lambda_2 \frac{(1-\lambda_3)}{2}, \infty\right) \text{ when } \lambda_4 > 0, \lambda_5 \leq 0,$$

$$\left(-\infty, \lambda_1 + \lambda_2 \frac{(1+\lambda_3)}{2}\right) \text{ when } \lambda_4 \leq 0,\ \lambda_5 > 0.$$

In the case of non-negative random variables, the condition

$$\lambda_1 - \frac{\lambda_2}{2\lambda_4}(1-\lambda_3) \geq 0$$

would become necessary. The density function may be unimodal with or without truncated tails, U-shaped, S-shaped or monotone. The family also includes the exponential distribution when $\lambda_4 \to \infty$ and $\lambda_5 \to 0$, the generalized Pareto distribution when $\lambda_4 \to \infty$ and $|\lambda_5| < \infty$, and the power distribution when $\lambda_5 \to 0$ and $|\lambda_4| < \infty$. Some typical shapes of the distribution are displayed in Fig. 3.4. Tarsitano [564] has provided close approximations to various symmetric and asymmetric distributions using (3.63) and went on to recommend the usage of the model when a particular distributional form cannot be suggested from the physical situation under consideration. Setting $Z = \frac{2(X-\lambda_1)}{\lambda_2}$, Tarsitano [564] expressed the raw moments in the form

$$E(Z^r) = \sum_{j=0}^{r}(-1)^j \binom{r}{j} \left(\frac{1-\lambda_3}{\lambda_4}\right)^{r-j} \left(\frac{1+\lambda_3}{\lambda_5}\right)^{j} B(1+(r-j)\lambda_4, 1+j\lambda_5)$$

provided λ_4 and λ_5 are greater than $\frac{1}{r}$, where $B(\cdot,\cdot)$ is the complete beta function, as before. The mean and variance are deduced from the above expression as

$$\mu = \lambda_1 - \frac{\lambda_2(1-\lambda_3)}{2(1+\lambda_4)} + \frac{\lambda_2(1+\lambda_3)}{2(1+\lambda_5)},$$

$$\sigma^2 = \frac{(1-\lambda_3)^2}{\lambda_4^2(2\lambda_4+1)} + \frac{(\lambda_3+1)^2}{\lambda_5^2(2\lambda_5+1)} - \frac{2(1-\lambda_3)}{\lambda_4\lambda_5} B(\lambda_4+1, \lambda_5+1).$$

The L-moments take on simpler expressions in this case, and the first four are as follows:

$$L_1 = \mu,$$

$$L_2 = \frac{\lambda_2(1-\lambda_3)}{2(\lambda_4+1)_{(2)}} + \frac{\lambda_2(1+\lambda_3)}{2(\lambda_5+1)_{(2)}},$$

$$L_3 = \frac{\lambda_2(1-\lambda_3)(\lambda_4-1)}{2(\lambda_4+1)_{(3)}} - \frac{\lambda_2(1+\lambda_3)(\lambda_5-1)}{2(\lambda_5+1)_{(3)}},$$

$$L_4 = \frac{\lambda_2(1-\lambda_3)(\lambda_4-1)^{(2)}}{2(\lambda_4+1)_{(4)}} - \frac{\lambda_2(1+\lambda_3)(\lambda_5-1)^{(2)}}{2(\lambda_5+1)_{(4)}}.$$

Percentile-based measures of location, spread, skewness and kurtosis can also be presented, but they involve rather cumbersome expressions. For example, the median is given by

$$M = \lambda_1 - \frac{\lambda_2(1-\lambda_3)}{2\lambda_4} + \frac{\lambda_2(1+\lambda_3)}{2\lambda_5} + \frac{\lambda_2(1-\lambda_3)}{\lambda_4}\left(\frac{1}{2}\right)^{\lambda_4+1} - \frac{\lambda_2(1-\lambda_3)}{\lambda_5}\left(\frac{1}{2}\right)^{\lambda_5+1}.$$

The means of order statistics are as follows:

$$E(X_{r:n}) = \lambda_1 + \frac{\lambda_2}{2}\left[\frac{1+\lambda_3}{\lambda_5} - \frac{1-\lambda_3}{\lambda_4}\right] + \frac{\lambda_2(1-\lambda_3)}{2\lambda_4}\frac{B(\lambda_4+r-1, n-r+1)}{B(r, n-r+1)}$$
$$- \frac{\lambda_2(1+\lambda_3)}{2\lambda_5}\frac{B(r, \lambda_1+n-r+1)}{B(r, n-r+1)}, \quad r = 1, \ldots, n, \tag{3.64}$$

$$E(X_{1:n}) = \lambda_1 + \frac{\lambda_2}{2}\left[\frac{1+\lambda_3}{\lambda_5} - \frac{1-\lambda_3}{\lambda_4}\right] + \frac{n!}{(\lambda_4)_{(n)}} - n\frac{\lambda_2}{2\lambda_5}\frac{(1+\lambda_3)}{n+\lambda_5},$$

$$E(X_{n:n}) = \lambda_1 + \frac{\lambda_2}{2}\left[\frac{1+\lambda_3}{\lambda_5} - \frac{1-\lambda_3}{\lambda_4}\right] + \frac{n\lambda_2(1-\lambda_3)}{2\lambda_4(\lambda_4+n-1)} - \frac{\lambda_2(1+\lambda_3)n!}{2(\lambda_5+1)_{(n)}}.$$

Tarsitano [564] discussed the estimation problem through nonlinear least-squares and least absolute deviation approaches. For a random sample $X_1, X_2, \ldots, X_n$ of size n from (3.59), under the least-squares approach, we consider

$$X_{r:n} = E(X_{r:n}) + \varepsilon_r, \quad r = 1, 2, \ldots, n,$$

and then seek the parameter values that minimize

$$\sum_{r=1}^{n} [X_{r:n} - E(X_{r:n})]^2. \tag{3.65}$$

In terms of expectations of order statistics in (3.60), realize that $X_{r:n}$ is an estimate of the expectation in (3.64), which incidentally is linear in λ_1, λ_2 and λ_3 and nonlinear in λ_4 and λ_5. So, as in the case of Osturk and Dale method discussed earlier, we may fix (λ_4, λ_5) and determine λ_1, λ_2 and λ_3. Then, (λ_4, λ_5) can be found such that (3.65) is minimized. In the least absolute deviation procedure, the objective function to be minimized is

$$\sum_{r=1}^{n} |X_{r:n} - Q(u_r^*)|,$$

where

$$u_r^* = Q(B_u^{-1}(r, n-r+1)).$$

3.3 Power-Pareto Distribution

As seen earlier in Table 1.1, the quantile function of the power distribution is of the form

$$Q_1(u) = \alpha u^{\lambda_1}, \quad 0 \le u \le 1;\ \alpha, \lambda_1 > 0,$$

while that of the Pareto distribution is

$$Q_2(u) = \sigma(1-u)^{-\lambda_2}, \quad 0 \le u \le 1;\ \sigma, \lambda_2 > 0.$$

A new quantile function can then be formed by taking the product of these two as

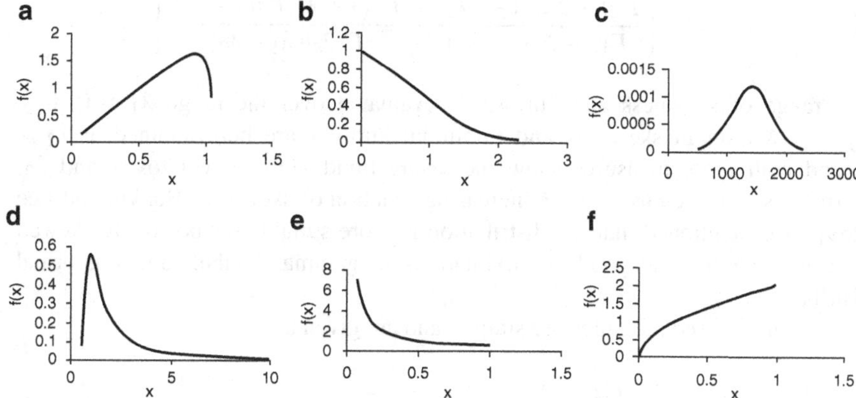

Fig. 3.5 Density plots of power-Pareto distribution for some choices of (C,λ_1,λ_2). (**a**) (1,0.5,0.01); (**b**) (1,1,0.2); (**c**) (1,0.2,0.1); (**d**) (1,0.1,1); (**e**) (1,0.5,0.001); (**f**) (1,2,0.001)

$$Q(u) = \frac{Cu^{\lambda_1}}{(1-u)^{\lambda_2}}, \quad 0 \le u \le 1, \tag{3.66}$$

where $C > 0$, $\lambda_1, \lambda_2 > 0$ and one of the λ's may be equal to zero. The distribution of a random variable X with (3.66) as its quantile function is called the power-Pareto distribution. Gilchrist [215] and Hankin and Lee [259] have studied the properties of this distribution. It has the quantile density function as

$$q(u) = \frac{Cu^{\lambda_1-1}}{(1-u)^{\lambda_2+1}}[\lambda_1(1-u)+\lambda_2 u]. \tag{3.67}$$

In (3.66), C is a scale parameter, λ_1 and λ_2 are shape parameters, with λ_1 controlling the left tail and λ_2 the right tail. The shape of the density function is displayed in Fig. 3.5 for some parameter values when the scale parameter $C = 1$.

Conventional moments of (3.66) are given by

$$E(X^r) = \int_0^1 \left[\frac{Cu^{\lambda_1}}{(1-u)^{\lambda_2}}\right]^r du = C^r B(1+r\lambda_1, 1-r\lambda_2)$$

which exists whenever $\lambda_2 < \frac{1}{r}$. From this, the mean and variance are obtained as

$$\mu = \frac{C\Gamma(1+\lambda_1)\Gamma(1-\lambda_2)}{\Gamma(2+\lambda_1-\lambda_2)} \tag{3.68}$$

and

$$\sigma^2 = C^2 \left\{ \frac{\Gamma(1+2\lambda_1, 1-2\lambda_1)}{\Gamma(2+2\lambda_1-2\lambda_2)} - \frac{\Gamma^2(1+\lambda_1)\Gamma^2(1-\lambda_2)}{\Gamma(2+\lambda_1-\lambda_2)} \right\}.$$

The range of skewness and kurtosis is evaluated over the range $\lambda_1 > 0$, $0 \le \lambda_2 < \frac{1}{4}$. Minimum skewness and minimum kurtosis are both attained at $\lambda_2 = 0$, and both these measures show increasing trend with respect to λ_1 and λ_2. Kurtosis is also seen to be as an increasing function of skewness. Hankin and Lee [259] have mentioned that the distribution is more suitable for positively skewed data and can provide good approximations to gamma, Weibull and lognormal distributions.

Percentile-based measures are simpler and are given by

$$M = C2^{\lambda_2-\lambda_1},$$

$$\text{IQR} = C4^{\lambda_2-\lambda_1}(3^{\lambda_1} - 3^{-\lambda_2}),$$

$$S = \frac{3^{\lambda_1} + 3^{-\lambda_2} - 2^{\lambda_2-\lambda_1+1}}{3^{\lambda_1} - 3^{-\lambda_2}},$$

$$T = \frac{2^{\lambda_2-\lambda_1}(7^{\lambda_1} - 5^{\lambda_1}3^{-\lambda_2} + 3^{\lambda_1}5^{-\lambda_2} - 7^{-\lambda_2})}{3^{\lambda_1} - 3^{-\lambda_2}}.$$

Further, the first four L-moments are as follows:

$$L_1 = \mu,$$

$$L_2 = \frac{C(\lambda_1+\lambda_2)}{\lambda_1 - \lambda_2 + 2} B(\lambda_1+1, 1-\lambda_2),$$

$$L_3 = \frac{C(\lambda_1^2 + \lambda_2^2 + 4\lambda_1\lambda_2 + \lambda_2 - \lambda_1)B((\lambda_1+1, 1-\lambda_2))}{(\lambda_1 - \lambda_2 + 2)_{(2)}},$$

$$L_4 = \frac{C(\lambda_1+\lambda_2)(\lambda_1^2 + \lambda_2^2 + 8\lambda_1\lambda_2 - 3\lambda_1 + 3\lambda_2 + 2)}{(\lambda_1 - \lambda_2 + 2)_{(3)}} B(\lambda_1+1, 1-\lambda_2),$$

where $B(\cdot,\cdot)$ is the complete beta function. From these expressions, we readily obtain the L-skewness and L-kurtosis measures as

$$\tau_3 = \frac{\lambda_1^2 - \lambda_2^2 + 4\lambda_1\lambda_2 + \lambda_2 - \lambda_1}{(\lambda_1 - \lambda_2 + 3)}$$

and

$$\tau_4 = \frac{\lambda_1^2 - \lambda_2^2 + 8\lambda_1\lambda_2 - 3\lambda_1 + 3\lambda_2 + 2}{(\lambda_1 - \lambda_2 + 3)(\lambda_1 - \lambda_2 + 4)}.$$

The expected value of the rth order statistic is

$$E(X_{r:n}) = C\frac{B(\lambda_1 + r, n - \lambda_2 - r)}{B(r, n - r + 1)} \quad n > \lambda_2 + r, \quad r = 1, 2, \ldots, n,$$

while the quantile functions of $X_{1:n}$ and $X_{n:n}$ are given by

$$Q_1(u) = C\frac{\{1 - (1 - u^{1/n})\}^{\lambda_1}}{(1 - u)^{n/\lambda_2}}$$

and

$$Q_n(u) = C\frac{u^{\frac{\lambda_1}{n}}}{(1 - u^{\frac{1}{n}})^{\lambda_2}}.$$

It is easily seen that the hazard quantile function is

$$H(u) = (1 - u)^{\lambda_2}\{Cu^{\lambda_1 - 1}(\lambda_1(1 - u) + \lambda_2 u)\}^{-1}, \tag{3.69}$$

the mean residual quantile function is

$$M(u) = \frac{1}{1 - u}B_u(1 - \lambda_2, 1 + \lambda_1) - \frac{Cu^{\lambda_1}}{(1 - u)^{\lambda_2}}, \tag{3.70}$$

the reversed hazard quantile function is

$$\Lambda(u) = (1 - u)^{\lambda_2 + 1}[Cu^{\lambda_1}(\lambda_1(1 - u) + \lambda_2 u)]^{-1}, \tag{3.71}$$

and the reversed mean residual quantile function is

$$R(u) = \frac{Cu^{\lambda_1}}{(1 - u)^{\lambda_2}} - \frac{1}{u}B_u(\lambda_1 + 1, 1 - \lambda_2). \tag{3.72}$$

Next, upon denoting the quantile function of the distribution by $Q(u; C, \lambda_1, \lambda_2)$, we have the following characterization results for this family of distributions (Nair and Sankaran [443]).

Theorem 3.1. *A non-negative variable X is distributed as $Q(u; C, \lambda_1, 0)$ if and only if*

(i) $H(u) = k_1 u[(1 - u)Q(u)]^{-1}$, $k_1 > 0$;
(ii) $M(u) = k_1[Q(u)]^{-1}$;
(iii) $R(u) = k_2 Q(u)$, $k_2 < 1$;
(iv) $\Lambda(u)R(u) = k_3$, $k_2 < 1$,

where k_i's are constants.

Theorem 3.2. *A non-negative random variable X is distributed as $Q_1(u;C,0,\lambda_2)$ if and only if*

1. $H(u) = A_1[Q_1(u)]^{-1}$, $A_1 > 0$;
2. $\Lambda(u) = A_1(1-u)[uQ_1(u)]^{-1}$;
3. $M(u) = A_2 Q_1(u)$;
4. $M(u)H(u) = A_3$, $A_3 > 0$,

where A_1, A_2 and A_3 are constants.

An interesting special case of (3.66) arises when $\lambda_1 = \lambda_2 = \lambda > 0$ in which case it becomes the loglogistic distribution (see Table 1.1). A detailed analysis of theloglogistic model and its applications in reliability studies have been made by Cox and Oakes [158] and Gupta et al. [237]. In this case, we deduce the following characterizations from the above.

Theorem 3.3. *A non-negative random variable X has loglogistic distribution with*

$$Q(u) = C\left(\frac{u}{1-u}\right)^{\lambda}, \quad C, \lambda > 0,$$

if and only if one of the following conditions hold:

(i) $H(u) = \frac{ku}{Q(u)}$;

(ii) $\Lambda(u) = \frac{k(1-u)}{Q(u)}$.

Hankin and Lee [259] proposed two methods of estimation—the least-squares and the maximum likelihood. In the least-squares method, they use

$$E(\log X_{r:n}) = \log C + \lambda_1 E(\log U_{r:n}) - \lambda_2 E(\log(1-U_{r:n})), \tag{3.73}$$

since $\log X_{r:n}$ and $\log Q(U_{r:n})$ have the same distribution, where $U_{r:n}$ is the rth order statistic from a sample of size n from the uniform distribution. Thus, from (3.36), we have

$$\begin{aligned} E(\log U_{r:n}) &= \frac{1}{B(r,n-r+1)} = \int_0^1 (\log u)u^{r-1}(1-u)^{n-r}du \\ &= -\left(\frac{1}{r} + \frac{1}{r+1} + \cdots + \frac{1}{n}\right) \end{aligned} \tag{3.74}$$

and

$$E(\log(1-U_{r:n})) = -\left(\frac{1}{n-r} + \frac{1}{n-r+1} + \cdots + \frac{1}{n}\right). \tag{3.75}$$

Then, the model parameters estimated by minimizing

$$\sum[\log X_{r:n} - E(\log X_{r:n})]^2.$$

Substituting (3.74) and (3.75) into the expression of $E(\log X_{r:n})$ in (3.73), we have an ordinary linear regression problem and is solved by standard programs available for the purpose. Maximum likelihood estimates are calculated as described earlier in Sect. 3.2.2 by following the steps described in Hankin and Lee [259]. In a comparison of the two methods by means of simulated variances, Hankin and Lee [259] found the least-squares method to be better for small samples when the parameters λ_1 and λ_2 are roughly equal and the maximum likelihood method to be better otherwise.

3.4 Govindarajulu's Distribution

Govindarajulu's [224] model is the earliest attempt to introduce a quantile function, not having an explicit form of distribution function, for modelling data on failure times. He considered the quantile function

$$Q(u) = \theta + \sigma\left\{(\beta+1)u^{\beta} - \beta u^{\beta+1}\right\}, \quad 0 \le u \le 1;\ \sigma,\beta > 0. \tag{3.76}$$

He used it to model the data on the failure times of a set of 20 refrigerators that were run to destruction under advanced stress conditions. Even though the validity of the model and its application to nonparametric inference were studied by him, the properties of the distribution were not explored. We now present a detailed study of its properties and applications.

The support of the distribution in (3.76) is $(\theta, \theta+\sigma)$. Since we treat it as a lifetime model, θ is set to be zero so that (3.76) reduces to

$$Q(u) = \sigma\left\{(\beta+1)u^{\beta} - \beta u^{\beta+1}\right\}, \quad 0 \le u \le 1. \tag{3.77}$$

Note that there is no loss of generality in studying the properties of this distribution based on (3.77) since the transformation $Y = X + \theta$, where X has its quantile function $Q(u)$ as in (3.77), will provide the corresponding results for (3.76). From (3.77), the quantile density function is

$$q(u) = \sigma\beta(\beta+1)u^{\beta-1}(1-u). \tag{3.78}$$

Equation (3.78) yields the density function of X as

$$f(x) = [\sigma\beta(\beta+1)]^{-1}F^{1-\beta}(x)(1-F(x))^{-1}. \tag{3.79}$$

Thus, this model belongs to the class of distributions, possessing density function explicitly in terms of the distribution function, discussed by Jones [307]. Further, by differentiating (3.78), we get

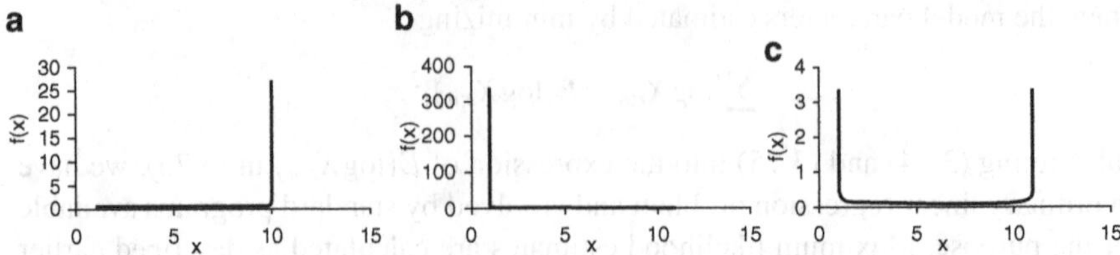

Fig. 3.6 Density plots of Govindarajulu's distribution for some choices of β. **(a)** $\beta = 3$; **(b)** $\beta = 0.5$; **(c)** $\beta = 2$

$$q'(u) = \sigma\beta(\beta+1)u^{\beta-2}[(\beta-1)-\beta u]$$

from which we observe that the density function is monotone decreasing for $\beta \le 1$, and $q'(u) = 0$ gives $u_0 = \beta^{-1}(\beta-1)$. Thus, when $\beta > 1$, there is an antimode at u_0. Figure 3.6 shows the shapes of the density function for some choices of β.

The raw moments are given by

$$E(X^r) = \int_0^1 [Q(p)]^r dp = \sigma^r \sum_{j=0}^{r} (-1)^j \binom{r}{j} (\beta+1)^{r-j}\beta^j/(\beta r + j + 1).$$

In particular, the mean and variance are

$$\mu = 2\sigma(\beta+2)^{-1},$$

$$\text{var} = \frac{\beta^2(5\beta+7)\sigma^2}{(2\beta+1)(2\beta+2)(\beta+2)^2}.$$

Moreover, we have the median as

$$M = \sigma 2^{-(\beta+1)}(\beta+2),$$

the interquartile range as

$$\text{IQR} = \sigma 4^{-(\beta+1)}[3^\beta(\beta+4) - (3\beta+4)],$$

the skewness as

$$S = \frac{\sigma[(\beta+1)\frac{3^\beta+1}{4^\beta} - \frac{\beta(3^{\beta+1}+1)}{4^\beta} + \frac{\beta+2}{2^\beta}]}{\text{IQR}}$$

and the kurtosis as

$$T = \frac{\sigma\left[(\beta+1)\frac{(7^\beta-5^\beta+3^\beta-1)}{8^\beta} - \frac{\beta(7^{\beta+1}-5^{\beta+1}+3^{\beta+1}-1)}{8^{\beta+1}}\right]}{\text{IQR}}$$

as percentile-based descriptive measures. Much simpler expressions are available for the L-moments as follows:

$$L_1 = \mu,$$
$$L_2 = \frac{2\beta\sigma}{(\beta+2)(\beta+3)},$$
$$L_3 = \frac{2\beta(\beta-2)\sigma}{(\beta+2)_{(3)}},$$
$$L_4 = \frac{2\beta^3 - 12\beta^2 + 10\beta)\sigma}{(\beta+2)_{(4)}}.$$

Consequently, we have

$$\tau_3 = \frac{\beta-2}{\beta+4} \quad \text{and} \quad \tau_4 = \frac{(\beta-5)(\beta-1)}{(\beta+4)(\beta+5)}.$$

With τ_3 being an increasing function of β, its limits are obtained as $\beta \to 0$ and $\beta \to \infty$. These limits show that τ_3 lies between $(-\frac{1}{2}, 1)$, and so it does not cover the entire range $(-1, 1)$. But the distribution has negatively skewed, symmetric (at $\beta = 2$) and positively skewed members. The L-kurtosis τ_4 is nonmonotone, decreasing initially, reaching a minimum in the symmetric case, and then increasing to unity.

A particularly interesting property of Govindarajulu's distribution is the distribution of its order statistics. The density function of $X_{r:n}$ is

$$\begin{aligned} f_r(x) &= \frac{1}{B(r,n-r+1)} f(x) F^{r-1}(x)(1-F(x))^{n-r} \\ &= \frac{1}{\sigma\beta(\beta+1)B(r,n-r+1)} F^{r-\beta}(x)(1-F(x))^{n-r-1}, \end{aligned} \tag{3.80}$$

upon using (3.79). So, we have

$$\begin{aligned} E(X_{r:n}) &= \frac{1}{B(r,n-r+1)} \int_0^1 Q(u) u^{r-1} (1-u)^{n-r} du \\ &= \frac{n!\Gamma(\beta+r)\sigma}{(r-1)!\Gamma(n+\beta+2)} \{(n+1)(\beta+1) - \beta(r-1)\}. \end{aligned} \tag{3.81}$$

In particular,

$$E(X_{1:n}) = \frac{(n+1)!\sigma\Gamma(\beta+2)}{\Gamma(n+\beta+2)} = \frac{(n+1)!\sigma}{(\beta+2)_{(n)}}$$

and

$$E(X_{n:n}) = \frac{n(n+2\beta+1)\sigma}{(n+\beta)(n+\beta+1)}.$$

The quantile functions of $X_{1:n}$ and $X_{n:n}$ are

$$Q_1(u) = \sigma[(1-(1-u)^{\frac{1}{n}})]^{\beta}[1+\beta(1-u)^{\frac{1}{n}}]$$

and

$$Q_n(u) = \sigma[(\beta+1)u^{\frac{\beta}{n}} - \beta u^{\frac{\beta+1}{n}}].$$

All the reliability functions also have tractable forms. The hazard quantile function is given by

$$H(u) = [\sigma\beta(\beta+1)u^{\beta-1}(1-u)^2]^{-1}$$

and the mean residual quantile function is

$$M(u) = [2-(\beta+1)(\beta+2)u^{\beta} + \beta(\beta+2)u^{\beta+1} - \beta(\beta+1)u^{\beta+2}] \\ \times[(\beta+2)(1-u)]^{-1}\sigma.$$

From the expression of the quantile function, it is evident that the parameter β largely controls the left tail and therefore facilitates in modelling reliability concepts in reversed time. Accordingly, the reversed hazard and reversed mean residual quantile functions are given by

$$\Lambda(u) = [\sigma\beta(\beta+1)u^{\beta}(1-u)]^{-1},$$
$$R(u) = \frac{\beta\sigma}{\beta+2}[\beta+2-(\beta+1)u]u^{\beta},$$

respectively. The reversed variance residual quantile function has the expression

$$D(u) = u^{-1}\int_0^u R^2(p)dp$$
$$= \frac{\sigma^2\beta^2u^{2\beta}}{(\beta+2)^2}\left\{\frac{(\beta+1)^2u^2}{2\beta+3} - (\beta+2)u + \frac{(\beta+1)^2}{2\beta+1}\right\}.$$

We further note that the function

$$R(u)\Lambda(u) = \frac{(\beta+1)^{-1} - (\beta+2)^{-1}u}{1-u} \quad (3.82)$$

is a homographic function of u.

Characterization problems of life distributions by the relationship between the reversed hazard rate and the reversed mean residual life in the distribution function approach have been discussed in literature; see Chandra and Roy [135]. In this spirit, from (3.82), we have the following theorem.

Theorem 3.4. *For a non-negative random variable X, the relationship*

$$R(u)\Lambda(u) = \frac{a+bu}{1-u} \tag{3.83}$$

holds for all $0<u<1$ *if and only if*

$$Q(u) = K\left(\frac{a}{1-a}u^{\frac{1}{a}-1} - au^{\frac{1}{a}}\right) \tag{3.84}$$

provided that a and b are real numbers satisfying

$$\frac{1}{a}+\frac{1}{b} = -1. \tag{3.85}$$

Proof. Suppose (3.83) holds. Then, we have

$$\left\{\frac{1}{u}\int_0^u pq(p)dp\right\}\left[uq(u)\right]^{-1} = \frac{a+bu}{1-u}. \tag{3.86}$$

Equation (3.86) simplifies to

$$\frac{uq(u)}{\int_0^u pq(p)dp} = \frac{1-u}{u(a+bu)} = \frac{1}{au} - \frac{a+b}{a(a+bu)}.$$

Upon integrating the above equation, we obtain

$$\log\int_0^u pq(p)dp = \frac{1}{a}\log u - \frac{a+b}{ab}\log(a+bu) + \log K,$$

or

$$\int_0^u pq(p)dp = Ku^{\frac{1}{a}}(a+bu)$$

with the use of (3.85). By differentiation, we then obtain

$$q(u) = Ku^{\frac{1}{a}-2}(1-u).$$

Integrating the last expression from 0 to u and setting $Q(0)=0$, we get (3.84). The converse part follows from the equations

$$\Lambda(u) = [Ku^{\frac{1}{a}-1}(1-u)]^{-1}$$

and

$$R(u) = \frac{Ka}{a+1} u^{\frac{1}{a-1}} (a+1-u).$$

Remark 3.2. Govindarajulu's distribution is secured when $a = (1+\beta)^{-1}$. The condition imposed on a and b in (3.85) can be relaxed to provide a more general family of quantile functions.

Regarding the estimation of the parameters σ and β, all the conventional methods like method of moments, percentiles, least-squares and maximum likelihood can be applied to the distribution quite easily. For example, in the method of moments, equating the mean and variance, we obtain

$$\bar{X} = \frac{2\sigma}{\beta+2} \quad \text{and} \quad S^2 = \frac{\beta^2(5\beta+7)\sigma^2}{(2\beta+1)(2\beta+2)(\beta+2)^2}. \tag{3.87}$$

Thus, we get

$$\frac{\bar{X}^2}{S^2} = \frac{4(2\beta+1)(2\beta+2)}{\beta^2(5\beta+7)}$$

which may be solved to get β. Then, by substituting it in (3.87), the estimate of σ can be found. There may be more than one solution for β and in this case a goodness of fit may then be applied to locate the best solution. The method of L-moments and some results comparing the different methods are presented in Sect. 3.6. Compared to the more flexible quantile functions discussed in the earlier sections, the estimation problem is easily resolvable in this case with no computational complexities. One of the major limitations of Govindarajulu's model, as mentioned earlier, is that it cannot cover the entire skewness range. In the admissible range, however, it provides good approximations to other distributions as will be seen in Sect. 3.6.

3.5 Generalized Weibull Family

This particular family of distributions is different from the distributions discussed so far in this chapter in the sense that it has a closed-form expression for the distribution function. So, all conventional methods of analysis are possible in this case. As a generalization of the Weibull distribution, the generalized Weibull family is defined by Mudholkar et al. [428] as

$$Q(u) = \begin{cases} \sigma\left[\frac{1-(1-u)^{\lambda}}{\lambda}\right]^{\alpha} & ,\ \lambda \neq 0 \\ \sigma(-\log(1-u))^{\alpha} & ,\ \lambda = 0 \end{cases}, \tag{3.88}$$

for $\alpha, \sigma > 0$ and real λ. The corresponding distribution function is

$$F(x) = 1 - \left\{1 - \lambda\left(\frac{x}{\sigma}\right)^{\frac{1}{\alpha}}\right\}^{\frac{1}{\lambda}}$$

with support $(0, \infty)$ for $\lambda \leq 0$ and $(0, \frac{\sigma}{\lambda^{\alpha}})$ for $\lambda > 0$. The quantile density function has the form

$$q(u) = \sigma\alpha\left[\frac{1-(1-u)^{\lambda}}{\lambda}\right]^{\alpha-1}(1-u)^{\lambda-1}. \tag{3.89}$$

The density function has a wide variety of shapes that include U-shaped, unimodal and monotone increasing or decreasing shapes. The raw moments are given by

$$E(X^r) = \begin{cases} \frac{B(\frac{1}{\lambda}, r\alpha+1)}{\lambda^{r\alpha+1}}\sigma^r & ,\ \lambda > 0 \\ \frac{B(-r\alpha-\frac{1}{\lambda}, r\alpha+1)}{(-\lambda)^{r\alpha+1}}\sigma^r & ,\ \lambda < 0 \end{cases}.$$

Moments of all orders exist for $\alpha > 0$, $\lambda > 0$. If $\lambda < 0$, then $E(X^r)$ exists if $\alpha\lambda > -r^{-1}$. The expressions for the percentile-based descriptive measures are as follows:

$$M = \left(\frac{1-(\frac{1}{2})^{\lambda}}{\lambda}\right)^{\alpha}\sigma,$$

$$\text{IQR} = \frac{\sigma}{\lambda^2}\left[\left(1-\left(\frac{1}{4}\right)^{\lambda}\right)^{\alpha} - \left(1-\left(\frac{3}{4}\right)^{\lambda}\right)^{\alpha}\right],$$

$$S = \frac{(1-(\frac{1}{4})^{\lambda})^{\alpha} + (1-(\frac{3}{4})^{\lambda})^{\alpha} - 2(1-(\frac{1}{2})^{\lambda})}{(1-(\frac{1}{4})^{\lambda})^{\alpha} - (1-(\frac{3}{4})^{\lambda})^{\alpha}},$$

$$T = \frac{(1-(\frac{1}{8})^{\lambda})^{\alpha} - (1-(\frac{3}{8})^{\lambda})^{\alpha} + (1-(\frac{5}{8})^{\lambda})^{\alpha} - (1-(\frac{7}{8})^{\lambda})^{\alpha}]}{(1-(\frac{1}{4})^{\lambda})^{\alpha} - (1-(\frac{3}{4})^{\lambda})^{\alpha}}.$$

For the calculation of L-moments, we use the result

$$\int_0^1 u^r Q(u)du = \sum_{y=0}^{r} \frac{(-1)^y\binom{r}{y}\sigma}{\lambda^{\alpha+1}} B\left(\frac{y+1}{\lambda}, \alpha+1\right)$$

in (1.34)–(1.37). Various reliability characteristics are determined as follows:

$$H(u) = \left[\sigma\alpha\left(\frac{1-(1-u)^{\lambda}}{\lambda}\right)^{\alpha-1}(1-u)^{\lambda}\right]^{-1},$$

$$M(u) = \frac{\sigma\alpha}{\lambda^{\alpha}} B_{(1-u)^{\lambda}} \Lambda(u),$$

$$\Lambda(u) = \left[u\sigma^{\alpha}\left(\frac{1-(1-u)^{\lambda}}{\lambda}\right)^{\alpha-1}(1-u)^{\lambda-1}\right]^{-1},$$

$$R(u) = \frac{\sigma\alpha}{u\lambda^{\alpha}}\left[\left(\frac{1-(1-u)^{\lambda}}{\lambda}\right)^{\alpha} - B\left(\frac{1}{\lambda}+1,\alpha\right) + B_{(1-u)^{\alpha}}\left(\alpha,\frac{1}{\lambda+1}\right)\right].$$

The parameters of the model are estimated by the method of maximum likelihood as discussed in Mudholkar et al. [428]. Due to the variety of shapes that the hazard functions can assume (see Chap. 4 for details), it is a useful model for survival data. This distribution has also appeared in some other discussions including assessment of tests of exponentiality (Mudholkar and Kollia [425]), approximations to sampling distributions, analysis of censored survival data (Mudholkar et al. [428]), and generating samples and approximating other distributions. Chi-squared goodness-of-fit tests for this family of distributions have been discussed by Voinov et al. [575].

3.6 Applications to Lifetime Data

In order to emphasize the applications of quantile function in reliability analysis, we demonstrate here that some of the models discussed in the preceding sections can serve as useful lifetime distributions. The conditions in the parameters that make the underlying random variables non-negative have been obtained. We now fit these models for some real data on failure times. Three representative models, one each from the lambda family, the power-Pareto and Govindarajulu's distributions, will be considered for this purpose. The first two examples have been discussed in Nair and Vineshkumar [452].

The four-parameter lambda distribution in (3.44) is applied to the data of 100 observations on failure times of aluminum coupons (*data source*: Birnbaum and Saunders [104] and quoted in Lai and Xie [368]). The last observation in the data is excluded from the analysis to extract equal frequencies in the bins. Distribute the data into ten classes, each containing ten observations in ascending order of magnitude. For estimating the parameters, we use the method of L-moments. The first four sample L-moments are $l_1 = 1,391.79$, $l_2 = 215.683$, $l_3 = 3.570$ and $l_4 = 20.7676$. Thus, the model parameters need to be solved from

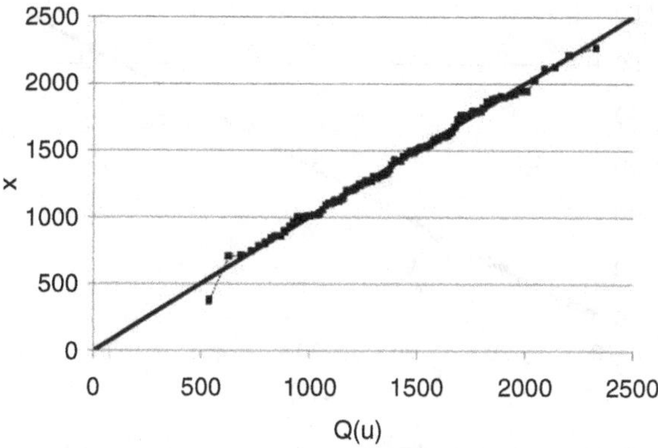

Fig. 3.7 *Q-Q* plot for the data on lifetimes of aluminum coupons

$$\lambda_1 + \frac{1}{\lambda_2}\left(\frac{1}{\lambda_4+1} - \frac{1}{\lambda_3+1}\right) = 1{,}391.79,$$

$$\frac{1}{\lambda_2}\left(\frac{1}{(\lambda_3+1)(\lambda_3+2)}) + \frac{1}{(\lambda_4+1)(\lambda_4+2)}\right) = 215.683,$$

$$\frac{1}{\lambda_2}\left(\frac{\lambda_3-1}{(\lambda_3+1)(\lambda_3+2)(\lambda_3+4)} - \frac{\lambda_4-1}{(\lambda_4+1)(\lambda_4+2)(\lambda_4+3)}\right) = 3.570,$$

$$\frac{1}{\lambda_2}\left(\frac{(\lambda_3-1)(\lambda_3-2)}{(\lambda_3+1)(\lambda_3+2)(\lambda_3+3)(\lambda_3+4)} - \frac{(\lambda_4-1)(\lambda_4-2)}{(\lambda_4+1)(\lambda_4+2)(\lambda_4+3)(\lambda_4+4)}\right) = 20.7676.$$

Among the solutions, the best fitting one, determined by the chi-square test (i.e., the parameter estimates that gave the least chi-square value), is

$$\hat{\lambda}_1 = 1{,}382.18, \quad \hat{\lambda}_2 = 0.0033, \quad \hat{\lambda}_3 = 0.2706 \quad \text{and} \quad \hat{\lambda}_4 = 0.2211. \tag{3.90}$$

Further, the upper limit of the support is 2,750.7, and thus the estimated support (256.28, 2,750.7) covers the range of observations (370, 2,240) in the data. Using (3.44) for $u = \frac{1}{10}, \frac{2}{10}, \cdots$, and the fact that if U has a uniform distribution on $[0,1]$ then X and $Q(u)$ have identical distributions, we find the observed frequencies in the classes to be 10, 10, 9, 12, 8, 11, 8, 12 and 10. Of course, under the uniform model, the expected frequencies are 10 in each class. Thus, the optimized chi-square value for the fit is $\chi^2 = 1.8$ which does not lead to rejection of the model in (3.44) for the given data. The *Q-Q* plot corresponding to the model is presented in Fig. 3.7.

The second example concerns the power-Pareto distribution in (3.66). To ascertain the potential of the model, we fit it to the data on the times to first failure of 20 electric carts, presented by Zimmer et al. [604], and also quoted in Lai and Xie [368]. Here again, the method of L-moments is adopted. The sample L-moments are

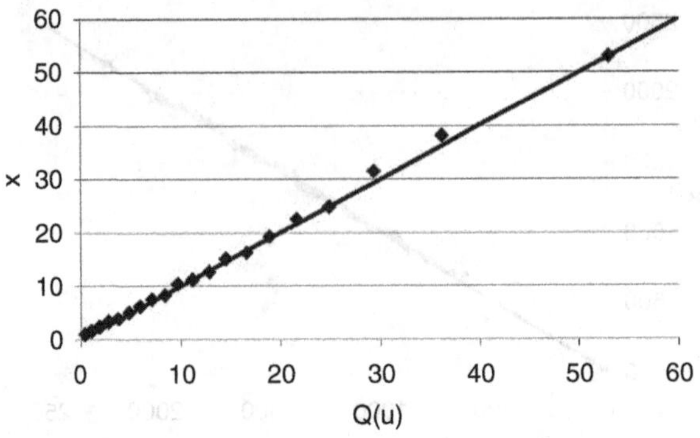

Fig. 3.8 *Q-Q* plot for the data on times to first failure of electric carts

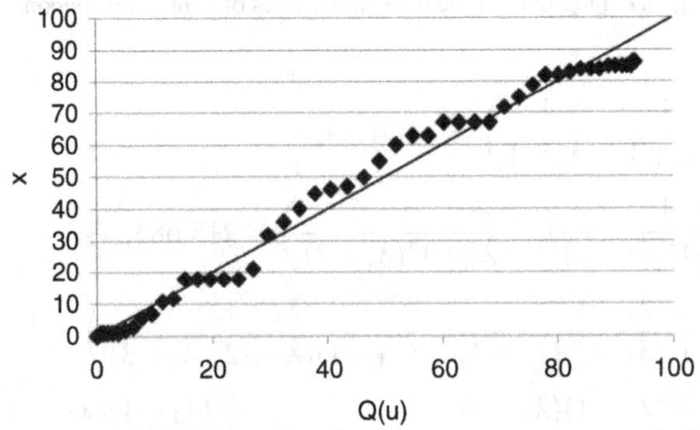

Fig. 3.9 *Q-Q* plot for the data on failure times of devices using Govindarajulu's model

$l_1 = 14.675, l_2 = 7.335$ and $l_3 = 2.4678$. Equating the population *L*-moments L_1, L_2 and L_3 presented in Sect. 3.3 to l_1, l_2 and l_3 and solving the resulting equations, we obtain

$$\hat{\lambda}_1 = 0.2346, \quad \hat{\lambda}_2 = 0.0967 \quad \text{and} \quad \hat{C} = 1{,}530.3.$$

The corresponding *Q-Q* plot is presented in Fig. 3.8.

Govindarajulu's distribution has already been shown as a suitable model for failure times in the original paper of Govindarajulu [224]. We reinforce this by fitting it to the data on the failure times of 50 devices, reported in Lai and Xie [368]. Equating the first two *L*-moments with those of the sample, the estimates of the model parameters are obtained as

$$\hat{\sigma} = 93.463 \quad \text{and} \quad \hat{\beta} = 2.0915.$$

Dividing the data into five groups of ten observations each, we find by proceeding as in the first example that the chi-square value is 1.8, which does not lead to the rejection of the considered model. Figure 3.9 presents the *Q-Q* plot of the fit obtained.

The objectives of these illustrations were limited to the purpose of demonstrating the use of quantile function models in reliability analysis. A full theoretical analysis and demonstration to real data situations of all the reliability functions *vis-a-vis* their ageing properties will be taken up subsequently in Chap. 4.

Chapter 4
Ageing Concepts

Abstract A considerable part of reliability theory is dedicated to the study of ageing concepts, their properties, implications and applications. In this chapter, we review some of the important results in this area and translate the basic definitions to make them amenable for a quantile-based analysis. Ageing represents the phenomenon by which the residual life of a unit is affected by its age in some probabilistic sense. It can be positive ageing, negative ageing or no ageing, according to whether the residual lifetime decreases, increases or remains the same as age advances. Generally, one investigates whether a given ageing concept preserves certain reliability operations such as formation of coherent structures, mixtures and convolutions. We first introduce the basic ideas behind convergence, mixtures, convolutions, shock models and equilibrium distributions. The ageing concepts are studied under three broad categories—based on hazard functions, residual life functions and survival functions. The IHR, IHR(2), IGHR, NBUHR, NBUHRA, SIHR, IHRA, DMTTF, IHRA* t_0 classes and their duals along with their properties come under ageing notions related to the hazard function. In the class of concepts based on residual life, we discuss DMRL, DMRLHA, UBA, UBAE, HUBAE, DRMRL, DVRL, DVRLA, NDMRL, NDVRL, IPRL-α, DMERL classes and their duals. Those defined in terms of the survival function include NBU, NBU-t_0, NBU* t_0, NBU(2), SNBU, NBUE, NBU(2)-t_0, NBUL, NBUP-α, NBUE, HNBUE, $\mathscr{L}$-class, $\mathscr{M}$-class and the renewal notions NBRU, RNBU, RNBUE, RNBRU and RNBRUE. A brief discussion is also made on classes of distributions possessing monotonic properties for reliability concepts in reversed time. The ageing properties of the quantile function models introduced in Chap. 3 are presented. Finally, some definitions and results on relative ageing are detailed.

4.1 Introduction

A considerable part of reliability theory is dedicated to the study of ageing concepts, their properties, implications and applications. We review here some important results in this area and translate the basic results to make them amenable for a

N.U. Nair et al., *Quantile-Based Reliability Analysis*, Statistics for Industry and Technology, DOI 10.1007/978-0-8176-8361-0_4,

quantile-based analysis. We recall that the lifetime of a unit is the time up to which the unit performs as it is required to do. The age of the unit is the time up to which the unit fails to function. By the term ageing of a mechanical or biological unit with a lifetime distribution, we mean that the residual life of the unit is affected by its age in some probabilistic sense. This description includes the cases in which a unit experiences no ageing, positive ageing or negative ageing. Positive ageing simply means that the residual lifetime decreases when the age of the unit increases and it reflects an adverse effect of age on the random lifetime of the unit. This is a common feature of equipments or mechanical systems that are subject to gradual wear and tear while in operation. Negative ageing, which is the dual concept of positive ageing, has a beneficial effect on the life of the unit as age progresses. A well-known example is that of human beings where the average remaining life increases after infancy. This is also the case with mechanical devices that suffer from designing or manufacturing errors or operation by inexperienced hands initially which improves their performance later, as well as systems that undergo preventive maintenance which reduces wear out failures. No ageing applies to units whose remaining life continues to be the same at all ages. This situation is identified by equating the lifetime distribution with its residual lifetime distribution, i.e.,

$$\frac{\bar{F}(x+t)}{\bar{F}(t)} = \bar{F}(x) \qquad \forall\, x,t. \tag{4.1}$$

It is known that the exponential distribution is the only lifetime model that satisfies (4.1). Sometimes, (4.1) is referred to as the lack of memory property of the exponential distribution, and it plays a key role in the definitions of various ageing criteria.

There are several advantages in studying ageing concepts. Primarily, they reveal in some sense the pattern in which a unit deteriorates or improves in its functioning with respect to age. This, of course, imparts valuable information about the reliability of a unit. Secondly, life distributions are classified on the basis of their ageing behaviour. In practice, a knowledge of this ageing behaviour could be used effectively for model selection. Nonparametric inferential methods generally make few assumptions about the population from which observations are drawn. Ageing behaviour is an advantage in such situations, especially when the ageing class has geometric properties like convexity, star shapedness, etc. Moreover, these classes have several desirable properties in their own right such as bounds on the reliability function, distributional properties like unimodality, inequalities among moments and preservation properties under various reliability operations. Further, the identification of the no ageing property of the exponential law becomes a base in the construction of tests of hypothesis of exponentiality against different ageing criteria. Often, reliability classes, with appropriate interpretations, form the basis of model selection in other disciplines. While discussing different ageing classes and criteria, we distinguish them as based on the hazard quantile function, residual quantile function and some other properties. The quantile-based definitions of the ageing concepts discussed here are available in Nair and Vineshkumar [454].

4.2 Reliability Operations

Before defining ageing concepts, we provide some preliminary details which are vital to reliability theory. It is customary to investigate whether a given life distribution possessing a specific ageing property preserves it under various reliability operations. Important among the operations are formation of coherent structures of independent components, addition of lifelengths or convolution and formation of mixtures of distributions.

4.2.1 Coherent Systems

Coherent systems, introduced by Birnbaum et al. [103], helped to a great extent in laying the foundation of reliability theory as a separate discipline. Here, we consider systems that can be in only one of two states, functioning or failed. A structure function ϕ of a coherent system of n components (or a coherent system of order n) is such that, with $x_i = 1$ if the ith component functions and $x_i = 0$ if it fails,

(i) $\phi(1,1,\ldots,1) = 1$,
(ii) $\phi(0,0,\ldots,0) = 0$,
(iii) $\phi(x_1,x_2,\ldots,x_n) \geq \phi(y_1,y_2,\ldots,y_n)$, whenever $x_i \geq y_i$, $i = 1,2,\ldots,n$.

For example, the structure function of series and parallel systems described in Chap. 1 are

$$\phi(x_1,x_2,\ldots x_n) = x_1 x_2 \ldots x_n$$

and

$$\phi(x_1,x_2,\ldots x_n) = 1-(1-x_1)(1-x_2)\ldots(1-x_n),$$

respectively. The states of the components of a coherent system are represented by Bernoulli random variables and the reliability function of the system is then

$$R(p_1,p_2,\ldots p_n) = E(\phi(X_1,X_2,\ldots,X_n)), \quad \text{with } p_i = P(X_i = 1),$$

where $X_1,X_2,\ldots,X_n$ are independent. Introducing a concept of time T now, let $X(t) = 0$ or 1 depending on whether the unit is failed or functioning at time t, we have

$$P(X(t) = 1) = \bar{F}(t) \tag{4.2}$$

as the survival function of T. For a coherent system of order n, $\phi(X_1(t),\dots,X_n(t))$ is the performance process and the system life T has survival function

$$\bar{F}(t) = R(\bar{F}_1(t),\dots,\bar{F}_n(t)),$$

where $\bar{F}_i$ is the survival function of the component lifelengths as defined in (4.2). For further details and properties, one may refer to Barlow and Proschan [69].

4.2.2 Convolution

Let X and Y be two independent random variables with distribution functions $F(x)$ and $G(x)$. The distribution of $X+Y$, specified by the distribution function

$$H_1(x) = \int_0^x F(x-t)dG(t), \tag{4.3}$$

is referred to as the convolution of X and Y, and is usually denoted by $H_1 = F * G$. In the reliability context, convolution is interpreted as the operation of adding lifelength X to Y. When a spare part is available as the replacement of the original part, in case of failure of the latter, the two together acts as a system of two components and the available total lifetime in this case is $X+Y$, the sum of the lengths of lives of the original and spare parts. Also, if the component with life Y fails at any time t preceding x, while component with life X fails during the remaining interval of time $x-t$, then (4.3) is the probability that the sum $X+Y$ does not exceed x.

4.2.3 Mixture

Assuming that $F(x|\theta)$, $\theta \in \Theta$, be a family of distributions and $G(\theta)$ be a distribution on Θ, the mixture of F with G is given by

$$H_2(x) = \int_\Theta F(x|\theta)\,dG(\theta). \tag{4.4}$$

The hazard rate of (4.4) is

$$h^*(x) = \frac{\int_\Theta f(x|\theta)dG(\theta)}{\int_\Theta \bar{F}(x|\theta)dG(\theta)}.$$

If the hazard rate of $F(x|\theta)$ is bounded by $h_1 < h(x|\theta) < h_2$ for all θ in the support of G, then $h_1 < h^*(x) < h_2$. A particular case of interest in practice is when the population consists of observations belonging to two different categories with proportions α and $1-\alpha$ having distribution functions F_1 and F_2. Then, the mixture distribution in (4.4) becomes

$$H_2(x) = \alpha F_1(x) + (1-\alpha)F_2(x),$$

which is commonly referred to as the two-component mixture model. In this case,

$$h^*(x) = \alpha(x)h_1(x) + (1-\alpha(x))h_2(x),$$

where $h_i(x)$ is the hazard rate corresponding to F_i, and

$$\alpha(x) = \frac{\alpha_1 \bar{F}_1(x)}{\alpha_1 \bar{F}_1(x) + \alpha_2 \bar{F}_2(x)};$$

furthermore, we have

$$\min(h_1(x), h_2(x)) \le h^*(x) \le \max(h_1(x), h_2(x)).$$

For an elaborate discussion on mixtures of life distributions, one may refer to Marshall and Olkin [412].

4.2.4 *Shock Models*

A unit may fail because of the changes within it or due to the changes in the environment. An approach to describe the state of the system over time is through a stochastic process, and the unit is deemed to have failed when the designated process surpasses a threshold level. For units that deteriorate over time, the failure occurs as a result of accumulated shocks received over time. Let $(N(t), t \ge 0)$ be the number of shocks received by the unit in the time interval $[0,t]$ and $\bar{P}_K = P(K > k)$, the probability that a unit survives k shocks, $k = 1, 2, \ldots$. Then, the survival function of the unit is given by

$$\bar{H}_3(x) = \sum_{k=0}^{\infty} [P(N(x) = k)]\bar{P}_K, \quad x \ge 0. \tag{4.5}$$

Different models for $N(t)$, like Poisson process, birth process and so on, have been considered in the literature. In the case of a Poisson process, (4.5) simplifies to

$$\bar{H}_3(x) = \sum_{k=0}^{\infty} \bar{P}_K e^{-\lambda x} \frac{(\lambda x)^k}{k!}. \tag{4.6}$$

Attention is usually given to the problem of ascertaining whether or not a concept of ageing admits a shock model interpretation; for details, see Esary et al. [189].

4.2.5 Equilibrium Distributions

A topic that is quite useful in the analysis of the ageing phenomenon is that of equilibrium distributions. Assume that we have a set of n units. We start working with a new unit at time zero, replace it upon failure by a second unit, and so on. If the failure times X_i, $i = 1,2,\dots,n$, of the units are independent and identically distributed, then the sequence of points (S_n), where $S_n = X_1 + X_2 + \cdots + X_n$ constitute a renewal process (Cox [157]). Let $F(x)$ be the common distribution function of X_i's, satisfying $F(0) = 0$ and $\mu = E(X_i) < \infty$. Upon denoting the age and remaining life of the unit in use at time T by U_T and V_T, respectively, the asymptotic distribution of U_T and V_T turn out to be

$$\bar{G}(x) = \mu^{-1} \int_x^\infty \bar{F}(t)dt, \tag{4.7}$$

which is called the equilibrium distribution corresponding to F. We shall denote by Z the random variable with survival function in (4.7).

For a non-negative and absolutely continuous random variable X with $E(X^n) < \infty$, Fagiouli and Pellerey [191] extended (4.7) by defining the nth order equilibrium recursively through the survival functions

$$S_n(x) = \mu_{n-1}^{-1} \int_x^\infty S_{n-1}(t)dt, \quad n = 1,2,3,\dots, \tag{4.8}$$

where $\mu_n = \int_0^\infty S_n(t)dt$. Notice that $S_1(x) = \bar{F}(x)$, $S_2(x) = \bar{G}(x)$ and $S_0(x) = \frac{f(x)}{f(0)}$, which is a survival function iff $f(x)$ is continuous and decreasing. If Z_n is the random variable corresponding to (4.8), we also have $Z_1 = X$ and $Z_2 = Z$ according to the earlier notation. The relationships between the survival functions $\bar{F}$ and S_n, hazard rates $h(x)$ and $h_n(x)$, and mean residual life functions $m(x)$ and $m_n(x)$ of X and Z_n, respectively, are as follows (Nair and Preeth [441]):

$$S_n(x) = \frac{E[(X-x)^{n-1}|X > x]}{E(X^{n-1})}\bar{F}(x),$$

$$h_n(x) = [m_{n-1}(x)]^{-1},$$

$$m_{n-1}(x) = \frac{m_n(x)}{1+m_n'(x)},$$

$$h_{n-1}(x) = h_n(x) - \frac{h_n'(x)}{h_n(x)} \quad n = 2,3,\dots,$$

$$m_n(x) = \frac{E[(X-x)^n|X > x]}{nE[(X-x)^{n-1}|X > x]}.$$

Nair and Preeth [441] also established that the generalized Pareto distribution (see Table 1.1) is characterized by any one of the relations

$$m_n(x) = C_n m(x),$$
$$h_n(x) = K_n h(x),$$
$$E[(X-x)^n | X > x] = A_n m^n(x),$$

where C_n, A_n and K_n are positive constants, and the two-component exponential mixture distribution

$$\bar{F}(x) = \alpha e^{-\lambda_1 x} + (1-\alpha)e^{-\lambda_2 x}, \quad x > 0;\ \alpha \geq 0,\ 0 < \lambda_1 < \lambda_2,$$

by

$$m_n(x) = \alpha_1 - \alpha_2 - \alpha_1 \alpha_2 h_n(x),$$

where

$$\alpha_i = \lambda_i^{-1}, \quad i = 1,2.$$

Various distributional properties and reliability aspects have been discussed by Gupta [233], Nanda et al. [460], Stein and Dattero [546] and Gupta [236].

Quantile-based analysis of equilibrium distributions is straightforward. Setting $x = Q(u)$ in (4.7), we have

$$G(Q_X(u)) = \mu^{-1} \int_0^u (1-p)q(p)dp, \tag{4.9}$$

where the integral

$$T_X(u) = \int_0^u (1-p)q(p)dp$$

is the well-known total time on test transform of the random variable X. The properties and reliability implications of $T_X(u)$ will be taken up later on in Chap. 5. Thus, from (4.9), we have

$$Q_X(u) = \mu^{-1} Q_Z(T_X(u))$$

or

$$Q_Z(u) = \mu Q_X(T_X^{-1}(u)).$$

Note that $T_X^{-1}(u)$ is a distribution function on $[0,1]$ since $T(u)$ is a quantile function.

Example 4.1. Let X be distributed as generalized Pareto with quantile function

$$Q(u) = \frac{b}{a}\left[(1-u)^{-\frac{a}{a+1}} - 1\right].$$

Then,

$$T(u) = \int_0^u (1-p)q(p)dp = b\left[1-(1-u)^{-\frac{1}{a+1}}\right]$$

and

$$\mu^{-1}T(u) = 1-(1-u)^{-\frac{1}{a+1}}.$$

Hence, the equilibrium distribution has its quantile function as

$$Q_Z(u) = Q[T^{-1}(u)] = Q\left[1-(1-u)^{-(a+1)}\right] = \frac{b}{a}\left[(1-u)^{-a}-1\right].$$

We note from (4.9) that

$$\begin{aligned}\bar{G}(Q_X(u)) &= 1-\mu^{-1}\int_0^u (1-p)q(p)dp \\ &= \mu^{-1}\left[\mu - \int_0^u (1-p)q(p)dp\right] \\ &= \mu^{-1}\int_u^1 (1-p)q(p)dp.\end{aligned}$$

Differentiating logarithmically, we obtain

$$\frac{g(Q_X(u))}{\bar{G}(Q_X(u))}q(u) = \frac{(1-u)q(u)}{\int_u^1 (1-p)q(p)dp}$$

or

$$H_Z(u) = [M_X(u)]^{-1}, \tag{4.10}$$

thus revealing that the hazard quantile function of the equilibrium random variable is simply the reciprocal of the mean residual quantile function of the baseline distribution. Since Z_n is in the equilibrium version of Z_{n-1}, we have

$$H_n(u) = [M_{n-1}(u)]^{-1}, \quad n = 2,3,\ldots, \tag{4.11}$$

and so

$$M_{n-1}(u) = M_n(u) - (1-u)M_n'(u). \tag{4.12}$$

Equation (4.10) can also be used to derive the quantile function of Z.

Example 4.2. The generalized lambda distribution with

$$Q_X(u) = \lambda_1 + \lambda_2^{-1}(u - (1-u)^{\lambda_4}), \quad \lambda_1 - \frac{1}{\lambda_2} \geq 0,$$

has the mean residual quantile function as

$$M_X(u) = \lambda_2^{-1}\left[\lambda_4(1+\lambda_4)^{-1}(1-u)^{\lambda_4} + \frac{1-u}{2}\right].$$

Hence,

$$H_Z(u) = \frac{\lambda_2}{\lambda_2(1+\lambda_4)^{-1}(1-u)^{\lambda_4} + \frac{1-u}{2}}.$$

The corresponding quantile function can then be obtained as

$$Q_Z(u) = \frac{1}{\lambda_2}\left[\frac{u}{2} + \frac{1}{1+\lambda_4}(1-(1-u)^{\lambda_4})\right].$$

4.3 Classes Based on Hazard Quantile Function

4.3.1 *Monotone Hazard Rate Classes*

These classes of life distributions are defined by the nature of the monotonicity of the hazard function, $h(x)$. In the sequel, we use the term increasing (decreasing) in the sense of non-decreasing (non-increasing). We say that the random variable X or its distribution has increasing hazard rate, or X is IHR in short, if and only if for all t such that $\bar{F}(t) > 0$,

$$\bar{F}_t(x) = \frac{\bar{F}(t+x)}{\bar{F}(t)}$$

is decreasing (increasing) in t for all $x \geq 0$. This means that the residual life distribution is stochastically decreasing (increasing) in t. It is immediate that this definition also implies that X is IHR if and only if the hazard rate $h(x) = \frac{f(x)}{\bar{F}(x)}$ is increasing. Similarly, X has decreasing hazard rate iff $h(x)$ is decreasing, and we say in this case that X is DHR. If $h(x)$ is differentiable, it follows that X is IHR (DHR) according as $h'(x) \geq (\leq) 0$. Since

$$h'(x) = \frac{dh(x)}{dx} = \frac{dh(Q(u))}{du}\frac{du}{dQ(u)} = H'(u)\frac{1}{q(u)}$$

and $q(u) > 0$, we can present the following definition in terms of hazard quantile functions.

Definition 4.1. A lifetime random variable is IHR (DHR) if and only if its hazard quantile function satisfies

$$H'(u) \geq (\leq)\, 0 \quad \text{for } 0 < u < 1.$$

Thus, all distributions specified in terms of $F(x)$ that are IHR (DHR) preserve the same property when specified by $Q(u)$ as well. We will retain the conventional nomenclature IHR (DHR) in the case of the quantile approach as well. Sometimes, it is easier to establish the IHR property by using any of the following equivalent conditions:

(i) $H(u_2) \geq H(u_1)$ for all $0 < u_1 < u_2 < 1$,
(ii) The quantile function of the residual life

$$Q(u_0 + (1-u_0)u) - Q(u_0) \text{ is a decreasing function of } u_0.$$

Property (i) is obvious. To prove Property (ii), we note that

$$\begin{aligned}
&Q(u_0 + (1-u)u_0) - Q(u_0) \text{ is decreasing}\\
&\quad\Rightarrow q(u_0 + (1-u_0)u)(2-u) - q(u_0) \leq 0\\
&\quad\Rightarrow \frac{1}{(1-u_0)q(u_0)} \leq \frac{1}{(1-(u_0+(1-u_0)u))(q(u_0+(1-u_0)u))}\\
&\quad\Rightarrow H(u_0) \leq H(u_0 + (1-u_0)u)\\
&\quad\Rightarrow \text{IHR by (i).}
\end{aligned}$$

The inverse implication also holds by taking $u_2 = u_1 + (1-u_1)u$ so that $u_2 \geq u_1$ for every $0 < u < 1$. Taking $u_1 = u_0$ and retracing the steps in the above proof, we get the required result.

Example 4.3. The generalized exponential law with quantile function (see Table 1.1)

$$Q(u) = \sigma[-\log(1-u^{1/\theta})]$$

has its hazard quantile function as

$$H(u) = \frac{\theta}{\sigma(1-u)}[u^{1-\frac{1}{\theta}} - u],$$

and

$$H'(u) = \frac{\theta}{\sigma(1-u)^2}\left[\frac{\theta-1+u}{\theta}u^{-\frac{1}{\theta}} - 1\right].$$

Hence, $H'(u) > 0$ for $\theta > 1$, $H'(u) < 0$ for $\theta < 1$, and $H(u)$ is constant for $\theta = 1$. Thus, X is IHR for $\theta \geq 1$ and DHR for $\theta \leq 1$.

Not all hazard quantile functions are monotone in nature. It may belong to some other categories like bathtub or upside bathtub-shape, periodic, roller-coaster shape and polynomial type. These alternative forms will be discussed now briefly.

Definition 4.2. The random variable X is said to have a bathtub-shaped hazard quantile function, or X is BT, if

$$H(u) = \begin{cases} H_1(u) & u \leq u_1 \\ c, & u_1 \leq u \leq u_2 \\ H_2(u) & u \geq u_2 \end{cases},$$

where c is a constant, $H_1(u)$ is strictly decreasing and $H_2(u)$ is strictly increasing. When $u_1, u_2 \to 0$, X is IHR and when $u_1, u_2 \to 1$, X is DHR. We also say that u_1 and u_2 are change points of $H(u)$. On the other hand, if $H_1(u)$ is strictly increasing and $H_2(u)$ is strictly decreasing, an upside-down bathtub-shaped (UBT) hazard quantile function results.

Often, many life distributions have only one change point in which case the following definition is more convenient.

Definition 4.3. Assuming differentiability of $H(u)$, X is BT (UBT) if and only if $H'(u) < (>)\ 0$ for u in $(0, u_0)$, $H'(u_0) = 0$ and $H'(u) > (<)\ 0$ in $(u_0, 1)$.

Bathtub-shaped curves arise in different scenarios. In some cases, the life of a unit is affected by a mixture of defects with varying intensities. Those with serious defects at the initial stages have a high rate of failure, but as the unit functions without failure, such defects no longer persist so that the hazard function decreases and later becomes steady with almost a constant rate. Finally, when the adverse effect of age surfaces, the hazard quantile exhibits increasing tendency until the unit fails to function. Other factors such as changes in the hazard conditions due to the unit or environment, wear out of items with flaws, introduction of tests, inspection or maintenance that limits the occurrence of failures can also give rise to BT or UBT distributions. There are cases in which a proportion of items come from an IHR distribution and the remaining come from a DHR distribution, producing BT-shaped hazard for the combined set of observations, as Kao [310] has demonstrated. Table 4.1 presents the behaviour of the hazard quantile functions of standard life distributions presented earlier in Table 2.1. When failures are caused by fatigue or corrosion, X follows unimodal or UBT distributions such as lognormal, inverse Gaussian and inverted gamma, which do not have tractable quantile functions. Further details can be seen in Jiang et al. [292].

Glaser [220] established a general theorem that facilitates the determination of whether X is IHR, DHR, BT or UBT. He made use of the function $\eta(x) = -\frac{f'(x)}{f(x)}$ which is identified with

Table 4.1 Nature of hazard quantile functions of distributions in Table 2.1

Distribution	Shape of $H(u)$
Exponential	Constant
Weibull	Increasing for $\lambda > 1$, constant for $\lambda = 1$, decreasing for $\lambda < 1$.
Pareto II	Decreasing
Rescaled beta	Increasing
Half-logistic	Increasing
Power	Increasing for $\beta \geq 1$ and BT for $\beta < 1$
Pareto I	Decreasing
Burr type XII	Decreasing for $c \leq 1$ and UBT for $c > 2$
Geompertz	Increasing for $C > 1$, decreasing for $C < 1$, and constant for $C = 1$.
Log logistic	Decreasing for $\beta \leq 1$ and UBT for $\beta > 1$
Exponential geometric	Decreasing
Exponentiated Weibull	$\lambda \leq 1, \lambda\theta \leq 1$ decreasing $\lambda \geq 1, \lambda\theta \geq 1$ increasing $\lambda < 1, \lambda\theta > 1$ UBT $\lambda > 1, \lambda\theta < 1$ BT
Generalized exponentiated	$\theta \leq 1$ decreasing $\theta \geq 1$ increasing
Extended Weibull	$\theta \geq 1\ \lambda \geq 1$ increasing $\theta \leq 1\ \lambda \leq 1$ decreasing $\lambda > 1\ (< 1)$ initially increasing (decreasing) and eventually increasing (decreasing) but there may be and interval in which decreasing (increasing)
Inverse Weibull	UBT
Generalized Pareto	$a < 0$ increasing $a > 0$ increasing $a \to 0$ constant
Exponential power	$\alpha \geq 1$ increasing, $\alpha < 1$ BT
Modified Weibull extension	$\alpha \geq 1$ increasing, $0 < \alpha < 1$ BT
Log Weibull	$0 \leq k \leq 1$ decreasing, $k > 1$ UBT
Dimitrakopoulou et al.	$\alpha > 1, \beta \geq 1, \alpha < 1, \beta \leq 1$ increasing $\alpha \geq 1, \beta < 1$ and $\alpha\beta \leq 1 (> 1)$, Increasing (BT) $\alpha\beta < 1$ unimodel, $\alpha \geq 1$ increasing
Geometric Weibull	$0 < \alpha \leq 1$ decreasing, $\alpha > 1$ increasing
Logistic exponential	$0 < k < 1$ BT, $k > 1$ UBT, $k = 1$ constant
Kus	Decreasing
Greenwich	UBT

$$J(u) = \frac{q'(u)}{q^2(u)}. \tag{4.13}$$

Parzen [484] refers to $J(u)$ as the score function, and it proves to be quite useful in classifying probability distributions according to tail length. In terms of $J(u)$, we have the following adaptation of Glaser's result.

Theorem 4.1. *(a) X is IHR (DHR) according to $J(u)$ being increasing (decreasing) for all u;*

(b) Let $J(u)$ be BT (UBT) in the sense of Definition 4.3. Then, if there exist a u_0 for which $J(u_0) = H(u_0)$, X is BT (UBT). If there is no such u_0, then X is IHR (DHR).

Theorem 4.2. *Let*

$$\lim_{u\to 0} \frac{1}{q(u)} = \alpha \quad \textit{and} \quad \lim_{u\to 0} \frac{J(u)}{H(u)} = \beta.$$

Then,

(a) if $J(u)$ is BT, X is IHR if either $\alpha = 0$ or $\beta < 1$, and X is BT if either $\alpha = \infty$ or $\beta > 1$;

(b) if $J(u)$ is UBT, X is DHR if $\alpha = \infty$ or $\beta > 1$, and UBT if $\alpha = 0$ or $\beta < 1$.

Remark 4.1. Notice that, in the above two theorems, $J(u)$ is increasing or decreasing according to

$$J'(u) = q(u)q''(u) - [q'(u)]^2$$

being $\geq$ or ≤ 0, and

$$\frac{J(u)}{H(u)} = \frac{(1-u)q'(u)}{q(u)},$$

so that the relevant quantities can be directly obtained from the quantile density function $q(u)$.

Example 4.4. Consider the inverse Gaussian law with probability density function

$$f(x) = \left(\frac{\lambda}{2\pi x^3}\right)^{\frac{1}{2}} \exp\left[-\frac{\lambda}{2\mu^2 x}(x-\mu)^2\right], \quad x > 0;\ \lambda, \mu > 0.$$

The distribution has no tractable quantile function. Yet, we can write

$$[q(u)]^{-1} = \left(\frac{\lambda}{2\pi}\right)^{\frac{1}{2}} \left\{Q^{-\frac{3}{2}}(u) \exp\left[-\frac{\lambda}{2\mu^2}\left(Q(u) - 2\mu + \frac{\mu^2}{Q(u)}\right)\right]\right\}, \quad 0 \leq u \leq 1, \tag{4.14}$$

and

$$\begin{aligned}J(u) &= \frac{d}{du}[q(u)]^{-1}\\&= \left(\frac{\lambda}{2\pi}\right)^{\frac{1}{2}} \exp\left[-\frac{(Q(u)-\mu)^2}{2\mu^2 Q(u)}\right] \frac{q(u)Q^{-\frac{5}{2}}(u)}{2\mu^2}\left(\frac{3\mu^2 Q(u)+\lambda(Q^2(u)-\mu^2)}{Q(u)}\right)\\&= \frac{3\mu^2 Q(u)+\lambda(Q^2(u)-\mu^2)}{2\mu^2 Q(u)},\end{aligned}$$

on using (4.14). Hence,

$$J'(u) = \frac{q(u)}{2Q^3(\mu)}(2\lambda - 3Q(u))$$

which is increasing for $Q(u) < \frac{2\lambda}{3}$ and decreasing for $Q(u) > \frac{2\lambda}{3}$. Hence, $J(u)$ is UBT with change point satisfying $Q(u) = \frac{2\lambda}{3}$. By Theorem 4.1 or Part (b) of Theorem 4.2, X is UBT. The same method works well for other distributions with $Q(u)$ being not of closed form.

Gupta and Warren [249] have extended Glaser's results to cover cases with more than one change point. Their result can be translated into quantile functions as below.

Theorem 4.3. *If $Q(u)$ is strictly increasing and $q(u)$ is twice differentiable and $J'(u) = 0$ has n solutions $0 < u_1 < u_2 \cdots < u_n < 1$, there exists at most one solution for $H'(u) = 0$ in $[u_{k-1}, u_k]$, $k = 1, 2, \ldots, n$.*

There is a vast literature on BT models including various formulations, methods of construction and applications. We will return to these issues in Chap. 8. In spite of the popularity of BT- and UBT-shaped hazard functions, there have been criticism against its indiscriminate usage and caution to the extent that they are more of a myth than reality; see, for example, Tabot [558] and Wong [586]. Klutke et al. [343] pointed out that a bimodal density function as a mixture of subpopulations does not yield a decreasing hazard function early in life. This means that a bathtub hazard rate cannot accommodate this feature of early failures. Further, if the hazard rate is decreasing in an interval, the density must also decrease in that interval. In a series of papers, Wong [587–589] advocated the concept of roller-coaster-shaped hazard functions citing the following physical characteristics for the hazard function to take such a shape. The shape is initially generated by basic failure mechanisms which lead to decreasing hazard rate and then humps are created by changing hazard conditions, wear out, distribution of flawed items, etc. Roller-coaster curves thus exhibit alternate monotonicities repeatedly.

Definition 4.4. If there exist points $0 < u_1 < u_2 \cdots < u_k < 1$ such that in the interval $[u_{k-1}, u_k]$, $1 \le k \le n+1$, $u_0 = 0$, $u_{n+1} = 1$, $H(u)$ is strictly monotone and it has opposite monotonicity in any two adjacent such intervals, we say that X has roller-coaster-shaped hazard quantile function with change points $u_1, u_2, \ldots, u_k$.

Another typical criterion of interest in this context is that of periodic hazard rate. A hazard function $h(x)$ is periodic if

$$B(x,t) = B(x+nc,t)$$

$n \geq 0$ is an integer and $C \geq 0$, where

$$B(x,t) = \frac{F(x+t) - F(x)}{1 - F(x)}.$$

Prakasa Rao [496] has shown that distributions with periodic hazard rates will be of the form

$$\bar{F}(x) = p(x)e^{-\alpha x}, \quad x \geq 0; \alpha > 0,$$

or exponential, where $p(\cdot)$ is a periodic function with period c and support contained in the set $(nc, n \geq 0)$. This follows from the fact that the hazard function is either a constant (in which case the definition is trivially satisfied since X is exponential) or has the form

$$h(x) = \frac{\alpha p(x) - p'(x)}{p(x)}$$

along with the fact that $p(x)$ being periodic, $p'(x)$ is also periodic. It can also be seen that (Chukova and Dimitrov [148] and Chukova et al. [149]) a non-negative random variable with continuous density function has periodic hazard rate if and only if it has almost lack of memory property. The almost lack of memory property means that there exists a sequence of distinct constants (a_n) such that

$$P(X \geq b + x | X \geq b) = P(X \geq x)$$

holds for any $b = a_n$, $n = 1, 2, 3, \ldots$ and all $x \geq 0$. Castillo and Sieworek [130] have considered the reliability of computing systems and showed that hard disk failures seem to follow the work load. The influence of this work load can be accounted for by addition of a periodic hazard rate. More details are present in Chukova and Dimitrov [148], Boyan [119], Dimitrov et al. [179] and Tanguy [561].

Apart from providing monotone hazard rates, the IHR and DHR classes of distributions possess some other important properties, which are as follows:

1. If X_1 and X_2 are IHR, their convolution is IHR, while DHR does not preserve convolution;
2. It is not true that a coherent system of independent IHR components is necessarily IHR. The DHR class is also not closed under the formation of

coherent systems. A parallel system of independent and identical IHR units is IHR while a series system of IHR units, not necessarily identical, is also IHR;

3. The operation of formation of mixtures is preserved under DHR class and is not preserved for the IHR class;
4. The harmonic mean of two IHR survival probabilities is an IHR survival probability;
5. If X is IHR (DHR), then $\log \bar{F}(x)$ is concave (convex);
6. If X is IHR (DHR) and ξ_p is the pth percentile, then

$$\begin{aligned}\bar{F}(x) &\geq (\leq)\, e^{-\alpha x},\ x \leq \xi_p \\ &\leq (\geq)\, e^{-\alpha x},\ x \geq \xi_p\end{aligned},$$

where $\alpha = -\frac{\log(1-p)}{\xi_p}$;

7. If X is IHR, then

$$\bar{F}(x) \geq \begin{cases} e^{-\frac{x}{\mu}}, & t < \mu \\ 0, & t \geq \mu \end{cases},$$

or equivalently

$$Q(u) \geq -\mu \log(1-u) \text{ for } t < \mu;$$

8. If X is IHR, then

$$-\mu \log(1-u) \leq \xi_u \leq -\mu \frac{\log(1-u)}{u}, \quad u \leq 1-e^{-1},$$

and

$$\mu \leq \xi_u \leq -\mu \frac{\log(1-u)}{u}, \quad \mu \geq 1-e^{-1};$$

9. If X is DHR, then

$$\begin{aligned}Q(u) &\leq -\mu \log(1-u), \quad Q(u) \leq \mu, \\ &\leq \frac{\mu e^{-1}}{(1-u)}, \quad Q(u) \geq \mu;\end{aligned}$$

10. If X_1 and X_2 are IHR with hazard rates $h_1(x)$ and $h_2(x)$, then the hazard rate of the convolution $h_c(x)$ satisfies $h_c(x) \leq \min(h_1(x), h_2(x))$;
11. The quantile density function of an IHR random variable is increasing;
12. We define the kth spacing between order statistics $X_{1:n}, \ldots, X_{n:n}$ by $X_{k:n} - X_{k-1:n}$. The order statistics of IHR distributions are IHR, but the spacings of IHR

distributions are not IHR. In the case of DHR class, the spacings are DHR, while order statistics are not;

13. In life testing experiments, some units may not fail at all during the course of the test. Hence, one cannot calculate the sample mean life for estimating μ. If M is the median, using Property 9 above, the mean can then be estimated as

$$\frac{M}{2\log 2} \leq \mu \leq \frac{M}{\log 2};$$

14. If X_1 and X_2 are identically distributed and IHR, then (Ahmad [21])

$$2^{\frac{(r+2)(r-1)}{2}} E(\min(X_1,X_2))^r \geq r!\mu^r, \quad r \geq 2,$$

and

$$E(\min(X_1,X_2))^{2r+2} \geq \binom{2r+2}{2r+1}\left(\frac{1}{2}\right)^{2r+2}(\mu'_{r+1})^2;$$

15. If X is IHR (DHR), its residual life X_t is also IHR (DHR);
16. If X is IHR (DHR), Z is also IHR (DHR). The converse is true if and only if the ratio of their densities $\frac{f(x)}{g(x)}$ is increasing (decreasing).

Results 1–12 are discussed in Barlow and Proschan [70] while 15 and 16 are presented in Gupta and Kirmani [242].

The equilibrium distribution discussed in Sect. 4.2 is a particular case of a more general class of distributions called weighted distributions. The random variable Y with density function

$$g(x) = \frac{w(x)f(x)}{Ew(X)}, \quad w(x) > 0,\ Ew(X) < \infty,$$

specifies the weighted distribution corresponding to the random variable X. Then,

$$h_Y(x) = \frac{w(x)}{E[w(X)|X > x]} h_X(x).$$

The equilibrium distribution results as a special case when $w(x) = \frac{1}{h(x)}$. For various identities connecting reliability functions of X and Y and conditions on the distribution of X for preserving DHR/IHR property of X, one may refer to Jain et al. [290], Gupta and Kirmani [242], Oluyede [473], Bartoszewicz and Skolimowska [77] and Misra et al. [417]. Returning to the quantile approach,

$$G(x) = \frac{\int_0^x w(x)f(x)dx}{\mu_w}, \quad \text{with } \mu_w = E[w(X)],$$

becomes

$$F^*(u) = G(Q(u)) = \frac{\int_0^u W(p)dp}{\mu_W}, \quad \text{with } W(p) = w(Q(p)),$$

which is a distribution function on $(0,1)$.

Blazej [107] has shown that if X is IHR and $C(u)$ is an increasing positive function on $(0,1)$ such that

$$\int_0^1 \frac{C(u)du}{1-u} = +\infty,$$

then Y is IHR. For a choice of $C(u)$, one can choose it as a constant, a positive increasing function on $(0,1)$, or a positive decreasing function ϕ such that $\phi(x) > a > 0$.

Lariviere and Porteus [375] refer to the ratio $t(x) = \frac{xf(x)}{\bar{F}(x)}$ as a generalized hazard rate and if $t(x)$ is non-decreasing, X is said to be increasing generalized hazard rate (IGHR). Lariviere [374] has shown that X is IGHR is equivalent to $\log X$ being IHR. Further, if $\log X$ is IHR (IGHR), then $\alpha + \beta \log X$ (αX^β) is IHR (IGHR). The concept of generalized hazard rate is primarily intended for use in operations management. Occasionally, the above results on IGHR becomes handy in verifying whether X is IHR.

4.3.2 Increasing Hazard Rate(2)

The notion of stochastic dominance plays a role in defining certain ageing classes. If X_1 and X_2 are two lifetime random variables with distribution functions F_1 and F_2, respectively, then X_1 has stochastic dominance over X_2 of the first order, D_1, if $F_1(x) \leq F_2(x)$. The dominance of order 2, D_2, is defined as

$$\int_0^x F_1(t)dt \leq \int_0^x F_2(t)dt$$

and $\bar{D}_2$ as

$$\int_x^\infty \bar{F}_1(t)dt \geq \int_x^\infty \bar{F}_2(t)dt,$$

while the third order stochastic dominance D_3 and $\bar{D}_3$ are

$$\int_0^y \int_0^x F_1(t)dtdx \leq \int_0^y \int_0^x F_2(t)dtdx$$

and

$$\int_y^\infty \int_x^\infty \bar{F}_1(t)dtdx \geq \int_y^\infty \int_x^\infty \bar{F}_2(t)dtdx,$$

respectively. Implications among these orders are as follows:

$$D_1 \Rightarrow D_2(\bar{D}_2) \Rightarrow D_3(\bar{D}_3).$$

Deshpande et al. [172] defined increasing hazard rate of order 2 (IHR(2)) by requiring the residual life X_{t_1} to have stochastic dominance D_2 over X_{t_2}. In other words, X is IHR(2) if and only if for every fixed $x \geq 0$, one of the following conditions are satisfied:

(a) $\int_0^x \frac{\bar{F}(t+s)}{\bar{F}(t)} ds$ is non-decreasing in s;

(b) $\int_0^x \frac{\bar{F}(t+s)}{\bar{F}(t)} dt \leq \int_0^x \frac{\bar{F}(t+v)}{\bar{F}(t)} dt \quad \forall x \geq 0,\ s \geq v.$

Definition 4.5. A lifetime random variable X is IHR(2) if and only if

$$\int_0^u [Q(t+(1-t)v) - Q(t)]dt \geq \int_0^u Q[(t+(1-t)s) - Q(t)]dt \tag{4.15}$$

for all $u \geq 0$ and $t \leq s$.

When X is IFR, we have

$$Q(t+(1-t)v) - Q(t) \geq Q(t+(1-t)s) - Q(t)$$

which implies (4.15) on integrating the last inequality over $(0,u)$. Hence, IHR $\Rightarrow$ IHR(2). Since $g_1(x) \geq g_2(x)$ does not always imply $\frac{dg_1}{dx} \geq \frac{dg_2}{dx}$, IHR(2) need not imply IHR, thus proving IHR(2) contains IHR. However, this class does not seem to have received much attention in practice.

4.3.3 New Better Than Used in Hazard Rate

This concept uses the idea that at the initial age the hazard rate will be less than that of a used one, indicating positive ageing. We say that X is new better than used in hazard rate (NBUHR) if $h(0) \leq h(x)$ for $x \geq 0$. The dual class is new worse than used in hazard rate (NWUHR) if $h(0) \geq h(x)$ for $x \geq 0$. When the term failure rate is used instead of hazard rate, the abbreviation becomes NBUFR, which is often used in the literature (Loh [404]).

Definition 4.6. A lifetime X is new better (worse) than used in hazard rate if and only if $H(0) \leq (\geq)\ H(u)$ for $u \geq 0$.

An associated concept is that of new better than used in hazard rate average (NBUHRA). Let $F(x)$ be such that $F(0) = 0$ and

$$\frac{\log \bar{F}(x)}{x} \leq \lim_{x \to 0} \frac{\log \bar{F}(x)}{x}.$$

Then, X is NBUHRA. If $\bar{F}(x)$ is continuously differentiable over $(0, \varepsilon)$ for some $\varepsilon > 0$, then the above condition is also equivalent to

$$h(0) \leq \frac{1}{x} \int_0^x h(t)dt, \quad x \geq 0. \tag{4.16}$$

In this case, the hazard rate of a new component is less than its average hazard rate in $(0, x)$ for all $x \geq 0$ (Loh [404]). Further, X is new worse than used in hazard rate average (NWUHRA) if the inequality in (4.16) is reversed. The NBUHRA and NWUHRA classes are also denoted by NBUFRA and NWUFRA in the literature. We now rewrite the definitions of the classes by making use of (4.16) in terms of hazard quantile function.

Definition 4.7. We say that X is NBUHRA (NWUHRA) if and only if the hazard quantile function $H(u)$ satisfies

$$H(0) \leq (\geq) \frac{-\log(1-u)}{Q(u)} \quad \text{for all } u. \tag{4.17}$$

Definition 4.7 can also be seen as

$$Q(u) \leq -\frac{1}{H(0)} \log(1-u)$$

meaning that the quantile function of X is less than that of the exponential distribution with the same hazard rate as X. In the first order stochastic dominance mentioned above, the lifetime X is worse than the exponential and hence X ages positively if it is NBUHRA. The process of averaging visible in (4.16) is also implicit in (4.17) if one rewrites it as

$$H(0) \leq (\geq) \frac{\int_0^u H(p)q(p)dp}{\int_0^u q(p)dp}.$$

The right side acts as a weighted average with weight $\frac{q(u)}{Q(u)}$. Evidently,

$$\text{IHR} \Rightarrow \text{NBUHR}$$

and furthermore

$$\text{NBUHR} \Rightarrow H(0) \leq H(u) \Rightarrow \int_0^u H(0)q(p)dp \leq \int_0^u H(p)q(p)dp$$
$$\Rightarrow \text{NBUHRA},$$

but not conversely. A closely related measure of positive ageing, given in Bryson and Siddiqui [122], based on the interval-average hazard rate defined by

$$A(t,s) = \frac{1}{t}\int_{s}^{s+t} h(x)dx,$$

is increasing interval average hazard rate when

$$A(t_2,s) \geq A(t_1,s), \quad t_2 \geq t_1 \geq 0,\ s \geq 0.$$

They have shown that this criterion is equivalent to IHR. The properties of the above two classes have been discussed in Abouammoh and Ahmed [5], Gohout and Kunhert [222] and El-Bassiouny et al. [185]. It has been shown that

1. NBUHR is not preserved under convolution,
2. NBUHR class is closed under the formation of mixtures,
3. NBUHR class is closed under formation of coherent systems with independent components,
4. $\mu_r' \geq (\leq) \frac{f(0)\mu_{r+1}'}{(r+1)}$ according to X being NBUHR (NWUHR)

4.3.4 Stochastically Increasing Hazard Rates

The monotonicity of hazard rates can also be evaluated in random intervals of time giving a further extension of IHR and DHR concepts.

Definition 4.8. Let $X_0 = 0, X_1, \ldots, X_k, \ldots$ be a sequence of independent and identically distributed exponential random variables each with mean μ, and Y be independent of X_k. Then, Y is said to have stochastically increasing hazard rate, or Y is SIHR, if and only if (Singh and Deshpande [542])

$$P\left[\sum_0^k X_i \leq Y < \sum_0^{k+1} X_i | Y \geq \sum_0^k X_i\right] \geq P\left[\sum_0^{k-1} X_i \leq Y < \sum_0^{k} X_i | Y \leq \sum_0^{k-1} X_i\right].$$

This means that the conditional probability of a unit with lifetime Y will not fail before the random time $\sum_1^k X_i$ given that it has not failed in $\sum_0^{k-1} X_i$, is decreasing in $k = 1,2,3,\ldots$. The dual class SDHR is defined similarly. It may be noted that

$$\text{IHR} \Rightarrow \text{SIHR}$$

and that both SIHR and SDHR hold if and only if Y is exponential. This concept has not been used much in reliability analysis.

4.3.5 Increasing Hazard Rate Average

Introduced by Birnbaum et al. [102], the increasing hazard rate average (IHRA) class and its dual decreasing hazard rate average (DHRA) class are among the basic classes of life distributions. A lifetime X is IHRA (DHRA) if and only if

$$-\frac{\log \bar{F}(x)}{x} \text{ is increasing (decreasing) in } x > 0.$$

Since $-\log \bar{F}(x) = \int_0^x h(t)dt$, X is IHRA means that the average hazard rate in $(0,x)$ defined by

$$\frac{1}{x}\int_0^x h(t)dt \quad \text{is increasing.}$$

A real valued function $\phi(x)$ on $[0,\infty)$ is star-shaped if $\phi(0) = 0$ and $\frac{\phi(x)}{x}$ is increasing in $x > 0$. Hence, X is IHRA if and only if $\log \bar{F}(x)$ is star-shaped.

Definition 4.9. We say that X is IHRA (DHRA) if and only if $\frac{Q(u)}{-\log(1-u)}$ is decreasing (increasing) in $0 < u < 1$.

Theorem 4.4. *The following conditions are equivalent:*

(i) X *is IHRA;*
(ii) $\frac{\int_0^u H(p)q(p)dp}{\int_0^u q(p)dp}$ *is increasing;*
(iii) $H(u) \geq \frac{Z(u)}{Q(u)}$, *with* $Z(u) = -\log(1-u)$.

Proof. (i) $\Leftrightarrow$ (ii)

$$X \text{ is IHRA} \Leftrightarrow -\frac{\log(1-u)}{Q(u)} \text{ is increasing}$$

$$\Leftrightarrow \text{(ii), since } -\log(1-u) = \int_0^u H(p)q(p)dp.$$

(ii) $\Leftrightarrow$ (iii). From (ii), we have

$$\frac{\int_0^u H(p)q(p)dp}{\int_0^u q(p)dp} \text{ is increasing} \Leftrightarrow Q(u)H(u)q(u) - q(u)\int_0^u H(p)q(p)dp \geq 0$$

$$\Leftrightarrow \frac{Q(u)}{1-u} - q(u)Z(u) \geq 0$$

$$\Leftrightarrow \frac{1}{(1-u)q(u)} \geq \frac{Z(u)}{Q(u)}$$

$$\Leftrightarrow H(u) \geq \frac{Z(u)}{Q(u)}.$$

We now list some key properties of the IHRA and DHRA classes.

1. IHR $\Rightarrow$ IHRA. The converse need not be true. Marshall and Olkin [412] have shown that
$$\bar{F}(x) = e^{-\lambda_1 x} + e^{-\lambda_2 x} - e^{-(\lambda_1+\lambda_2)x}$$
is IHRA, but X is UBT.
2. If X is IHRA, then $\frac{\bar{F}(x)}{\bar{F}_Z(x)}$ is decreasing, where as before Z is the equilibrium random variable.
3. If X belongs to the one-parameter exponential family of distributions and $E(Z) = E(X)$, then X has exponential distribution.
4. $h_Z(x) = c\,h_X(x)$, where $c > 0$ is a constant, if and only if X follows generalized Pareto distribution.
5. When X is IHRA, its survival function $\bar{F}(x)$ can cross the survival function of any exponential at most once and only from above.
6. An IHRA distribution has finite moments of all orders.
7. The class of IHRA distributions is closed under the formation of coherent systems. It is the smallest class containing the exponential law, while DHRA class is not closed. Both IHRA and DHRA preserve formation of series systems.
8. The IHRA class is closed under convolution and its dual DHRA class does not possess such a property.
9. Mixtures of IHRA distributions are not necessarily IHRA, while if each $F(x;\theta)$ is DHRA, then the mixture is also DHRA.
10. Every IHRA distribution can be obtained as a limit in distribution of a sequence of coherent systems of components having exponential or degenerate distributions.
11. IHRA distributions arise when shocks occur according to a Poisson process in time, each independently causing random damage to the unit. The unit fails when accumulated damages exceed a threshold level.
12. We have
$$\mu_r' \le (\ge)\, \Gamma(r+1)\mu^r, \quad 0 < r < 1$$
$$\mu_r' \ge (\le)\, \Gamma(r+1)\mu^r, \quad 1 < r < \infty.$$
13. We have
$$\bar{F}(x) \begin{cases} \ge (\le)\, e^{-\alpha x}, & 0 < x < \xi p \\ \le (\ge)\, e^{-\alpha x}, & x > \xi p,\ \alpha = -\frac{1}{\xi p}\log(1-p). \end{cases}$$
14. The coefficient of variation is $\le (\ge) 1$ according to X being IHRA (DHRA).
15. $E(X_1^{r+1}) \ge E[\min(\frac{X_1}{\alpha}, \frac{X_2}{1-\alpha})]^{r+1}$, where X_1 and X_2 are independent and identically distributed IHRA variables (with r being an integer); see Ahmad and Mugadi [26].

16. For fixed $x > 0$, when X is IHRA,

$$\bar{F}(x) \le \begin{cases} 1 & , x \le \mu \\ e^{-\omega x} & , x > \mu \end{cases},$$

where ω is a function of x satisfying $1 - \omega\mu = e^{-\omega x}$.
17. When $X_1, X_2, \dots, X_n$ are iid IHRA, for all integers $r \ge 0$, $k \ge 2$, we have (Ahmad and Mugadi [26])

$$E(\min(X_1, X_2, \dots, X_n))^r \ge \frac{\mu'_{r+1}}{r+1}.$$

18. Let X be IHRA (DHRA). Then, $[\frac{\mu'_r}{\Gamma(r+1)}]^{\frac{1}{r}}$ is decreasing (increasing) in $r \ge 0$.
19. If X is IHRA (DHRA) and $C(u)$ is an increasing (decreasing) function on $(0,1)$, then $G(x) = \frac{\int_0^x w(x) f(x)}{\mu_x}$ is IHRA (DHRA) (Blazej [107]).

Proofs and further details of many of the above results can be found in Barlow and Proschan [69]. For results concerning weighted distributions, see Misra et al. [417], Bartoszewicz and Skolimowska [77] and Oluyede [473].

4.3.6 Decreasing Mean Time to Failure

The IHR and IHRA arise in evolving maintenance strategies in reliability engineering. In order that a unit functions satisfactorily without failures or disruption, reliability engineers adopt several types of maintenance strategies. These strategies spell out schemes of replacement before failure. One such method is to resort to an age-replacement policy in which a unit is replaced either when it fails or at an age T whichever is earlier. The number of failures $N(x)$ in $(0,x)$ with no planned replacements is a renewal process and $N_A(x,T)$ and the number of in-service failures under the age-replacement policy is also a renewal process. If $\bar{F}(x)$ is the survival function of X_i, the length of time between the $(i-1)$th and ith failures in the process $N(x)$, the distribution of $X_{i,A}$, the length of time between the $(i-1)$th and ith failures in $N_A(x,T)$ for fixed $T > 0$, is specified by

$$S_T(x) = [\bar{F}(T)]^n \bar{F}(x - nT), \quad nT \le x < (n+1)T, \; n = 0, 1, 2, \dots \tag{4.18}$$

A yardstick for the effectiveness of the strategy is to study the properties of the mean time to failure (MTTF) derived as the mean of (4.18), viz.,

$$M(T) = \frac{1}{F(T)} \int_0^T \bar{F}(t)dt. \tag{4.19}$$

Equation (4.19) makes it clear that the behaviour of $M(T)$ is associated with ageing properties of F. Thus, the class of distributions for which $M(T)$ is decreasing (DMTTF) and increasing (IMTTF) have been studied in literature by many authors including Klefsjö [334], Knopik [344, 345] and Asha and Nair [39]. We can rewrite (4.19) as

$$\mu(u) = M(Q(u)) = u^{-1}\int_0^u (1-p)q(p)dp \tag{4.20}$$

which gives the average time to failure at the $100(1-u)\%$ point of the distribution.

Definition 4.10. We say that $S_T(x)$ is decreasing (increasing) MTTF according to $\mu'(u) \leq (\geq)\ 0$. Then, we have the identities

$$\mu(u) = u^{-1}[\mu - (1-u)R(u)],$$
$$u\mu'(u) + \mu(u) = [H(u)]^{-1}.$$

Thus, $\mu(u)$ is increasing or decreasing according to $\mu(u) \geq (H(u))^{-1}$ or $\mu(u) \leq [H(u)]^{-1}$, for all u in $(0,1)$. The MTTF is BT (UBT) when $H(u)\mu(u) = 1$ and $H'(u_0) \leq (\geq) 0$ at u_0, where u_0 is the solution of the equation $H(u)\mu(u) = 1$. Li and Xu [393] have shown that

$$\text{IHR} \Rightarrow \text{IHRA} \Rightarrow \text{DMTTF}.$$

Asha and Nair [39] have further proved that if $\mu(u)$ is decreasing and concave, then X is IHR and hence IHRA. Knopik [344, 345] has established the following closure properties:

1. If the lifetimes of the components are independent with absolutely continuous distributions which are DMTTF, then any parallel system is also DMTTF, and moreover if the components are identically distributed also, then a series system of DMTTF components is DMTTF;
2. The DMTTF family is closed under convolution and weak convergence of distributions.

Li and Li [398] introduced the IHRA* t_0 (DHRA* t_0) classes which imply that the average hazard rate begins to increase (decrease) after t_0.

Definition 4.11. A random life X is IHRA* t_0, for all $x \geq t_0 > 0$ and all $\frac{t_0}{x} \leq b < 1$, if and only if $\bar{F}(bx) \geq \bar{F}^b(x)$.

This class of IHRA* t_0 distributions satisfies the following properties:

1. IHRA $\Rightarrow$ IHRA* t_0;
2. Let F be a life distribution with strictly increasing hazard rate. Denote the cumulative hazard rate of F by $C_F(x) = \int_0^x h(t)dt$ and

$$C^*(x) = \begin{cases} C_F(x) & , 0 \le x \le t_0 \\ C_G(x) - C_G(t_0) + C_F(t_0) & , x \ge t_0 \end{cases} .$$

If $C_F(t_0) \le C_G(t_0)$, then the life distribution determined by C^* is IHRA* t_0, but not IHRA. For example, take $\bar{F}(x) = a^{x^{1/2}}, 0 < a < 1, \bar{G}(x) = e^{-x^2}, x \ge 0, t_0^{3/2} \ge -\log a$;

3. If $\bar{F}_i$ is the survival function of an IHRA $*t_i$ unit, $i = 1, 2, \ldots, n$, of a system, then the coherent system is IHRA $*t_0$, where $t_0 = \max_{1 \le i \le n} t_i$.

4.4 Classes Based on Residual Quantile Function

In this section, we discuss various classes of life distributions arising from the monotonic nature of the mean, variance and percentile residual functions. As in the case of the hazard rate notions, the classes are identical irrespective of whether the definitions based on distribution functions or quantile functions are implemented.

4.4.1 Decreasing Mean Residual Life Class

For defining increasing hazard function classes, we utilized the fact that the quantile function of the residual life is decreasing in u_0. A weaker class can be obtained if we consider instead the monotonicity of the mean of the distribution. Thus, we have the decreasing (increasing) mean residual life DMRL (IMRL) class if $m(x)$ is a decreasing (increasing) function in $x > 0$.

Definition 4.12. A random variable X is said to have decreasing (increasing) mean residual quantile function if

$$M(u_1) \le (\ge) M(u_2) \quad \text{for } 0 \le u_2 \le u_1 < 1,$$

or equivalently

$$\int_0^1 [Q(u + (1-u)p) - Q(u)]dp$$

is a decreasing (increasing) function of u.

Setting $v = u + (1-u)p$, we see that

$$\int_0^1 [Q(u + (1-u)p) - Q(u)]dp = (1-u)^{-1} \int_u^1 Q(v)dv - Q(u) = M(u).$$

If $M(u)$ is differentiable, then X is DMRL (IMRL) according to $M'(u) \le (\ge) 0$.

Example 4.5. Let

$$Q(u) = \sigma(1-(1-u)^{\frac{1}{\alpha}}), \quad \sigma, \alpha > 0.$$

Then, we have

$$M(u) = \frac{\sigma}{\alpha+1}(1-u)^{\frac{1}{\alpha}}$$

so that

$$M'(u) = -\frac{\sigma}{\alpha(\alpha+1)}(1-u)^{\frac{1}{\alpha}-1} < 0.$$

Hence, X is DMRL.

Abouammoh and El-Neweihi [4], Abouammoh et al. [6], Gupta and Kirmani [242], Abu-Youssef [16] and Ahmad and Mugadi [26] have all discussed various properties of the classes of distributions generated by monotonic mean residual life function. Some of these properties are as follows:

1. If X is IHR (DHR), then X is DMRL (IMRL). This follows from the implications

$$X \text{ is IHR} \Rightarrow Z \text{ is IHR} \Rightarrow H_z(u) = \frac{1}{M_X(u)} \text{ is increasing}$$
$$\Rightarrow M_X(u) \text{ is decreasing.}$$

   The converse need not be true. However, if X is DMRL and $m(x)$ is convex, then X is IHR.
2. X is DMRL if and only if $h_X(x) \le h_Z(x)$ or $H_X(u) \le H_Z(u)$. We note from (2.37) that

$$\begin{aligned} H_X(u) \le H_Z(u) &\Leftrightarrow H_X(u) \le \frac{1}{M_X(u)} \\ &\Leftrightarrow H_X(u)M_X(u) \le 1 \\ &\Leftrightarrow 1 - H_X(u)(1-u)M'_X(u) \le 1 \\ &\Leftrightarrow M'_X(u) < 0. \end{aligned}$$

3. From Property 1, it follows that X is DMRL if and only if Z is IHR.
4. X is DMRL (IMRL) if and only if $E\phi(X-x|X>x)$ is decreasing (increasing) for all convex increasing functions ϕ.
5. The mixture of IMRL distributions is IMRL, provided the mixture has a finite mean. Generally, DMRL distributions are not closed under the formation of mixtures. However, for the mixture of non-crossing life distributions, IMRL class is closed.

6. Both IMRL and DMRL classes are not closed under the formation of coherent systems.
7. The convolution of both IMRL and DMRL distributions need not be of the same class.
8. We have

$$(r+1)E[X_1(\min X_1,X_2)^r] \geq (\leq)(r+2)\nu_2^{(r+1)},$$

where X_1 and X_2 are independent and identically distributed, $\nu_r = E(\min X_1,X_2)^r$, and also

$$\nu_2 \geq (\leq)\ \frac{\mu^2}{2}$$

when F is DMRL (IMRL).
9. $M_Z(u) \leq (\geq) M_X(u)$ if and only if X is DMRL.
10. If $\theta(u)$ is increasing and X is DMRL, then X is IHR. Recall the definition of

$$\theta(u) = \frac{1}{1-u}\int_u^1 Q(p)dp$$

as the mean quantile function. Then, upon differentiation, we find

$$(1-u)\theta'(u) - \theta(u) = -Q(u)$$

and

$$M(u) + Q(u) = \theta(u),$$

which together yields

$$\frac{\theta'(u)}{M(u)} = \frac{1}{1-u} = q(u)H(u).$$

11. Let

$$A_1(x,y) = \frac{1}{\bar{F}(x)}\int_x^y w(t)f(t)dt, \quad x<y.$$

If X is DMRL (IMRL), $A_1(x,y)$ is increasing (decreasing) on the support of X, and log convex (log concave) for x satisfying $f(x) > 0$, then the weighted random variable X_W is DMRL (IMRL) (Misra et al. [417]).
12. If $m(x)$ is strictly convex (concave) on $[0,\infty)$ and decreasing (increasing) for $x \geq 0$, then $h(x)$ is strictly increasing (decreasing) on $[0,\infty)$ (Kupka and Loo [363]).

13. A random variable with $E(X^n) < \infty$, $n = 1, 2, \ldots$, is generalized Pareto if and only if $M_n(u) = C_n M(u)$, where $M_n(u)$ is as defined in (4.12).
14. IHRA does not imply DMRL.

When the mean residual life declines in the interval $(0, t_0)$ and thereafter never greater than what it was at age t_0 (Kulasekera and Park [356]), the class of distributions is called BMRL-t_0 (better mean residual life at age t_0).

Bryson and Siddiqui [122] introduced net decreasing mean residual lifetime of X if and only if $m(x) \leq m(0)$ for all $x \geq 0$. This translates into the following definition.

Definition 4.13. We say that X has net decreasing mean residual lifetime (NDMRL) if and only if $M(u) \leq M(0)$ for $0 \leq u < 1$.

The NDMRL has the following implications:

$$\text{IHR} \Rightarrow \text{DMRL} \Rightarrow \text{NDMRL},$$

$$\text{IHR} \Rightarrow \text{IHRA} \Rightarrow \text{NDMRL}.$$

The dual class is defined by reversing the inequality and the implications among corresponding negative ageing concepts follow. Another criterion based on mean residual quantile function takes the harmonic averages in $(0, x)$. A distribution is decreasing mean residual life in harmonic average (DMRLHA) if and only if (Deshpande et al. [172])

$$\left[\frac{1}{x}\int_0^x \frac{dt}{m(t)}\right]^{-1} \quad \text{is decreasing in } x.$$

Accordingly, we have the following definition in terms of quantile functions.

Definition 4.14. We say that X is DMRLHA (IMRLHA) if and only if

$$\frac{\int_0^u \frac{q(p)}{M(p)}dp}{\int_0^u q(p)dp}$$

is decreasing (increasing) in u. It is easy to see that

(i) DMRL $\Rightarrow$ DMRLHA $\Rightarrow$ NBUE;
(ii) Since $H_Z(u) = \frac{1}{M_X(u)}$, we have

$$\frac{\int_0^u \frac{q(p)}{M(p)}dp}{\int_0^u q(p)dp} = \frac{\int_0^u H_Z(p)q(p)dp}{\int_0^u q(p)dp}.$$

Thus, X is DMRLHA $\Leftrightarrow$ Z is IFRA.

Honfeng and Yi [275] compared the failure rates of X and Z in defining what is called the new better (worse) than equilibrium hazard rate, NBEHR (NWEHR), if and only if $h_X(x) \leq h_Z(x)$. It is obvious that when $E(X) < \infty$,

$$\begin{aligned}\text{NBEHR} &\Leftrightarrow h_X(x) \leq h_Z(x) \Leftrightarrow h_X(x) \leq \frac{1}{m_X(x)}\\ &\Leftrightarrow h_X m_X(x) \leq 1\\ &\Leftrightarrow 1 + m'(x) \leq 1\\ &\Leftrightarrow X \text{ is DMRL.}\end{aligned}$$

Like the IHRA notion comparison with the class of DMTTF distributions (see Definition 4.9), we see that DMRL $\not\Rightarrow$ DMTTF and DMTTF $\not\Rightarrow$ DMRL (Li and Xu [393]).

4.4.2 Used Better Than Aged Class

When a unit is working with unknown age, to assess its ageing behaviour, Alzaid [36] introduced the used better than aged (UBA) and its dual used worse than aged (UWA) classes of life distributions. Two induced classes from these are the UBAE (used better than aged in expectation) and UWAE (used worse than aged in expectation). When $E(X) < \infty$ and $0 < m(\infty) < \infty$, the UBA (UWA) class is specified by

$$\bar{F}(x+t) \geq (\leq) \bar{F}(t) e^{-\frac{x}{m(\infty)}}, \; x \geq 0, \; t \geq 0.$$

Accordingly, we have the following definitions.

Definition 4.15. The random variable X is UBA (UWA) if and only if

$$Q(v + (1-v)u) - Q(v) \geq (\leq) - M(1) \log(1-u) \tag{4.21}$$

for all $0 \leq u, v < 1$, provided $0 < M(1) < \infty$.

Definition 4.16. We say that X is UBAE (UWAE) if and only if $0 < M(1) < \infty$ and

$$M(u) \geq (\leq) M(1) \quad \text{for all } u. \tag{4.22}$$

The following properties hold for these four classes of life distributions:

1. IHR $\Rightarrow$ DMRL $\Rightarrow$ UBA $\Rightarrow$ UBAE (see Alzaid [36] and Willmot and Cai [581] for proofs).
2. X is UBAE (UWAE) $\Leftrightarrow \bar{F}_Z(x+t) \geq (\leq) \bar{F}_Z(t) e^{-\frac{x}{m(\infty)}}$.
3. X is UBAE (UWAE) if and only if Z is UBA (UWA).

4. If

$$\bar{G}(x) = \frac{\int_0^\infty e^{-\alpha t}\bar{A}(x+t)dt}{\int_0^\infty e^{-\alpha t}\bar{A}(t)dt}, \quad x \geq 0,\ -\infty < \alpha < \infty,$$

then $1-\bar{A}$ is UBA (UBAE) $\Rightarrow G$ is UBA (UBAE). It may be observed that $\bar{G}(x)$ given above is a generalization of the equilibrium distribution.

5. If X is such that $E(X^{r+s+2}) < \infty$, then

$$\frac{\mu'_{r+s+2}}{(r+s+2)!} \geq \frac{\mu'_{r+1}(m(\infty))^{s+1}}{(r+1)!} \quad \text{if } X \text{ is UBA.}$$

6. The moment generating function $\phi(t)$ of X satisfies

$$\phi(t) \leq 1 + \frac{\mu t}{1-tm(\infty)} \quad \text{if } X \text{ is UBA,}$$

$$\phi(t) \leq 1 + \frac{\mu t + t^2(\frac{\mu'_2}{2} - \mu m(\infty))}{1-tm(\infty)} \quad \text{if } X \text{ is UBAE.}$$

7. When X is UBA (UBAE) and $E(X) < \infty$ $(E(X^2) < \infty)$, then all moments of X exist and are finite.
8. When X is UBAE, we have

$$\frac{\mu'_{r+s+3}}{(r+s+3)!} \geq \frac{\mu'_{r+2}}{(r+2)!}(m(\infty))^{s+1}.$$

Properties 5–8 are taken from Ahmad [22] while Properties 1–4 are from Willmot and Lin [583].

A weaker condition than UBAE is given in Kotlyar [353] as NUABE defined by

$$\int_x^\infty \bar{F}(t)dt \geq \mu e^{-\frac{x}{m(\infty)}}.$$

Nair and Sankaran [445] introduced another version of mean residual life function which is the expected value of the asymptotic conditional distribution of residual life given age in a renewal process. This function

$$m^*(x) = \frac{\int_x^\infty (t-x)\bar{F}(t)dt}{\int_x^\infty \bar{F}(t)dt} = \frac{E((X-x)^2|X>x)}{2m(x)} \tag{4.23}$$

is called the renewal mean residual life function (RMRL). In the quantile formulation, (4.23) is equivalent to

$$e(u) = m^*(Q(v)) = \frac{\int_u^1 [Q(p)-Q(u)](1-p)q(p)dp}{\int_u^1 (1-p)q(p)dp}. \tag{4.24}$$

They then showed that $m^*(x)$ is similar to the conventional mean residual life function $m(x)$ and can be employed in all applications just as $m(x)$. Differentiating

$$e(u)\int_u^1 (1-p)q(p)dp = \int_u^1 [Q(p)-Q(u)](1-p)q(p)dp$$

and simplifying using

$$M(u) = \frac{1}{1-u}\int_u^1 (1-p)q(p)dp,$$

we get

$$M(u) = \frac{e(u)q(u)}{q(u)+e'(u)}$$

or

$$e'(u) = q(u)\left[\frac{e(u)-M(u)}{M(u)}\right]. \tag{4.25}$$

Definition 4.17. The random variable X belongs to the decreasing renewal mean residual life (DRMRL) or increasing renewal mean residual life (IRMRL) class according to $e(u)$ being decreasing (increasing) in u.

The DRMRL (IRMRL) class has the following properties:

1. X is DRMRL (IRMRL) if and only if $e(u) \leq (\geq)\, M(u)$;
2. If X is DMRL (IMRL), then X is DRMRL (IRMRL);
3. If X is DRMRL (IRMRL), then X is DVRL (IVRL);
4. The exponential distribution is characterized by a constant $e(u)$;
5. X is DRMRL (IRMRL) if and only if $C^*(u) \leq (\geq) 1$;
6. X is DRMRL (IRMRL) if and only if $M_z(u) = M_X(u)$;
7. If X_t and Z_t are residual lives of X and Z, respectively, then

$$M_{X_t}(u) \geq M_{Z_t}(u) \Leftrightarrow X \text{ is DRMRL}.$$

The closure properties with respect to various reliability operations in this case seems to be an open problem.

4.4.3 Decreasing Variance Residual Life

Recall from (2.10) and (2.42) that the variance residual life is given by

$$\sigma^2(x) = \frac{2}{\bar{F}(x)} \int_x^\infty \int_u^\infty \bar{F}(t)dtdu - m^2(x),$$

and the corresponding quantile definition is

$$\begin{aligned} V(u) &= (1-u)^{-1} \int_u^1 Q^2(p)dp - (M(u)+Q(u))^2 \\ &= (1-u)^{-1} \int_u^1 M^2(p)dp. \end{aligned}$$

When $\sigma^2(x)$ is decreasing (increasing), we say that X is decreasing variance residual life—DVRL (increasing variance residual life—IVRL).

Definition 4.18. A lifetime random variable X is DVRL (IVRL) if $V(u)$ is decreasing (increasing) in u.

From a practical point of view, the class of DVRL distributions is studied as it indicates positive ageing, and also the uncertainty in the system behaviour decreases with age. Some characteristics of these two ageing criteria are as follows:

1. $V(u)$ is increasing (decreasing) if and only if $C^{*2}(u) \geq (\leq 1)$;
2. DMRL $\Rightarrow$ DVRL.

   To prove this implication in terms of quantile functions, we note that

$$\begin{aligned} V(u) &= \frac{1}{1-u} \int_u^1 [Q(p)-Q(u)]^2 dp - M^2(u) \\ &= \frac{2}{1-u} \int_u^1 (1-p)(Q(p)-Q(u))q(p)dp - M^2(u) \\ &= \frac{2}{1-u} \int_u^1 \int_p^1 q(p)(1-t)q(t)dpdt - M^2(u) \\ &= \frac{2}{1-u} \int_u^1 (1-p)q(p)M(p)dp - M^2(u). \end{aligned}$$

   Hence,

$$V(u) - M^2(u) = \frac{2}{1-u} \int_u^1 (1-p)(M(p)-M(u))q(p)dp.$$

   Since X is DMRL, $M(p) - M(u) \leq 0$ for all $u \leq p$, and so we have

$$V(u) - M^2(u) \leq 1 \text{ or } C^{*2}(u) \leq 1,$$

   and thus X is DVRL.
3. When F is strictly increasing, Z_n is DVRL if and only if Z is DMRL. This follows from the fact that

$$m_Z(t) = \frac{E(X_t^2)}{E(X_t)},$$

which is a decreasing function of t.

4. We have

$$\frac{M_Z(u)}{M_X(u)} = \frac{1}{2}(1 + C^{*2}(u)),$$

so that if F is strictly increasing, then X is DVRL if and only if $M_Z(u) \leq M(u)$.

5. In general, the DVRL class is not closed under mixing.
6. A family of distributions F_θ obeys the non-crossing property if, for any α_1 and α_2, the graphs of F_{α_1} and F_{α_2} do not intersect on their common support. Stoyanov and Al-sadi [548] proved that if F_α is IVRL for each $\alpha > 0$ and obeys the non-crossing property, then their mixture is IVRL.
7. Both DVRL and IVRL distributions are not closed under convolution.
8. Both DVRL and IVRL do not preserve the formation of coherent systems.
9. If X_1 and X_2 are independent copies of X and $Y = \min(X_1, X_2)$, then Al-Zahrani and Stoyanov [32] established that

$$\mu_1' \mu_2' \leq 4E(X_2 Y^2) - \frac{8}{3}E(Y^3)$$

and

$$\mu_2'^2 \leq \frac{16}{3}E(X_2 Y^3) - 4E(Y^4)$$

with equality sign holding in the two cases if and only if X is exponential.

10. We have

$$\bar{F}(x) \leq \frac{\mu}{\sigma(x) + x}, \quad x \geq \mu - \sigma(x)$$

for DVRL distributions, and

$$\bar{F}(x) \geq \frac{\mu - x}{\sigma(x)}$$

for IVRL distributions (Launer [377]).

11. If X is DVRL (IVRL), then

$$E(X^2|X \geq x) \leq (\geq) E^2(X|X \geq x) + [E(X|X > x) - x]^2.$$

For a comprehensive account of the above results and other properties of monotone variance residual life classes, we refer to Gupta [234], Gupta et al. [246], Abouammoh et al. [8] and Gupta and Kirmani [244].

Just as in the case of the mean residual quantile function, other concepts can be formulated for the variance as well. From Launer [377] and Abouammoh et al. [8], we have the following new classes.

Definition 4.19. A lifetime random variable X is net decreasing in variance residual quantile function (NDVRL) or net increasing in variance residual quantile function (NIVRL) according to $V(u) \leq (\geq)\ \sigma^2$.

Definition 4.20. We say that X has decreasing (increasing) variance residual life average DVRLA (IVRLA) if and only if

$$\frac{\int_0^u V(p)q(p)dp}{\int_0^u q(p)dp}$$

is decreasing (increasing). Further, X is new worse (better) than average variance residual life if and only if

$$\frac{\int_0^u V(p)q(p)dp}{\int_0^u q(p)dp} \leq (\geq)\ \sigma^2.$$

We have the implications DVRL $\Rightarrow$ NDVRL and NBUE $\Rightarrow$ NDVRL (see Definition 4.29 for NBUE).

4.4.4 Decreasing Percentile Residual Life Functions

Haines and Singpurwalla [258] discussed classes of life distributions based on the monotonic behaviour of the percentile residual life function $p_\alpha(x)$ defined in (2.19).

Definition 4.21. The random variable X has decreasing (increasing) percentile life DPRL (α) (IPRL (α)) if $F(0) = 0$ and $p_\alpha(x)$ is decreasing (increasing) in x. In terms of quantile functions, the conditions become $Q(0) = 0$ or $P_\alpha(u)$ defined in (2.19) is decreasing (increasing) in u. If X has a constant $p_\alpha(x)$ $(P_\alpha(u))$, it is both DPRL (α) and IPRL (α). Joe and Proschan [301] studied the distinguishing features of these two classes of distributions.

Recalling from (2.49) that

$$P_\alpha(u) = Q(\alpha + (1-\alpha)u) - Q(u),$$

we have

$$P'_\alpha(u) = [(1-\alpha)q(\alpha + (1-\alpha)u) - q(u)]q(u).$$

Hence,

$$\begin{aligned} X \text{ is DPRL}(\alpha) &\Leftrightarrow (1-\alpha)q(\alpha+(1-\alpha)u) \le q(u) \\ &\Leftrightarrow \frac{1}{q(\alpha+(1-\alpha)u)} \ge \frac{1-\alpha}{q(u)} \\ &\Leftrightarrow \frac{1-\alpha-u+\alpha u}{1-(\alpha+u-\alpha u)q(\alpha+u-\alpha u)} \ge \frac{(1-\alpha)(1-u)}{(1-u)q(u)} \\ &\Leftrightarrow H(u+\alpha(1-u)) \ge H(u) \\ &\Leftrightarrow X \text{ is IHR.} \end{aligned}$$

We see that X is IHR when X is DPRL (α) for all α in $(0,1)$. However, if X is DPRL (α) for some α, it is not necessary that X is IHR. Moreover, it is not necessary that DPRL (α) implies DPRL (β) for $\beta > \alpha$. Haines and Singpurwalla [258] showed that DPRL (α) class is not closed under formation of coherent systems or convolution or mixture of distributions.

A particular case of the α-percentile residual life is the median percentile life when $\alpha = \frac{1}{2}$. Lillo [399] pointed out that this measure is preferred over the mean residual life in view of its robustness and use in regression models; see also Kottas and Gelfand [354] and Csorgo and Csorgo [160]. Denoting decreasing median residual life function by DMERL and increasing median residual life by IMERL, we have the following properties:

1. IHR implies DMERL;
2. DMRL does not imply DMERL and DMERL does not imply DMRL;
3. IMRL and IMERL also have no mutual implications either.

4.5 Concepts Based on Survival Functions

There are several criteria available in the literature based on the comparison of survival functions (quantile functions) or their integral versions, and we describe them in this section.

4.5.1 New Better Than Used

A distribution with $F(0) = 0$ is said to be new better than used (NBU) if

$$\bar{F}(x+t) \le \bar{F}(x)\bar{F}(t) \tag{4.26}$$

for all $x, t > 0$. When the inequality in (4.26) is reversed, we say that the distribution is new worse than used (NWU). Here, we are comparing the residual life X_t and X of a unit and the definition says that a new unit has stochastically larger (smaller) life than one of age t, and therefore NBU (NWU) represents positive (negative) ageing. The equality in (4.26) holds if and only if

$$\bar{F}(x+t) = \bar{F}(x)\bar{F}(t)$$

which is a Cauchy functional equation with the only continuous solution of the form $\bar{F}(x) = e^{-\lambda x}$, or X is exponential. One may refer to Rao and Shanbhag [506] for a thorough discussion on characterizations of distributions based on such functional equations.

Definition 4.22. A random variable X with $Q(0) = 0$ is said to be NBU (NWU) if and only if

$$Q(u+v-uv) - Q(v) \leq Q(u) \tag{4.27}$$

for $0 \leq u < v < 1$.

The equality in (4.27) holds when

$$Q(u+v-uv) = Q(u) + Q(v)$$

or

$$Q(1-(1-u)(1-v)) = Q(1-(1-u)) + Q(1-(1-v)),$$

which reduces to the form

$$Q(1-x_1y_1) = Q(1-x_1) + Q(1-y_1).$$

The last equation has the continuous solution as

$$Q(u) = -k\log(1-u),$$

which means X is exponential.

The NBU property has several other implications as listed below:

1. IHRA (DHRA) $\Rightarrow$ NBU (NWU).
   To see this result in the quantile form, we take $0 < u < v < 1$ and express

$$X \text{ is IHRA} \Rightarrow \frac{\int_0^u H(p)q(p)dp}{Q(u)} \text{ is increasing in } u$$

$$\Rightarrow \frac{\int_0^{u+v(1-u)} H(p)q(p)dp}{Q(u+v(1-u))} \geq \frac{\int_0^u H(p)q(p)dp}{Q(u)}$$

$$\Rightarrow \frac{Q(u+v-uv)}{Q(u)} \leq \frac{\int_0^{u+v(1-u)}(1-p)^{-1}dp}{\int_0^u (1-p)^{-1}dp} = 1 + \frac{\log(1-u)}{\log(1-v)}$$

$$\Rightarrow \frac{Q(u+v-uv)}{Q(u)} - 1 \le \frac{\int_0^u H(p)q(p)dp}{\int_0^v H(p)q(p)dp}$$

$$\le \frac{Q(v)}{Q(u)} \left[\frac{\int_0^u H(p)q(p)dp}{Q(u)} \Big/ \frac{\int_0^v H(p)q(p)dp}{Q(v)} \right]$$

$$\le \frac{Q(v)}{Q(u)}.$$

Thus,

$$Q(u+v-u) \le Q(u) + Q(v) \Rightarrow X \text{ is NBU.}$$

2. An equivalent condition for X to be NBU is

$$\int g(\alpha x)h[(1-\alpha)x]dF(x) \le \int g(x)dF(x) \int h(x)dF(x)$$

for all non-negative increasing functions g and h and all $0 < \alpha < 1$ (Block et al. [111]).
3. If X is NBU, it has finite moments of all positive orders, which is a stronger result than those for IHR and IHRA classes.
4. If X has a density, the NBU (NWU) implies NBUHR (NWUHR):

$$X \text{ is NBU} \Leftrightarrow Q(u+v-uv) \le Q(u) + Q(v)$$

$$\Leftrightarrow \frac{Q(u+v-uv) - Q(u)}{v(1-u)} \le \frac{Q(v)}{v(1-u)}$$

$$\Rightarrow Q'(u) \le \frac{Q'(0)}{1-u} \text{ on taking limits as } v \to 0$$

$$\Rightarrow (1-u)q(u) \le q(0) \Rightarrow H(u) \ge H(0)$$

$$\Leftrightarrow X \text{ is NBUHR.}$$

5. If each component of a coherent system is NBU, then the system life is also NBU.
6. Convolution of two NBU distributions is NBU.
7. The residual life of an NBU distribution is not NBU. A necessary and sufficient condition for this to hold is that X is IHR.
8. The mixture of two NBU distributions is NBU, provided that the distributions of the components do not cross.
9. For a sequence (X_n) of independent lifetime random variables with NBU distributions, $S_N = X_1 + X_2 + \cdots + X_N$, where N is a positive integer-valued random variable, is also NBU.
10. When X is NBU, we have

$$\begin{aligned}\bar{F}(x) &\geq [\bar{F}(t)]^{\frac{1}{k}}, \quad \frac{t}{k+1} < x < \frac{t}{k}, \ k = 0,1,2,\ldots \\ &\leq [\bar{F}(t)]^{k}, \quad kt < x < (k+1)t,\end{aligned}$$

and when X is NWU,

$$\begin{aligned}\bar{F}(x) &\leq [\bar{F}(t)]^{\frac{1}{k+1}}, \quad \frac{t}{k+1} < x < \frac{t}{k}, \\ &\geq [\bar{F}(t)]^{k+1}, \quad kt < x < (k+1)t.\end{aligned}$$

11. When X is NBU (NWU), we have

$$\frac{\mu'_{r+s+2}}{\Gamma(r+s+3)} \geq (\leq) \frac{\mu'_{r+1}}{\Gamma(r+2)} \frac{\mu'_{s}}{\Gamma(s+2)} \quad r,s \geq 0.$$

12. NBU does not imply DMTTF.
13. If X is NBU and $C^*(u)$ is increasing, then Z is NBU.
14. If a coherent system from independent NBU components has exponential life, then it is essentially a series system with exponential components. We refer the readers to Shaked [530], Abouammoh and El-Neweihi [4] and Barlow and Proschan [70] for some further details in this regard.

There are several variants of the NBU concept presented in the literature. In situations wherein a unit or system deteriorates over time, say, up to an instant t_0, to make the system more effective, replacement or repairs are often thought of. But, by this operation, the system may not revert to the same effectiveness as at t_0. An ageing concept that is relevant in such a situation is to assume that the system lifelength is smaller from t_0 onwards compared to a new one. This idea gives rise to the NBU-t_0 (NWU-t_0) class of life distributions that satisfy (Hollander et al. [274])

$$\bar{F}(t_0 + x) \leq (\geq) \bar{F}(t_0)\bar{F}(x), \quad x \geq 0.$$

Definition 4.23. We say that X is NBU-u_0 (NWU-u_0) if and only if, for some $0 \leq u_0 < 1$, we have

$$Q(u + u_0 - uu_0) \leq (\geq) \ Q(u) + Q(u_0) \quad \text{for all } 0 \leq u < 1.$$

The class of distributions that are both NBU-t_0 and NWU-t_0 is not confined to exponential distributions as in the case of other ageing notions. Along with exponential laws, all distributions with periodic hazard quantile functions and those distributions whose quantile functions are specified by $Q(u) = Q_1(u)$, $0 \leq u \leq u_0$, where $Q_1(0) = 0$, are also both NBU-t_0 and NWU-t_0. Some other important properties of these two classes are as follows:

1. If $H_X(u) \le H_Y(u), 0 \le u \le v_0, H_X(u) = H_Y(u)$ in $(v_0, 1)$, and $H_X(u)$ is decreasing in $[0, v_1], 0 < v_1 < v_0$, then X is NBU-v_0, but not NBU. In general, NBU $\Rightarrow$ NBU-u_0;
2. NBU-t_0 property is preserved under the formation of coherent systems, but NWU-t_0 is not;
3. Both NBU-t_0 and NWU-t_0 are not preserved under convolution;
4. NBU-t_0 is preserved under mixtures of non-crossing distributions, but not for arbitrary mixtures. NWU-t_0 is not closed with respect to formation of mixtures; see Park [483] for more details.

Kayid [318] has presented a generalization of the NBU and NBU-t_0 classes. If A denotes the set of functions $a(u)$ satisfying $a(u) > 0$ in $(0, 1)$ and $a(u) = 0$ otherwise, X is said to be NBU with respect to $a(u)$, denoted by NBU$_{(a)}$, if and only if

$$\int_0^{F_t^{-1}(u)} a(F_t(x))dx \le \int_0^{F^{-1}(p)} a(F(x))dx,$$

where F_t is the usual residual life distribution of $X_t = X - t|(X > t)$. When $a(\cdot)$ is a constant, it is evident that NBU$_{(a)}$ reduces to NBU, and when the time is fixed as t_0, NBU$_{(a)} \Leftrightarrow$ NBU-t_0. For a non-negative X with continuous F, if X is NBU (NBU$_{(a)}$) and $a(u)$ is decreasing, then X is NBU$_{(a)}$ (NBU).

A slightly different concept is NBU* t_0 (NWU* t_0), defined by Li and Li [398] through the relationship

$$\bar{F}(x+y) \le (\ge) \bar{F}(x)\bar{F}(y)$$

for all $x \ge 0$, $y \ge t_0 > 0$. The difference between NBU-t_0 and NBU*t_0 is that in the former t_0 is a fixed time while in the latter it extends beyond t_0. From the above, we have the following definition.

Definition 4.24. We say that X is NBU*u_0 (NWU*u_0) if and only if the quantile function satisfies

$$Q(u+v-uv) \le (\ge) Q(u) + Q(v) \quad \text{for all } 0 \le u < 1 \text{ and } v \ge u_0.$$

The two classes NBU*u_0 and NWU*u_0 possess the following properties:

1. NBU $\Rightarrow$ NBU * u_0 $\Rightarrow$ NBU * u_1, $u_1 \ge u_0$;
2. IHRA * u_0 need not imply NBU * u_0;
3. If $X_1, X_2, \ldots, X_n$ are independent and NBU * t_i, $i = 1, 2, \ldots, n$, then the life of the coherent system with X_i as lifetimes of the components is also NBU * t_0, where $t_0 = \max(t_1, t_2, \ldots, t_n)$. As a consequence, a coherent system with n independent components each of which is NBU * t_0 is also NBU * t_0;
4. If X_1 and X_2 are independent NBU * t_0 lifetimes, their convolution is also NBU * t_0;
5. If a life distribution is NBU * t_0, it is also NBU-t_0. Thus, we have

$$\text{NBU} \Rightarrow \text{NBU} * t_0 \Rightarrow \text{NBU -}t_0.$$

Another generalization of the NBU class has been provided by Deshpande et al. [172] using second order stochastic dominance, called the new better (worse) than used in second order dominance, NBU(2) (NWU(2)).

Definition 4.25. A lifetime random variable X is said to be NBU(2) (NWU(2)) if and only if

$$\int_0^x \bar{F}(y)dy \geq \int_0^x \frac{\bar{F}(t+y)}{\bar{F}(t)}dy$$

for all $t, x \geq 0$, or equivalently,

$$\int_0^u (1-p)q(p)dp \geq \frac{1}{1-v}\int_0^u [1-Q^{-1}(Q(p)+Q(v))]q(p)dp$$

for all $0 \leq u, v < 1$.

Obviously,

$$\text{NBU (NWU)} \Rightarrow \text{NBU (2)(NWU(2))}.$$

Li and Kochar [389] have shown that NBU(2) class is closed under the formation of series systems and convolution. Li [396, 397] further established that NBU(2) class is closed with respect to formation of mixtures and parallel systems. The convolution of X_1 and X_2 which are NBU(2) is also NBU(2) (Hu and Xie [287]). Some limited converse results on the closure properties have been discussed in Li and Yam [394]. If a system possesses a particular ageing property, the problem is to examine whether components satisfy the same property. Li and Yam [394] have shown that if parallel and series systems consisting of independent and identically distributed components are NWU (2), then the components are also NWU (2).

A stochastic version of the NBU property has been discussed by Singh and Deshpande [542] along the lines of stochastically increasing hazard rates presented earlier.

Definition 4.26. A lifetime X is said to be stochastically new better than used (SNBU) if

$$P\left(X \geq \sum_{i=0}^{k+1} Y_i | X \geq \sum_{i=0}^{k} Y_i\right) \leq P(X \geq Y_{k+1}),$$

where $Y_0, Y_1, \ldots, Y_n, \ldots$, with $Y_0 = 0$, is a sequence of independent and identically distributed exponential random variables each with mean μ, and X is independent of the Y_i's. It has been established that

$$\text{SIHR} \Rightarrow \text{SNBU} \quad \text{and} \quad \text{NBU} \Rightarrow \text{SNBU}.$$

Yet another extension of the NBU and NWU property of ageing systems in the context of comparison of the reliability of new and used systems by the use of dynamic signatures has been provided by Samaniego et al. [514]. Recall that the signature of a system with n independent and identically distributed components is an n-dimensional vector whose ith component is the probability that the ith ordered component failure is fatal to the system. System signatures have found key applications in the study and comparison of engineered systems; see, for example, Samaniego [513]. Now, when a working used system is inspected at time t and it is observed that precisely k failures have occurred by that time, then the $(n-k)$-dimensional vector whose jth element is the probability that the $(k+j)$th ordered component failure is fatal to the system has been termed the dynamic signature by Samaniego et al. [514]. It is indeed a distribution-free measure of the design of the residual system. With such a notion of dynamic signature, these authors presented the following dynamic versions of NBU (NWU).

Definition 4.27. Let T denote the lifetime of a coherent system with n components whose lifetimes $X_1,\dots,X_n$ are independent and identically distributed with a continuous distribution function F over $(0,\infty)$. Let $X_{1:n} \le \cdots \le X_{n:n}$ denote the order statistics of $X_1,\dots,X_n$, and E_i be the event that $\{X_{1:n} \le t < X_{i+1:n}\}$, with $X_{0:n} \equiv 0$. Then, for fixed $i \in \{0,1,\dots,n-1\}$, T is said to be conditionally NBU, given i failed components, denoted by i-NBU, if for all $t > 0$, either $P(E_i \cap \{T > t\}) = 0$, or

$$P(T > x) \ge P(T > x+t \mid E_i \cap \{T > t\}) \quad \text{for all } x > 0.$$

Definition 4.28. A n-component is said to be uniformly new better than used, denoted by UNBU, if it is i-NBU for $i \in \{0,1,\dots,n-1\}$.

Samaniego et al. [514] have illustrated the use of these concepts in the performance evaluation of burn-in systems.

4.5.2 New Better Than Used in Convex Order

Cao and Wang [127] discussed a new class of distributions called new better than used in convex order (NBUC) and its dual new worse than used in convex order (NWUC). The NBUC (NWUC) class satisfies

$$\int_x^{\infty} \bar{F}_y(t)dy \le (\ge) \int_x^{\infty} \bar{F}(t)dt.$$

In terms of quantiles, we have the following definition.

Definition 4.29. We say that X is NBUC (NWUC) if and only if

$$\frac{1}{1-v}\int_u^1 [1-Q^{-1}(Q(p)+Q(v))]q(p)dp \le (\ge) \int_u^1 (1-p)q(p)dp.$$

These two classes possess the following properties:

1. NBU (NWU) $\Rightarrow$ NBUC (NWUC), as NBUC is the integrated version of NBU;
2. A parallel system of independent and identically distributed NBUC components is NBUC (Hendi et al. [269] and Li et al. [390]). Even when the components are independent and non-identical, NBUC class is preserved under the formation of parallel systems (Cai and Wu [124]);
3. The convolution of two independent NBUC variables is NBUC (Hu and Xie [287]);
4. The NBUC property is preserved under monotonic antistar-shaped transformation and under nonhomogeneous Poisson shock models (Li and Qiu [391]);
5. Under the formation of mixtures, the NBUC class is preserved (Li [397]);
6. If X is NBUC, then (Ahmad and Mugadi [26])

$$(r+2)!(s+1)!E(X^{r+s+3}) \le (r+s+3)!E(X^{r+2})E(X^{s+1}).$$

As an application of the concept, Belzunce et al. [87] compared the age replacement (block) policies and a renewal process with no planned replacements when the lifetime of the unit is NBUC.

A further extension of the NBUC class is the NBUCA class defined by

$$\int_0^\infty \int_x^\infty \bar{F}(u+t)dudx \le \bar{F}(t)\int_0^\infty \int_x^\infty \bar{F}(u)dudx \quad \text{for all } t \ge 0.$$

For properties and further details, we refer the readers to Ahmad and Mugadi [26].

Elabatal [186] studied the extensions of NBU(2) and NBUC classes at a specific age t_0, called NBU(2)-t_0 and NBUC-t_0, which can be defined as follows.

Definition 4.30. The NBU(2)-v_0 class of distributions is one that satisfies

$$\frac{1}{1-v_0}\int_0^u [1-Q^{-1}(Q(p)+Q(v_0))]q(p)dp \le \int_0^u (1-p)q(p)dp$$

for some $0 \le v_0 < 1$.

Definition 4.31. X is said to be NBUC-v_0 if, for some $0 \le v_0 < 1$, we have

$$\frac{1}{1-v_0}\int_u^1 [1-Q^{-1}(Q(p)+Q(v_0))]q(p)dp \le \int_u^1 (1-p)q(p)dp.$$

It is known that if X_1 and X_2 are independent NBUC-t_0 variables, then the convolution is also NBUC-t_0. The class is also closed under the formation of a parallel system of iid components which are NBUC-t_0. As Poisson shock model interpretation $\bar{H}(x)$ is NBUC-t_0 if (P_k) has discrete NBUC-t_0 property that satisfies

$$\sum_{j=k}^{\infty} \bar{P}_{i+j} \leq \bar{P}_i \sum_{j=k}^{\infty} \bar{P}_j, \quad \bar{P}_K = 1 - P_K.$$

Based on survival function, Hendi [268] introduced the increasing cumulative (decreasing) survival class, denoted by ICSS (DCSS), through the property

$$\int_0^x \bar{F}_{t_1}(y)du \leq (\geq) \int_0^x \bar{F}_{t_2}(y)dy$$

for all $x > 0$, $0 \leq t_1 \leq t_2 < \infty$. It can be seen that the ICSS (DCSS) class is equivalent to the IHR(2) (DHR(2)). However, Hendi [268] proved that DCSS is preserved under convolution, while ICSS is not closed under the formation of convolution and coherent structures. These properties could be read in conjunction with those of the IHR(2) class discussed earlier.

Yet another variant of the NBU distributions is the new better (worse) than used in Laplace order, denoted by NBUL (NWUL).

Definition 4.32. Yue and Cao [598] defined the NBUL (NWUL) class as one that satisfies the inequality

$$\int_0^{\infty} e^{-sx}\bar{F}(t+x)dx \leq (\geq) \int_0^{\infty} e^{-sx}\bar{F}(x)dx.$$

This concept has different interpretations in the context of ageing. One of them is by considering the mean life of a series system of two independent components, one having exponential survival function and the other having survival function $\bar{F}$. In two such systems A and B, if A has used component of age t while B has a used component with survival function $\bar{F}$, then F is NBUL means that the mean life of A is not larger than that of B. These classes have the following properties:

1. NBU $\Rightarrow$ NBU(2) $\Rightarrow$ NBUL;
2. Let X and Y be independent random variables with survival functions $\bar{F}$ and $e^{-\lambda x}$, respectively, and $W = \min(X,Y)$. Then, X is NBUL (NWUL) if and only if W is NBUE (NWUE), and some details of NBUE (NWUE) classes are presented in the next section;
3. NBUL is not closed under the formation of series systems. However, if the component survival functions are completely monotone, then the closure property holds. The NBUL concept is used in connection with replacement policies. For a detailed study of the properties of the classes and their applications, we refer to Yue and Cao [598], Al-Wasel [31], and Li and Qiu [391].

Joe and Proschan [301] have provided a classification of life distributions based on percentiles, which are as follows.

A lifetime random variable X is new better (worse) than used with respect to the α-percentile, denoted by NBUP-α (NWUP-α), if $F(0) = 0$ and $p_{\alpha}(0) \geq (\leq) p_{\alpha}(x)$ for all $x \geq 0$.

They then established the following properties:

1. NBU $\Leftrightarrow$ NBUP-α for all $0 < \alpha < 1$;
2. DPRL-$\alpha \Rightarrow$ NBUP-α for any $0 < \alpha < 1$;
3. If X is NBUP-α, then $\bar{F}(x) \le (1-\alpha)^n$, $np_\alpha(0) \le x < (n+1)p_\alpha(0)$ for $n = 0,1,2,\ldots$, and if F is continuous, $\bar{F}(x) \le (1-\alpha)^{n+1}, np_\alpha(0) \le x < (n+1)p_\alpha(0)$;
4. An NBUP-α distribution has a finite mean that is bounded above by $\frac{p_\alpha(0)}{\alpha}$. If F is continuous, F has a mean (possibly infinite) that is bounded from below by $\frac{(1-\alpha)p_\alpha(0)}{\alpha}$;
5. An NBUP-α distribution has finite moment of order $r > 0$;
6. The closure properties with respect to formation of coherent systems, convolution and mixtures do not hold for NBUP-α and NWUP-α distributions.

4.5.3 New Better Than Used in Expectation

Instead of comparing a life distribution with its residual life distribution, a weaker concept results when expectations are considered for this comparison. This leads to new better than used in expectation (NBUE) and its dual new worse than used in expectation. If $E(X) < \infty$, X is said to be NBUE (NWUE) if and only if

$$\mu \ge (\le) \int_0^\infty \frac{\bar{F}(x+t)}{\bar{F}(t)} dx = m(x)$$

for all $t \ge 0$ for which $\bar{F}(t) > 0$. This says that a used unit of any age has a smaller mean residual life than a new unit with the same life distribution.

Definition 4.33. We say that a lifetime X is NBUE if and only if

$$\frac{1}{1-v}\int_0^1 \{1 - Q^{-1}(Q(p)+Q(v))\} q(p)dp \le \mu = \int_0^1 (1-p)q(p)dp,$$

or

$$\frac{1}{1-u}\int_u^1 (1-p)q(p)dp \le \mu.$$

The NBUE and NWUE classes have the following properties:

1. NBU (NWU) $\Rightarrow$ NBUC (NWUC) $\Rightarrow$ NBUE (NWUE);
2. NBU(2) (NWU(2)) $\Rightarrow$ NBUE (NWUE);
3. DMRL (IMRL) $\Rightarrow$ NBUE (NWUE);
4. NBUE (NWUE) $\Rightarrow$ NDVRL (NIVRL);
5. NBUE (NWUE) $\Rightarrow M(u) \le (\ge) M(0)$. The last inequality is equivalent to $H_Z(u) \ge (\le) H_X(0)$, and so

$$X \text{ is NBUE (NWUE)} \Rightarrow Z \text{ is NBUHR (NWUHR)};$$

6. Both NBUE and NWUE classes are not closed under the formation of coherent systems;
7. The convolution of two NBUE distributions is NBUE, but this preservation property is not true for NWUE;
8. The mixture of two NBUE (NWUE) life distributions is not in general NBUE (NWUE), while the mixture of NWUE distributions, no two of which cross, is again NWUE. This property is not shared by NBUE class. For proofs of Properties 6–8, see Marshall and Proschan [413];
9. If X is NBUE (NWUE), then $\int_x^\infty \bar{F}(t)dt \le \mu e^{-\frac{x}{\mu}}$;
10. When X is NBUE, we have

$$\bar{F}(x) \ge \begin{cases} 1-\frac{x}{\mu} & , x \le \mu \\ 0 & , x \ge \mu \end{cases},$$

and when X is NWUE, we have

$$\bar{F}(x) \le \frac{\mu}{\mu+x}, \quad x \ge 0 \quad \text{(Haines and Singpurwalla [258])};$$

11. When X is NBUE, we have

$$F(x) \ge \frac{\sigma^2+\mu^2-x^2}{\sigma^2+(\mu+x)^2-x^2}, \quad x \le (\mu_2')^{\frac{1}{2}},$$

and in the case of NWUE, we have

$$\begin{aligned} F(x) &\le \frac{\sigma^2}{\sigma^2+(\mu+x)^2}, \quad 0 < x < \frac{2\sigma^2}{\mu} \\ &\ge \frac{\sigma^2}{\sigma^2+x^2}, \quad x \ge \frac{2\sigma^2}{\mu} \quad \text{(Launer [377])}; \end{aligned}$$

12. In the case of NBUE (NWUE) distributions, we have

$$\frac{\mu'_{r+1}}{\Gamma(r+2)} \le (\ge) \frac{\mu'_r}{\Gamma(r+1)}\mu \quad \text{(Barlow and Proschan [70])};$$

13. X is NBU does not imply Z is NBUE, nor X is NBUE implies Z is NBUE;
14. Neither DVRL $\Rightarrow$ NBUE nor NBUE $\Rightarrow$ DVRL. A common property shared by the two concepts is that the coefficient of variation of X is ≤ 1, provided F is strictly increasing;
15. If X is NBUE (NWUE), then $\bar{F}_Z(x) \le (\ge) \bar{F}_X(x)$;

16. If X is NBUE and $E(X) = E(Z)$, then X is exponential and converse is true as well. For details on Properties 14–16, see Gupta [233];
17. DMTTF $\Rightarrow$ NBUE.

4.5.4 Harmonically New Better Than Used

The harmonically new better (worse) than used in expectation HNBUE (HNWUE) class of life distribution, introduced by Rolski [512], consists of distributions for which

$$\int_x^\infty \bar{F}(t)dt \leq (\geq)\ \mu e^{-\frac{x}{\mu}}, \quad x \geq 0. \tag{4.28}$$

An equivalent definition is presented below.

Definition 4.34. A lifetime random variable X is HNBUE (HNWUE) if and only if one of the following conditions are satisfied:

(i)

$$\int_u^1 (1-p)q(p)dp \leq (\geq)\ \mu e^{-\frac{Q(u)}{\mu}};$$

(ii)

$$\left\{ \frac{\int_0^u \frac{q(p)dp}{M(p)}}{\int_0^u q(p)dp} \right\}^{-1} \leq (\geq)\ \mu.$$

The first definition follows directly from (4.28). To prove the equivalence of (i) and (ii), we observe that

$$\begin{aligned}
\text{(ii)} &\Leftrightarrow \int_0^u q(p)\left(\frac{1}{1-p}\int_p^1 (1-t)q(t)dt\right)^{-1} dp \geq \frac{Q(u)}{\mu} \\
&\Leftrightarrow \int_0^u \left[\frac{q(p)(1-p)}{\int_p^1 (1-t)q(t)dt}\right] dp \geq \frac{Q(u)}{\mu} \\
&\Leftrightarrow \log\mu - \log\int_0^u (1-p)q(p)dp) \geq \frac{Q(u)}{\mu} \\
&\Leftrightarrow \text{(i)}.
\end{aligned}$$

Thus, the HNBUE concept says that the harmonic mean of the mean residual hazard quantile function of a unit of age x is not grater than the harmonic mean life of a new unit. The two classes HNBUE and HNWUE enjoy the following properties:

1. NBUE (NWUE) $\Rightarrow$ HNBUE (HNWUE), which follows from

$$\begin{aligned} \text{NBUE } &\Rightarrow M(u) \le M(0) \\ &\Rightarrow \frac{q(u)}{M(u)} \ge \frac{q(u)}{M(0)} \\ &\Rightarrow \frac{\int_0^u \frac{q(p)}{M(p)}dp}{Q(u)} \ge \frac{\int_0^u q(p)dp}{M(0)Q(u)} \Rightarrow \text{HNBUE;} \end{aligned}$$

2. A necessary and sufficient condition that X is HNBUE (HNWUE) is that

$$E\phi(X) \le (\ge)\, E\phi(X^*)$$

   for all non-decreasing convex functions ϕ on $(0,\infty)$ with $\phi(0+) = 0$, where X^* is exponential with the same mean μ as X;
3. X is HNBUE if and only if $Q_z(u) \le Q_{X^*}(u)$;
4. The HNBUE class is closed under the operation of forming non-negative linear combination of HNBUE random variables;
5. Both classes are not preserved under the formation of coherent structures;
6. The HNWUE class is preserved under mixing, but HNWUE is not;
7. We have

$$\bar{F}(x) \le \begin{cases} 1 & , x < \mu \\ e^{\frac{\mu - x}{x}}, & , x > \mu \end{cases},$$

   when X is HNBUE;
8. $\mu^{r+3} \ge \frac{\mu_{r+3}}{(r+3)!}$ if X is HNBUE;
9. $\bar{H}(t) = \sum_{k=0}^{\infty} P(N(t) = k)\, \bar{P}_k$ is HNBUE (HNWUE), where $N(t)$ is a counting process governing the shocks and the interarrival times of shocks are independent HNBUE (HNWUE). For further details, one may refer to Klefsjö [332], Bhattacharjee and Kandar [98], Al-Ruzaize et al. [30], Basu and Bhattacharjee [79] and Cheng and Lam [144]. The kth order HNBUE has been studied by Basu and Ebrahimi [80].

Definition 4.35. A lifetime random variable is (k-HNBUE) if

$$\frac{1}{x}\int_0^x m^{-k}(t)dt \le (\ge)\mu^k \quad \text{for all } x > 0,$$

or

$$\frac{\int_0^u M^{-k}(p)q(p)dp}{\int_0^u q(p)dp} \le (\ge)\mu^k \quad \text{for } 0 < u < 1,$$

where $k = 1$ corresponds to the usual HNBUE. It is known that whenever X is $(k+1)$-HNBUE, it is also k-HNBUE.

Using stochastic dominance of order three, the HNBUE concept can be generalized, which results in HNBUE(3) (HNWUE(3)) defined by

$$\int_0^\infty \int_t^\infty \bar{F}(u)dudt \leq (\geq) \mu^2 e^{-\frac{x}{\mu}} \quad \text{for all } x,t \geq 0.$$

It is evident that HNBUE $\Rightarrow$ HNBUE(3).

4.5.5 $\mathscr{L}$ and $\mathscr{M}$ Classes

A still larger class than the HNBUE can be constructed using transforms. Klefsjö [337] introduced the $\mathscr{L}$-class by considering the Laplace transform of the survival function.

Definition 4.36. We say that a random variable X with finite mean μ belongs to the $\mathscr{L}$-class ($\bar{\mathscr{L}}$-class) if and only if

$$\int_0^\infty e^{-sx} \bar{F}(x)dx \geq (\leq) \frac{\mu}{1+s\mu}.$$

Chaudhury [138] found that for X in the $\mathscr{L}$-class, the coefficient of variation is ≤ 1. However, the exponential distribution is not characterized by the property that the coefficient of variation is unity. Further,

$$X \text{ is HNBUE (HNWUE)} \Rightarrow X \text{ is } \mathscr{L}(\bar{\mathscr{L}}).$$

Chaudhury [139] also established that if (F_n) is a sequence of life distributions in $\mathscr{L}$ and F_n converges weakly to F, then F also belongs to $\mathscr{L}$. Consider a sequence $X_1, X_2, \ldots,$ of independent and identically distributed random variables, and N as a geometric random variable over the set of positive integers. If N is independent of the X_i's, then the sum $S = \sum_{i=1}^N X_i$ is called a geometric compound. Lin and Hu [403] established the preservation of the $\mathscr{L}$ class under geometric compounding. Several other interesting properties are presented in Bhattacharjee and Sengupta [97] Lin [400], and Nanda [457].

Klar [330] has given an example of a distribution that belongs to $\mathscr{L}$ with the property that its hazard rate tends to zero and mean residual life tends to infinity, which led to some doubts about the $\mathscr{L}$-class representing positive ageing. To overcome this limitation, Klar and Muller [331] presented an ageing class in which the Laplace transform is replaced by the moment generating function, and referred to it as the $\mathscr{M}$ class.

Definition 4.37. We say that X belongs to the $\mathscr{M}$ class if

$$\int_0^\infty e^{tx} dF(t) \leq \frac{1}{1-\mu x}, \quad 0 \leq x < \frac{1}{\mu},$$

or

$$\int_0^\infty \bar{F}(t) e^{tx} dx \leq \frac{\mu}{1-\mu x},$$

where $\frac{1}{1-\mu x}$ is the moment generating function of the exponential distribution.

Notice that since $E(e^{tX})$ has to be finite, $x < \frac{1}{\mu}$, and so the case $h(x) \to 0$ as $x \to \infty$ does not arise. We have the following properties for the $\mathscr{M}$ class:

1. The $\mathscr{M}$ class contains all HNBUE distributions.
2. Distributions in the $\mathscr{M}$ class are closed under convolution of independent random variables;
3. The $\mathscr{M}$ class contains all random variables with $P(a < X < b) = 1$ and $E(X) \geq \frac{a+b}{2}$, $0 \leq a < b$, and also all symmetric distributions;
4. Let $X_Y = X - Y | (X > Y)$. Then, X is in $\mathscr{L}$ if $E(X_Y) < \mu$ for all Y independent of X, and have a density function of the form

$$g_t(x) = \frac{e^{tx} \bar{F}(x)}{\int e^{tx} \bar{F}(x) dx}.$$

   On the other hand, X is in $\mathscr{M}$ if $E(X_Y) < Y$ for all Y independent of X, and have density $g_t(x)$ above for some $0 < t < \mu^{-1}$;
5. X is in $\mathscr{M} \Rightarrow X$ is in $\mathscr{L}$, and

$$X_i \text{ is in } \mathscr{M} \Rightarrow \sum \alpha_i X_i \text{ is in } \mathscr{M}, \quad \text{with } \alpha_i \geq 0, \sum \alpha_i = 1,$$

   and the X_i's have a common mean.

4.5.6 Renewal Ageing Notions

The renewal ageing concepts essentially compare the reliability functions of four random variables—X, its residual life X_t, the equilibrium random variable Z, and the corresponding residual life Z_t. Results in this direction are due to Abouammoh et al. [7], Bon and Illayk [116], Abouammoh and Qamber [10] and Abdel-Aziz [2].

Definition 4.38. We say that X is new better (worse) than renewal used, denoted by NBRU (NWRU), if and only if, for all $t \geq 0$,

$$\int_{x+t}^\infty \frac{\bar{F}(u)du}{\bar{F}(x)} \leq (\geq) \int_x^\infty \bar{F}(u)du.$$

Definition 4.39. Renewal new is better (worse) than used, denoted by RNBU (RNWU), if and only if, for all t,

$$\frac{\bar{F}(x+t)}{\bar{F}(t)} \leq (\geq) \frac{1}{\mu} \int_x^{\infty} \bar{F}(u)du.$$

Definition 4.40. Renewal new is better (worse) than used in expectation, denoted by RNBUE (RNWUE), if and only if

$$2\mu \int_x^{\infty} \bar{F}(u)du \leq \mu_2' \bar{F}(x) \quad \text{or} \quad E(X_t) \leq E(Z).$$

Definition 4.41. Renewal new is better (worse) than renewal used, denoted by RNBRU (RNWRU), if and only if

$$\mu \int_{x+t}^{\infty} \bar{F}(u)du \leq \left(\int_x^{\infty} \bar{F}(u)du \right) \left(\int_t^{\infty} \bar{F}(u)du \right).$$

Definition 4.42. Renewal new is better (worse) than renewal used, denoted by RNBRUE (RNWRUE), if and only if $E(Z_t) \leq E(Z)$, or

$$2\mu \int_x^{\infty} \int_t^{\infty} \bar{F}(u)dudt \leq \mu_2' \int_x^{\infty} \bar{F}(u)du.$$

Definition 4.43. The random variable X has generalized increasing mean residual life (GIMRL) property if and only if, for all $x \geq 0$,

$$\frac{\int_t^{\infty} \bar{F}(u)du}{\bar{F}(t+x)} \text{ is increasing in } t.$$

Definition 4.44. Harmonically new renewal better than used in expectation, denoted by HRNBUE, if

$$\bar{F}_Z(t) \leq e^{-\frac{t}{\mu_Z}}.$$

The implications among these classes are as follows:

$$\text{GIMRL} \Rightarrow \text{RNBU} \Rightarrow \text{RNBUE} \Rightarrow \text{HRNBUE}, \tag{4.29}$$

$$\text{NBU} \Rightarrow \text{NBRU}.$$

By comparing the above definitions, we see that NBRU property is the same as NBUC discussed earlier. Bon and Illayk [116] established that if the first two moments of X are finite and if X has HRNBUE property, then X has an exponential distribution. Thus, all classes implied in (4.29) are gathered in the exponential class. The conversion of Definitions 4.38–4.44 can be accomplished in the same manner as in the earlier cases.

Since we have discussed a large number of classes based on ageing concepts and given separate implications, a consolidated diagram showing all the classes and mutual implications is presented in Fig. 4.1 for a quick reference.

4.6 Classes Based on Concepts in Reversed Time

Parallel developments have been attempted to generate life distributions based on the monotonicity properties of the reversed hazard rate function, reversed mean residual life function, and so on. However, a special feature of such criteria is that for lifetime random variables, they have monotonicity in only one direction (either decreasing or increasing). Hence, they fail to distinguish life distributions and are therefore of limited use in representing different types of ageing characteristics. But, other properties possessed by these classes could be of advantage in the analysis of data.

Definition 4.45. A lifetime random variable is decreasing reversed hazard rate (DRHR) if and only if $\Lambda(u)(\lambda(x))$ is decreasing for all $0 < u < 1$ $(x > 0)$.

The quantile function corresponding to the reversed residual life, $t - x|(X \leq t)$, is

$$Q_{u_0}(u) = Q(u) - Q((1-u)u_0) = \int_{u_0(1-u)}^{u_0} q(p)dp.$$

Hence, Definition 4.45 is equivalent to saying that X is DRHR if and only if

$$Q_{u_0}(u) \leq Q_{u_1}(u), \quad 0 < u_1 \leq u_2 < 1.$$

Block et al. [111] have proved that there does not exist a non-negative random variable that has increasing reversed hazard rate function. A large class of distributions including those that are IHR like the Weibull, gamma, and Pareto are DRHR. The DRHR distributions are closed under the formation of coherent systems (Nanda et al. [462]).

Definition 4.46. A life distribution has increasing reversed mean residual life time (increasing mean inactivity time) IMIT if and only if $r(x)$ $(R(u))$ is increasing in $x(u)$.

There is no non-negative random variable which has decreasing MIT over the entire domain $(0,\infty)$. Further, DRHR $\Rightarrow$ IMIT (Nanda et al. [462]). Similarly, the monotonicity of the reversed variance residual life $D(u)$ can be studied.

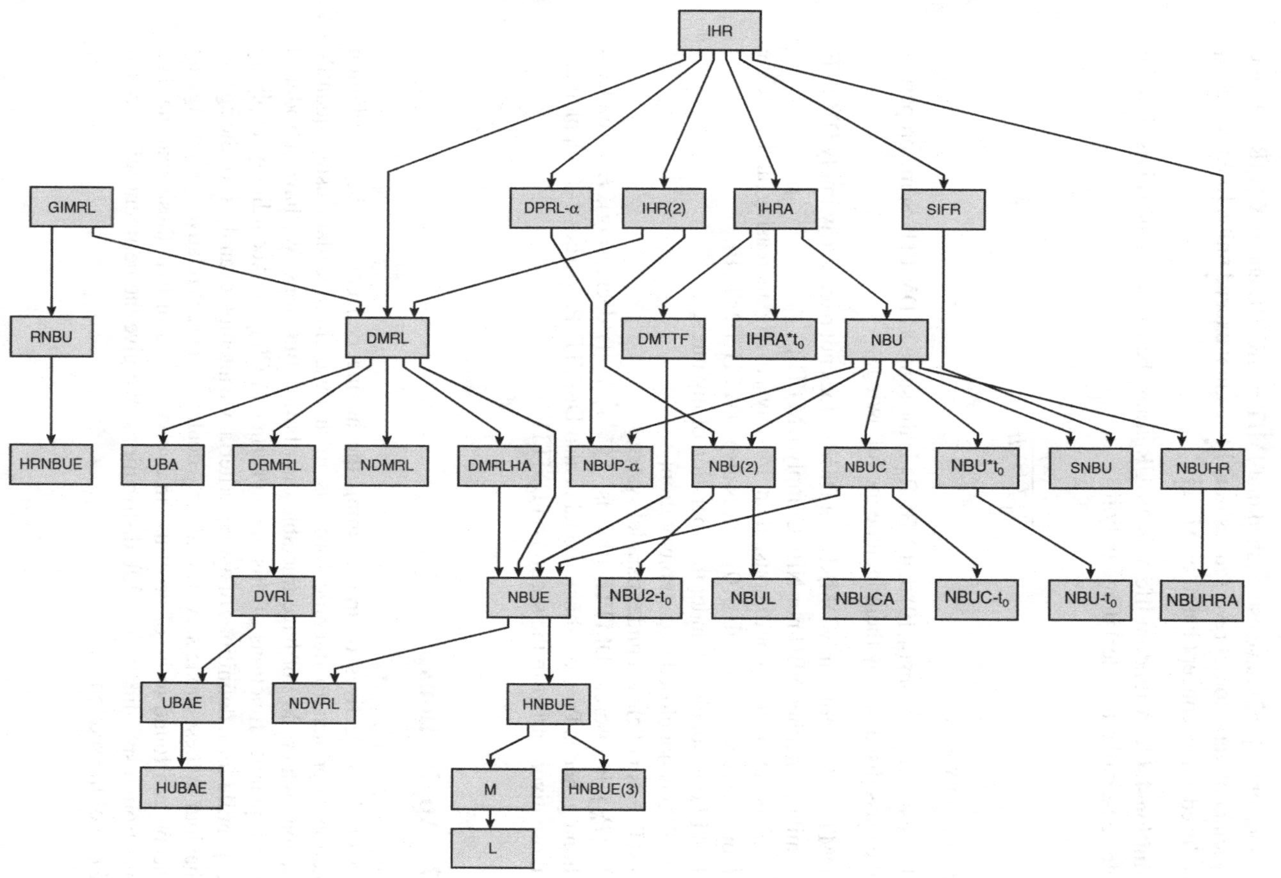

Fig. 4.1 Implications among different ageing concepts

Definition 4.47. We say that X is increasing reversed variance residual life (IRVRL) if and only if $v(x)$ $(D(u))$ is increasing in $x(u)$.

Nanda et al. [462] have established that IMIT $\Rightarrow$ IRVRL and if X is IRVRL, the coefficient of variation of reversed residual life cannot exceed unity. Li an Xu [393] introduced a new concept based on MTTF.

Definition 4.48. A random life X is NBUR$_h$ (new better than renewal used in the reversed hazard rate order) if and only if

$$\frac{F(x)}{\int_0^x \bar{F}(t)dt}$$

increases in $x \geq 0$.

It is easy to see from (4.19) that NBUR$_h$ is the same as DMTTF. Some properties of the class discussed by them include the following:

1. DMTTF does not imply DMRL and DMRL does not necessarily imply DMTTF. Similarly, neither NBU nor NBUC imply DMTTF;
2. If X is absolutely continuous, then for any strictly increasing and concave (convex) function ϕ with $\phi(0) = 0$, is also DMTTF (IMTTF);
3. DMTTF is not closed under the operation of mixtures;
4. IMTTF is not closed under convolution;
5. IMTTF is not closed under parallel systems;
6. If $\{P_K\}$ is discrete DMTTF (that is, $\sum_0^{k-1} \bar{P}_i/P_k$ is decreasing), $\bar{H}(t)$ under a homogeneous Poisson shock model is also DMTTF. Properties 1–6 supplement those given earlier in the section on DMTTF.

4.7 Applications

One of the objectives of transforming the ageing concepts in the distribution function approach to quantile forms is to analyse lifetime data using quantile functions which do not have tractable distribution functions. We have introduced several quantile functions of this nature earlier in Chap. 3. Accordingly, applying the quantile form definitions of ageing criteria, we attempt an analysis of the ageing behaviour in these models. A second topic dealt with here is relative ageing. Ageing concepts are found to be of great use in evolving tests of hypothesis that the data come from a specific class of life distributions. We give the pertinent references at the end of the section.

4.7.1 Analysis of Quantile Functions

Govindarajulu's distribution in (3.81) with

$$Q(u) = \sigma[(\beta+1)u^{\beta} - \beta u^{\beta+1}]$$

has hazard quantile function as

$$H(u) = [\sigma\beta(\beta+1u^{\beta-1}(1-u)^2]^{-1}.$$

Accordingly,

$$H'(u) = \frac{u(1+\beta)+(1-\beta)}{\sigma\beta(\beta+1)u^{\beta}(1-u)^3}. \tag{4.30}$$

It is evident from (3.33) that $H(u)$ is increasing for $\beta \leq 1$ and for $\beta > 1$, $H'(u) = 0$ at $u_0 = \frac{\beta-1}{\beta+1}$. Hence, X is IHR for $0 < \beta < 1$, and BT for $\beta > 1$, with change point u_0.

The mean residual quantile function from Sect. 3.4 is

$$M(u) = \frac{\sigma}{(\beta+2)(1-u)}\left[2-(\beta+1)(\beta+2)u^{\beta}+2\beta(\beta+2)u^{\beta+1}-\beta(\beta+1)u^{\beta+2}\right]$$

which is decreasing for $\beta < 1$. At $\beta = 1$,

$$M(u) = \frac{2\sigma(1-u)^2}{3}$$

again decreases. But, at $\beta = 2$,

$$M(u) = \frac{\sigma}{2}(1-u)^2(1+3u)$$

and so $M'(u) = 0$ at $u = \frac{1}{9}$. We see that $M(u)$ is nonmonotone, being increasing in $(0, \frac{1}{9})$ and then decreasing in $(\frac{1}{9}, 1)$ with change point $u_0 = \frac{1}{9}$. Thus, $M(u)$ is of UBT shape. Notice that at $\beta = 2$, the change point of the failure rate is $u_0 = \frac{1}{3}$ and at this value $M(u)$ is decreasing. In the case of the refrigerator failure data studied in Sect. 3.4, the parameters are $\sigma = 1$ and $\hat{\beta} = 2.94$. We have $M(u)$ initially increasing and then decreasing with change point $u \doteq 0.2673$, while the hazard quantile function is BT shaped with change point $u_0 \doteq 0.493$. Thus, the change point occurs earlier for the mean residual quantile function. Figure 4.2 presents the shapes of the hazard quantile function for selected values of the parameters.

Consider the power-Pareto distribution with its hazard quantile function as

$$H(u) = (1-u)^{\lambda_2}[cu^{\lambda_1-1}\{\lambda_1(1-u)+\lambda_2 u\}]^{-1}.$$

The nature of the hazard rate function for some values of C_1, λ_1 and λ_2 is exhibited in Fig. 4.3. Differentiating $H(u)$, we see that the sign of $H(u)$ depends on

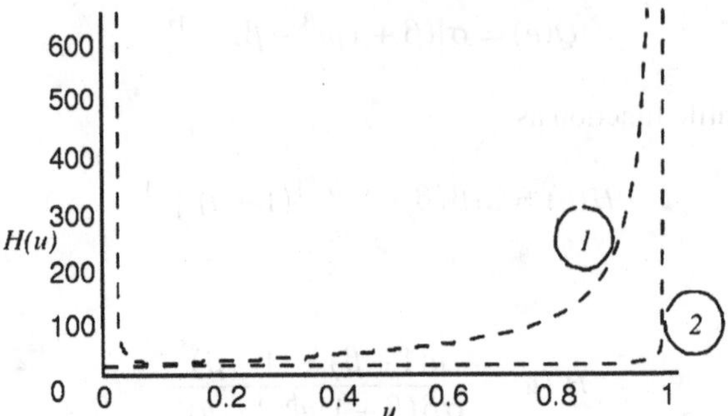

Fig. 4.2 Plots of hazard quantile function when (1) $\beta = 0.1$, $\sigma = 1$ and (2) $\beta = 2$, $\sigma = 1$ for Govindarajalu's distribution

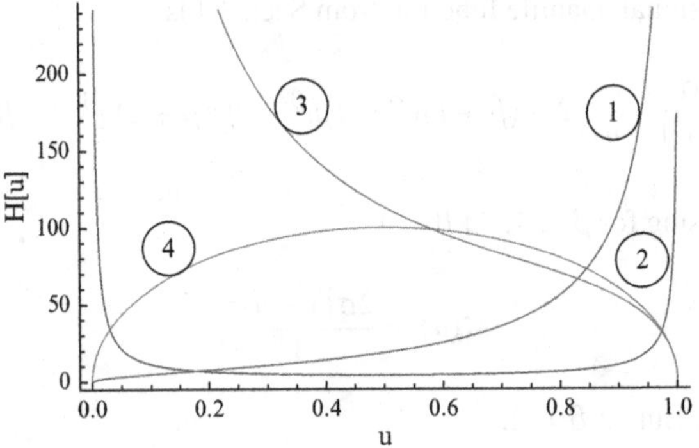

Fig. 4.3 Plots of hazard quantile function when (1) $C = 0.1$, $\lambda_1 = 0.5$, $\lambda_2 = 0.01$; (2) $C = 0.5$, $\lambda_1 = 2$, $\lambda_2 = 0.01$; (3) $C = 0.01$, $\lambda_1 = 2$, $\lambda_2 = 0.5$; (4) $C = 0.01$, $\lambda_1 = 0.5$, $\lambda_2 = 0.5$, for the power-Pareto distribution

$$g(u) = -[(\lambda_1 - \lambda_2)^2 u^2 + (\lambda_1 - 2\lambda_1^2 + 2\lambda_1\lambda_2)u + \lambda_1(\lambda_1 - 1)].$$

Denoting the admissible roots of $g(u) = 0$ by u_1 and u_2 with $u_1 > u_2$, we see that $H(u)$ is decreasing when

$$\lambda_1(1 - 4\lambda_2) + 4\lambda_2^2 \leq 0 \quad \text{or} \quad \lambda_1 = 0$$

for all u. Further, $H(u)$ decreases when

$$\lambda_1(1 - 4\lambda_2) + 4\lambda_2^2 > 0 \text{ for all } u \text{ outside the interval } (u_2, u_1),$$

and increases within (u_2, u_1). If there is only one root for $H'(u) = 0$, that is, $u_1 = u_2 = u_0$, then $H(u)$ is decreasing. For $\lambda_2 = 0$, $H(u)$ is increasing. Summarizing the shape of $H(u)$, we have

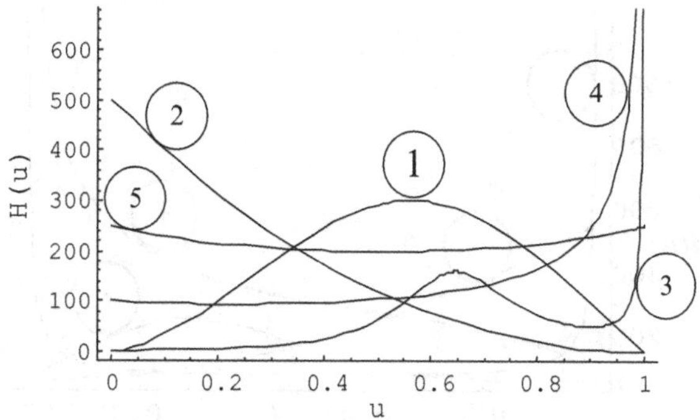

Fig. 4.4 Plots of hazard quantile function when (1) $\lambda_1 = 0$, $\lambda_2 = 100$, $\lambda_3 = -0.5$, $\lambda_4 = -0.1$; (2) $\lambda_1 = 0$, $\lambda_2 = 500$, $\lambda_3 = 3$, $\lambda_4 = 2$; (3) $\lambda_1 = 0$, $\lambda_2 = 2$, $\lambda_3 = 10$, $\lambda_4 = 5$; (4) $\lambda_1 = 0$, $\lambda_2 = 100$, $\lambda_3 = 2$, $\lambda_4 = 0.5$; (5) $\lambda_1 = 0$, $\lambda_2 = 250$, $\lambda_3 = 2$, $\lambda_4 = 0.001$, for the generalized Tukey lambda distribution

$$X \text{ is DHR for } \lambda_1 = 0 \text{ or } \lambda_1(1-4\lambda_2)+4\lambda_2^2 \le 0,$$

$$X \text{ is IHR for } \lambda_2 = 0,$$

and $H(u)$ has opposite monotonicities to that in (u_2, u_1) where it is increasing. It can be verified that X is IHR when $\lambda_1 = 2$, $\lambda_2 = 0$, DHR when $\lambda_1 = 3$, $\lambda_2 = 2$, and non-monotonic when $\lambda_1 = \lambda_2 = \frac{1}{2}$. For an application to real data, we return to Sect. 3.6, where the power-Pareto distribution did provide a good fit for the data on the failure times of 20 electric carts. The hazard quantile function is

$$H(u) = (1-u)^{0.0967}[1530.53u^{-0.7654}(0.2346(1-u)+0.0967u)]^{-1}.$$

Here, $\lambda_1(1-4\lambda_2)+4\lambda_2^2 > 0$ and so the hazard curve is initially increasing and then becomes BT shaped.

The generalized Tukey lambda distribution of Freimer et al. [203] has its hazard quantile function as

$$H(u) = \lambda_2[(1-u)^{\lambda_4} + u^{\lambda_3-1}(1-u)]^{-1}.$$

The sign of $H'(u)$ depends on the function

$$g(u) = \lambda_2[\lambda_4(1-u)^{\lambda_4-1} + \lambda_3 u^{\lambda_3-1} + (1-\lambda_3)u^{\lambda_3-2}].$$

The hazard quantile function can take on a wide variety of shapes as can be seen in Fig. 4.4. It is easy to see that X is IHR when $\lambda_2 > 0$, $\lambda_4 > 0$ and $0 < \lambda_3 < 1$, subject to the condition $\lambda_1 - \frac{1}{\lambda_2\lambda_3} \ge 0$ which is required for X to have a life distribution. When $\lambda_4 = 0$,

$$g(u) = \lambda_2[u^{\lambda_3-2}(\lambda_3 u + 1 - \lambda_3)]$$

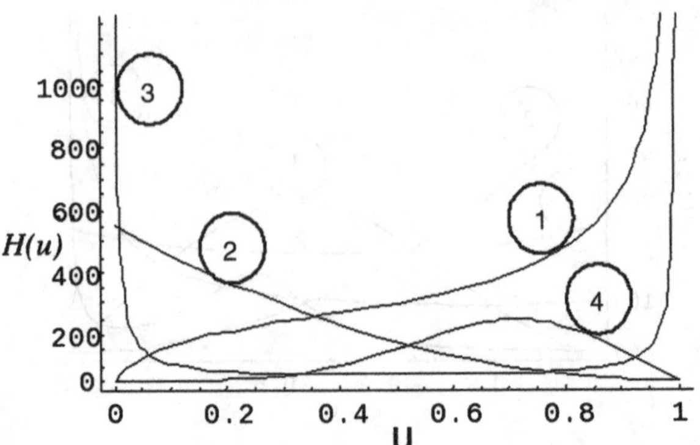

Fig. 4.5 Plots of hazard quantile function when (1) $\lambda_1 = 1$, $\lambda_2 = 100$, $\lambda_3 = 0.05$, $\lambda_4 = 0.5$; (2) $\lambda_1 = 0$, $\lambda_2 = -1000$, $\lambda_3 = 0$, $\lambda_4 = -2$; (3) $\lambda_1 = 1$, $\lambda_2 = 10$, $\lambda_3 = 2$, $\lambda_4 = 0$; (4) $\lambda_1 = 0$, $\lambda_2 = -1000$, $\lambda_3 = -2$, $\lambda_4 = -1$, for the generalized lambda distribution

so that $H'(u) = 0$ has a solution $u = \frac{\lambda_3 - 1}{\lambda_3}$. In this case, $H(u)$ is BT-shaped. An exhaustive analysis using $g(u)$ given above is difficult and for this reason we have presented above only some illustrative cases that exhibits the flexibility of $H(u)$ to adopt to different kinds of ageing behaviour.

The generalized lambda distribution, like the Freimer et al. [203] model, has quite a flexible hazard quantile function. Recall from (3.5) that the distribution has

$$H(u) = \lambda_2[(1-u)(\lambda_3 u^{\lambda_3 - 1} + \lambda_4 (1-u)^{\lambda_4 - 1})]^{-1}.$$

We shall now take some special cases. When $\lambda_3 = 0$, $\lambda_4 > 0$,

$$H(u) = \frac{\lambda_2}{\lambda_4 (1-u)^{\lambda_4}}$$

and so X is IHR if $\lambda_2 > 0$, and DHR when $\lambda_2 < 0$. Setting $\lambda_4 = 0$,

$$H(u) = \frac{\lambda_2}{(1-u)u^{\lambda_3 - 1}}$$

showing that X is IHR for $0 < \lambda_3 < 1$, and BT for $\lambda_3 > 1$ with change point $u_0 = \frac{\lambda_3 - 1}{\lambda_3}$. Finally, when $\lambda_3 = 2$, $\lambda_4 = 1$,

$$H(u) = \frac{\lambda_2}{(1-u)[2u+1]}$$

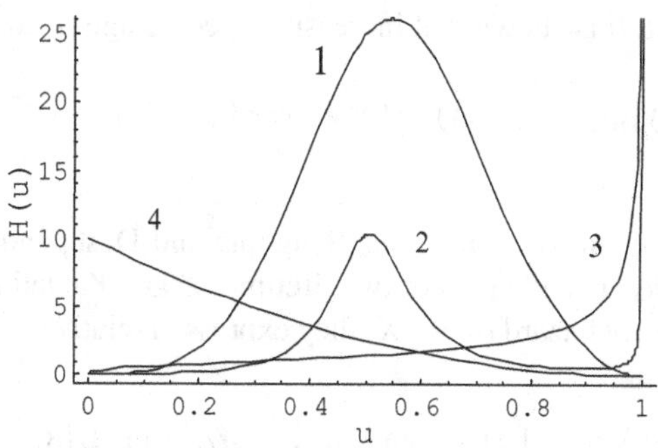

Fig. 4.6 Plots of hazard quantile function when (1) $\lambda_1 = 0$, $\lambda_2 = 0.01$, $\lambda_3 = 0.5$, $\lambda_4 = -2$; (2) $\lambda_1 = 0$, $\lambda_2 = 100$, $\lambda_3 = 0.5$, $\lambda_4 = 10$; (3) $\lambda_1 = 0$, $\lambda_2 = 1$, $\lambda_3 = 0.6$, $\lambda_4 = 0.5$; (4) $\lambda_1 = 0$, $\lambda_2 = 0.1$, $\lambda_3 = 1$, $\lambda_4 = -5$, for the van Staden–Loots model

so that when $\lambda_2 > 0$, $H'(u) = 0$ at $u = \frac{1}{4}$ and X is UBT with change point $u_0 = \frac{1}{4}$. The shapes of the hazard function can also be seen from Fig. 4.5 for some selected choices of the four parameters.

Finally, the van Staden–Loots model defined in (3.26), characterized by

$$H(u) = \lambda_2(1-u)[(1-\lambda_3)u^{\lambda_4 - 1} + \lambda_3(1-u)^{\lambda_4 - 1}],$$

also possesses different shapes of hazard quantile functions. Figure 4.6 presents the plot of $H(u)$ for some selected parameter values showing different shapes.

4.7.2 Relative Ageing

The role of relative ageing concepts is either to compare the ageing patterns of two units at a fixed time or to ascertain whether the same unit is ageing more positively or more negatively at different points of time.

Consider two units whose lifetimes follow the same distribution $F(x)$, and let y be the chronological age of one unit and the other is new. Bryson and Siddiqui [122] then argued that $\frac{\bar{F}(y+x)}{\bar{F}(y)}$ is the probability that the older system will survive the same duration x given its survival up to time y as the new one with survival probability $\bar{F}(x)$. They define the specific ageing factor as

$$A(x,y) = \frac{\bar{F}(y)\bar{F}(x)}{\bar{F}(x+y)}, \quad x,y > 0,$$

which compares the two survival probabilities. Note that $A(x,y) > (< 1)$ will mean that the older system has aged in the sense that it has less (more) probability of survival than a new unit. It is shown that increasing specific age factor

$$A(y_1,x) \geq A(y_2,x) \quad \text{for all } x \geq 0,\ y_2 \geq y_1$$

is equivalent to IHR.

Instead of comparing survival functions, Sengupta and Deshpande [526] made use of the failure rates $h_X(x)$ and $h_Y(x)$ of two lifetimes X and Y. Defining $\mathscr{H}_G(x) = \int_0^x h(t)dt$ as the cumulative hazard rate of X, they expressed relative ageing concepts as follows:

1. The random variable X ages faster than Y if $Z = \mathscr{H}_G(X)$ is IHR;
2. X is ageing faster than Y in average if Z is IHRA;
3. The random variable X is ageing faster than Y in quantile if Z is NBU.

They also obtained bounds and inequalities on $\bar{F}(x)$ in all three cases. Jiang et al. [292] dealt with unimodal hazard rates and defined the ageing intensity function as

$$L(x) = \frac{xh(x)}{\mathscr{H}(x)}$$

which, in terms of quantile functions, becomes

$$l(u) = L(Q(u)) = \frac{Q(u)H(u)}{\int_0^u H(p)q(p)dp}.$$

When X is IHR,

$$H(u) \geq \frac{\int_0^u H(p)q(p)dp}{\int_0^u q(p)dp}$$

and hence $l(u) > 1$. Thus, the value of $l(u)$ quantifies the intensity of ageing. In a different setting, Abraham and Nair [13] denoted the remaining life of an old unit of lifelength X which has the same probability of survival as a new unit of age y by $g(x,y)$. They then showed that

$$g(x,y) = \mathscr{H}^{-1}(\mathscr{H}(x) + \mathscr{H}(y)) - x$$

is a necessary and sufficient condition for

$$P(X > g(x,y) + x | X > x) = P(X > y).$$

That is, if y is the αth quantile of X, then $g(x,y)$ is the αth quantile of the residual life distribution of X. A relative ageing factor is

$$B(x,y) = y^{-1}g(x,y)$$

which reveals the rate at which an old unit is losing or gaining life in relation to a new unit with identical life distribution. They showed that X is IHR if and only if $B(x,y)$ is decreasing in x for every $y > 0$, and $g(x,y) < y$ if and only if X is NBU. Theories on the quantification of ageing are still at the formative stage, but that they have interesting relationships with ageing concepts is evident from the above discussion.

For the application of the ageing concept in a real situation, it is desirable to have a test procedure. The tests are usually performed by assuming the null hypothesis that the population distribution is exponential against the alternative that it belongs to some specific ageing class, excluding exponential. As a basis for the test, one often uses moment inequalities, inequalities for survival functions, total time on test transformation (see next chapter), order statistics, stochastic orders (Chap. 8) and so on. A comprehensive survey of various tests available for this purpose has been provided by Lai and Xie [368].

The various ageing properties discussed in this chapter will be revisited in Chap. 8 in terms of stochastic orders. Some general theorems and properties established there will shed additional light on their relevance in reliability theory and other applied disciplines.

Chapter 5
Total Time on Test Transforms

Abstract The total time on test transform is essentially a quantile-based concept developed in the early 1970s. Apart from its applications in reliability problems, it has also been found useful in other areas like stochastic modelling, maintenance scheduling, risk assessment of strategies and energy sales. When several units are tested for studying their life lengths, some of the units would fail while others may survive the test period. The sum of all observed and incomplete life lengths is the total time on test statistic. As the number of units on test tends to infinity, the limit of this statistic is called the total time on test transform (TTT). The definitions and properties of these two concepts are discussed and the functional forms of TTT for several life distributions are presented in Table 5.1. We discuss the Lorenz curve, Bonferroni curve and the Leimkuhler curve which are closely related to the TTT. Identities connecting various curves, characterizations of distributions in terms of these curves and their relationships with various reliability functions are detailed subsequently. In view of the ageing classes in the quantile set-up introduced in Chap. 4, it is possible to characterize these classes in terms of TTT. Accordingly, we give necessary and sufficient conditions for IHR, IHRA, DMRL, NBU, NBUE, HNBUE, NBUHR, NBUHRA, IFHA*t_0, UBAE, DMRLHA, DVRL, and NBU-t_0 classes in terms of the total time on test transform. Another interesting property of the TTT is that it uniquely determines the lifetime distribution. There have been several generalizations of the TTT. We discuss these extensions and their properties, with special reference to the TTT of order n. Relationships between the reliability functions of the baseline model and those of the TTT of order n (which is also a quantile function) are described and then utilized to describe the pattern of ageing of the transformed distributions. Some life distributions are characterized. The discussion of the applications of TTT in modelling includes derivation of the L-moments and other descriptive measures of the original distribution. Some of the areas in reliability engineering that widely use TTT are preventive maintenance, availability, replacement problems and burn-in strategies.

N.U. Nair et al., *Quantile-Based Reliability Analysis*, Statistics for Industry and Technology, DOI 10.1007/978-0-8176-8361-0_5,

5.1 Introduction

The concept of total time on test transform (TTT) was studied in the early 1970s; see, e.g., Barlow and Doksum [67] and Barlow et al. [65]. When several units are tested for studying their life lengths, some of the units would fail while others may survive the test duration. The sum of all observed and incomplete life lengths is generally visualized as the total time on test statistic. When the number of items placed on test tends to infinity, the limit of this statistic is called the total time on test transform. A formal definition of these two concepts will be introduced in the next section. The TTT is essentially a quantile-based concept, although it is discussed often in the literature in terms of $F(x)$.

Many papers on TTT concentrate on reliability and its engineering applications. This include analysis of life lengths and new classes of ageing; see Abouammoh and Khalique [9], Ahmad et al. [25] and Kayid [318]. A special characteristic of TTT is that the basic ageing properties can be interpreted and determined through it. The works of Barlow and Campo [66], Bergman [89], Klefsjö [334], Abouammoh and Khalique [9] and Perez-Ocon et al. [492] are all of this nature. Properties of TTT were used for construction of bathtub-shaped distributions by Haupt and Schabe [266] and Nair et al. [447]. Much of the literature has focused on developing test procedures, most of which are for exponentially against alternatives like IHR, IHRA, NBUE, DMRL and HNBUE. For this, one may refer to Bergman [90], Klefsjö [335, 336], Kochar and Deshpande [348], Aarset [1], Xie [592, 593], Bergman and Klefsjö [96], Wei [579] and Ahmed et al. [25].

Applications of TTT can be found in a variety of fields. Of these, the role of TTT in reliability engineering will be taken up separately in Sect. 5.5. The optimal quantum of energy that may be sold under long-term contracts using TTT is discussed in Campo [125] and risk assessment of strategies in Zhao et al. [601]. TTT plotting of censored data (Westberg and Klefsjö [578]), problem of repairable limits (Dohi et al. [180]), normalized TTT plots and spacings (Ebrahimi and Spizzichino [183]), maintenance scheduling (Kumar and Westberg [357], Klefsjö and Westberg [340]), estimation in stationary observations (Csorgo and Yu [161]) and stochastic modelling (Vera and Lynch [573]) are some of the other topics discussed in the context of total time on test.

5.2 Definitions and Properties

We now give formal definitions of various concepts based on total time on test.

Definition 5.1. Suppose n items are under test and successive failures are observed at $X_{1:n} \leq X_{2:n} \leq \cdots \leq X_{n:n}$, and let $X_{r:n} < t \leq X_{r+1:n}$, where $X_{r:n}$'s are order statistics from the distribution of a lifetime random variable X with absolutely continuous distribution function $F(x)$. Then, the total time on test statistic during $(0,t)$ is defined as

$$\tau(t) = nX_{1:n} + (n-1)(X_{2:n} - X_{1:n}) + \cdots + (n-r+1)(X_{r:n} - X_{r-1:n}) \\ + (n-r)(t - X_{r:n}). \tag{5.1}$$

The above expression is arrived at by noting that the test time observed between 0 and $X_{1:n}$ is $nX_{1:n}$, that between $X_{1:n}$ and $X_{2:n}$ is $(n-1)(X_{2:n} - X_{1:n})$ and so on, and finally that between $X_{r:n}$ and t is $(n-r)(t - X_{r:n})$. Also, the total time up to the rth failure is

$$\tau(X_{r:n}) = nX_{1:n} + (n-1)(X_{2:n} - X_{1:n}) + \cdots + (n-r+1)(X_{r:n} - X_{r-1:n}). \tag{5.2}$$

It may also be noted that (5.1) is equivalent to

$$\tau(t) = X_{1:n} + X_{2:n} + \cdots + X_{r:n} + (n-r)t.$$

Definition 5.2. The quantity

$$\phi_{r:n} = \frac{\tau(X_{r:n})}{\tau(X_{n:n})} = \frac{\sum_{j=1}^{r}(n-j+1)(X_{j:n} - X_{j-1:n})}{\sum_{j=1}^{n}(n-j+1)(X_{j:n} - X_{j-1:n})}, \quad \text{with } X_{0:n} = 0, \tag{5.3}$$

is called the scaled total time on test statistic (scaled TTT statistic).

Noting that $\bar{X}_n = \frac{1}{n}(X_{1:n} + \cdots + X_{n:n})$ is the sample mean of the n order statistics, we have $\phi_{r:n} = \frac{\tau(X_{r:n})}{n\bar{X}_n}$. The empirical distribution function defined in terms of the order statistics is

$$F_n(t) = \begin{cases} 0, & t < X_{1:n}, \\ \frac{r}{n}, & X_{r:n} \le t < X_{r+1:n},\ r = 1,2,\ldots,n-1, \\ 1, & t \ge X_{n:n}. \end{cases}$$

If there exists an inverse function

$$F_n^{-1}(t) = \inf[x \ge 0 | F_n(x) > t],$$

we can verify that

$$\int_0^{F_n^{-1}(\frac{r}{n})} \bar{F}_n(t)dt = \sum_{j=1}^{r}\left(1 - \frac{j-1}{n}\right)(X_{j:n} - X_{j-1:n}) = \frac{\tau(X_{r:n})}{n}$$

and

$$\lim_{n\to\infty} \lim_{\frac{r}{n}\to u} \int_0^{F_n^{-1}(\frac{r}{n})} \bar{F}_n(t)dt = \int_0^{F^{-1}(u)} \bar{F}(t)dt \tag{5.4}$$

uniformly in u belonging to $[0,1]$. The expression on the right side of (5.4), viz.,

$$\int_0^{F^{-1}(u)} \bar{F}(t)dt = H_F^{-1}(u), \tag{5.5}$$

is called the total time on test transform. Accordingly we have, with a slightly different notation $T(u)$ for $H_F^{-1}(u)$, the following definition.

Definition 5.3. The TTT of a lifetime random variable X is defined as

$$T(u) = \int_0^u (1-p)q(p)dp. \tag{5.6}$$

Example 5.1. The linear hazard quantile function family of distributions specified by

$$Q(u) = \log\left(\frac{a+bu}{a(1-u)}\right)^{\frac{1}{a+b}}$$

(see Chap. 2) has

$$q(u) = [(1-u)(a+bu)]^{-1},$$

and so, from (5.6), we find

$$T(u) = \frac{1}{b}\log\left(\frac{a+bu}{a}\right).$$

The expressions for TTT for some specific life distributions are presented in Table 5.1.

Some important properties of the TTT in (5.6) are the following:

1. $T(0)=0$, $T(1)=\mu$. $T(u)$ is an increasing function if and only if F is continuous. In this case, $T(u)$ is a quantile function and the corresponding distribution is called the transformed distribution;
2. The baseline distribution F is uniquely determined by $T(u)$. To see this, we differentiate (5.6) to get

$$T'(u) = (1-u)q(u), \tag{5.7}$$

and thence

$$Q(u) = \int_0^u \frac{T'(p)}{1-p}dp;$$

3. From Table 5.1, we see that the graph of the TTT of the exponential distribution is the diagonal line in the unit square;

Table 5.1 Total time on test transforms for some specific life distributions

Distribution	$T(u)$
Exponential	$\lambda^{-1}u$
Pareto II	$\frac{\alpha}{c-1}[1-(1-u)^{\frac{c-1}{c}}]$
Rescaled beta	$\frac{R}{c+1}[1-(1-u)^{\frac{c+1}{c}}]$
Weibull	$\frac{\sigma}{\lambda n^{1/\lambda}} I_{-n\log(1-u)}(\frac{1}{\lambda})$
Half-logistic	$2\sigma\log(1+u)$
Power	$\frac{\alpha u^{\frac{1}{\beta}}}{(1+\beta)}(1+\beta-u)$
Govindarajulu	$\sigma(\beta+1)u^{\beta}[(1-u)^2+2(\beta+1)^{-1}u(1-u)$ $+((\beta+1)(\beta+2))^{-1}u^2]$
Generalized lambda	$\lambda_2^{-1}\left\{\frac{u^{\lambda_3}}{\lambda_3+1}(1+\lambda_3(1-u))+\frac{\lambda_4}{\lambda_4+1}(1-(1-u)^{\lambda_4+1})\right\}$
Generalized Tukey lambda	$\lambda_2^{-1}\left\{\frac{u^{\lambda_3}}{\lambda_3(\lambda_3+1)}(1+\lambda_3(1-u))+\frac{1}{\lambda_4+1}(1-(1-u)^{\lambda_4+1})\right\}$
van Staden–Loots	$\lambda_2^{-1}\left\{\frac{(1-\lambda_3)u^{\lambda_4}}{\lambda_4}\frac{(1+\lambda_4(1-u))}{(\lambda_4+1)}+\frac{\lambda_3}{\lambda_4+1}(1-(1-u)^{\lambda_4})\right\}$
Power-Pareto	$C[\lambda_1 B_u(\lambda_1,2-\lambda_2)+\lambda_2 B_u(\lambda_1+1,1-\lambda_2)]$

4. Many identities exist between $T(u)$ and the basic reliability functions introduced earlier in Sects. 2.3–2.6. Directly from (5.7) and (2.30), we have

$$T'(u)=\frac{1}{H(u)}. \tag{5.8}$$

Again from (2.35), we find

$$T(u)=\mu-\int_u^1(1-p)q(p)dp=\mu-(1-u)M(u)$$

and consequently

$$M(u)=\frac{\mu-T(u)}{1-u}, \tag{5.9}$$

which relates TTT and the mean residual quantile function. On the other hand, from (2.46), we find

$$V(u)=\frac{1}{1-u}\int_u^1 M^2(p)dp$$

and hence

$$V(u) = \frac{1}{1-u}\int_u^1 \left(\frac{\mu - T(p)}{1-p}\right)^2 dp, \tag{5.10}$$

or equivalently

$$T(u) = \mu - (1-u)[(1-u)V'(u) - V(u)]^{\frac{1}{2}}. \tag{5.11}$$

Next, with regard to functions in reversed time, we have

$$T(u) = Q(u) - \int_0^u pq(p)dp$$

or

$$uq(u) = [Q(u) - T(u)]',$$

and so

$$\Lambda(u) = \frac{1}{[Q(u) - T(u)]'}. \tag{5.12}$$

Also from (2.50), the reversed mean residual quantile function satisfies

$$uR(u) = \int_0^u pq(p)dp$$

and

$$uR(u) = Q(u) - T(u),$$

and consequently

$$T(u) = Q(u) - uR(u). \tag{5.13}$$

Finally, we use the reversed variance quantile function

$$D(u) = \frac{1}{u}\int_0^u R^2(p)dp$$

to write

$$D(u) = \frac{1}{u}\int_0^u \frac{Q(p) - T(p)}{p^2}dp.$$

These relationships are used in the next section to characterize the ageing properties in terms of total time on test transform.

Definition 5.4. We say that

$$\phi(u) = \frac{\int_0^u (1-p)q(p)dp}{\int_0^1 (1-p)q(p)dp} = \frac{T(u)}{\mu} \tag{5.14}$$

is the scaled total time on test transform, or scaled transform in short, of the random variable X.

Definition 5.5. The plot of the points $(\frac{r}{n}, \phi_{r,n})$, $r = 1, 2, \ldots, n$, when connected by consecutive straight lines, is called the TTT-plot.

The statistic $\frac{1}{n}\tau(X_{r:n})$ converges uniformly in u to the TTT as $n \to \infty$ and $\frac{r}{n} \to u$. Now, we present the asymptotic distribution, which is due to Barlow and Campo [66]. Let $\phi_{r,n} = \left\{\phi_n(p) = \frac{H_n^{-1}(p)}{H_n^{-1}(1)},\ 0 \le p \le 1\right\}$ be the scaled TTT process. Define

$$S_n(p) = \sqrt{n}\left\{\frac{H_n^{-1}(p)}{\sum_1^n X_{j:n}} - \phi(p)\right\}$$

for $\frac{j-1}{n} \le p \le \frac{j}{n}$ and $1 \le j \le n$, with $S_n(0) = S_n(1) = 0$. Upon using

$$\phi(u) = \frac{1}{\mu}\left\{(1-u)Q(u) + \int_0^u Q(p)dp\right\},$$

we see that

$$\frac{H_n^{-1}(\frac{j}{n})}{H_n^{-1}(1)} = \int_0^{j/n} \frac{F_n^{-1}(u)}{\sum X_{j:n}} d\nu_n(u)du + \left(1 - \frac{j}{n}\right)\frac{X_{j:n}}{\sum X_{j:n}}$$

converges to

$$\int_0^u \frac{Q(p)}{\mu}dp + \frac{(1-u)}{\mu}Q(u)$$

with probability one and uniformly in $0 \le u \le 1$ as $n \to \infty$, where $\nu_n(u)$ puts mass $\frac{1}{n}$ at $u = \frac{j}{n}$. Next,

$$\begin{aligned} S_n\left(\frac{j}{n}\right) &= \sqrt{n}\left(\frac{H_n^{-1}(\frac{j}{n})}{\sum X_{j:n}} - \phi(u)\right) \\ &\doteq \int_0^{\frac{j}{n}} \sqrt{n}\left(\frac{X_{[nu]:n}}{\sum X_{j:n}} - \frac{Q(u)}{\mu}\right) d\nu_n(u) + \left(1 - \frac{j}{n}\right)\left(\frac{X_{j:n}}{\sum X_{j:n}} - \frac{F^{-1}(\frac{j}{n})}{\mu}\right), \end{aligned}$$

where $[t]$ denotes the greatest integer contained in t. Then,

$$\lim_{n\to\infty}\sqrt{n}\left(\frac{H_n^{-1}(p)}{\sum X_{j:n}}-\phi(u)\right)=\int_0^u\theta(p)dp+(1-u)\theta(u),$$

with

$$\theta(u)=-\frac{q(u)}{\mu}A(u)+\frac{Q(u)}{\mu^2}\int_0^1 A(p)q(p)dp$$

and

$$\lim_{n\to\infty}\sqrt{n}\left\{\frac{H_n^{-1}(u)}{\sum X_{j:n}}-\phi(u)\right\}=A(u).$$

In the above, $\{A(u), 0\leq u\leq 1\}$ is the Brownian bridge process.

5.3 Relationships with Other Curves

The similarity between the Lorenz curve used in economics and the TTT and the corresponding results have been discussed by Chandra and Singpurwalla [134] and Pham and Turkkan [493]. If X is a non-negative random variable with finite mean, the Lorenz curve is defined as

$$L(u)=\frac{1}{\mu}\int_0^u Q(p)dp, \tag{5.15}$$

which is itself a continuous distribution function with $L(0)=0$ and $L(1)=1$. It is a bow-shaped curve below the diagonal of the unit square. Used as a measure of inequality in economics, we note that as the bow is more bent, the amount of inequality increases. Also $L(u)$ is convex, increasing and is such that $L(u)\leq u$, $0\leq u\leq 1$. The Lorenz curve determines the distribution of F up to a scale. Two well-known measures of inequality that are related to the Lorenz curve are the Gini index and the Pietra index. There are many analytic expressions for calculating the Gini index, including

$$G=2\int_0^1(u-L(u))du=1-2\int_0^1 L(u)du. \tag{5.16}$$

In addition,

$$G=1-2\mu^{-1}\int_0^1\int_0^u Q(p)dydu=1-\mu^{-1}E(X_{1:2}),$$

where $X_{1:2}$ is the smallest of a sample of size 2 from the population.

Next, the Pietra index is obtained from the maximum vertical deviation between $L(u)$ and the line $L(u) = u$, given by

$$P = \mu^{-1} \int_0^{F(\mu)} (\mu - Q(p))dp = F(\mu) - L(F(\mu)). \tag{5.17}$$

It can be seen that P is $\frac{1}{2\mu} \int_0^\infty |x - \mu| f(x)dx$, half the relative mean deviation. A detailed account of the results concerning $L(u)$ and G can be found in Kleiber and Kotz [341].

The cumulative Lorenz curve of X is given by

$$CL(u) = \int_0^1 L(u)du = \frac{1}{\mu} \int_0^1 \int_0^u Q(p)dudp. \tag{5.18}$$

Chandra and Singpurwalla [133] observed that both $L(u)$ and $L^{-1}(u)$ are distribution functions, L is convex, L^{-1} is concave, and that $L(u)$ is related to the mean residual life function $m(x)$. In the quantile set-up, the Lorenz curve can be related to all the basic reliability functions. For example, we have from (2.34) and (5.15) that

$$\begin{aligned} Q(u) + M(u) &= \frac{1}{1-u} \int_u^1 Q(p)dp, \\ \mu - \int_0^u Q(p)dp &= (1-u)(Q(u) + M(u)), \\ \mu[1 - L(u)] &= (1-u)(Q(u) + M(u)), \end{aligned}$$

and so

$$M(u) = \frac{\mu(1 - L(u))}{1-u} - Q(u) = \mu \left[\frac{1 - L(u)}{1-u} - L'(u) \right].$$

Now, $H(u)$ is recovered from (2.37) and $V(u)$ from (2.46), after substituting for $M(u)$. A much simpler expression results for the reversed mean residual quantile function $R(u)$ as

$$R(u) = Q(u) - \frac{\mu L(u)}{u} = \mu[L'(u) - u^{-1}L(u)].$$

Also,

$$[\Lambda(u)]^{-1} = R(u) + uR'(u)$$

and

$$D(u) = \frac{1}{u} \int_0^u R^2(p)dp.$$

Example 5.2. The Pareto distribution is one of the basic distributions used in modelling income data and it plays a role similar to the exponential distribution in reliability. Its quantile function is (Table 1.1)

$$Q(u) = \sigma(1-u)^{-\frac{1}{\alpha}}$$

and so we obtain the following expressions:

$$L(u) = 1-(1-u)^{1-\frac{1}{\alpha}} \quad \text{since } \mu = \frac{\sigma\alpha}{\alpha-1},\ \alpha > 1,$$

$$M(u) = \frac{\sigma\alpha[(1-u)^{1-\frac{1}{\alpha}}]}{(\alpha-1)1-u} - \sigma(1-u)^{-\frac{1}{\alpha}} = \frac{\sigma}{\alpha-1}(1-u)^{-\frac{1}{\alpha}},$$

$$H(u) = [M(u)-(1-u)M'(u)]^{-1} = \frac{\alpha(1-u)^{\frac{1}{\alpha}}}{\sigma},$$

$$V(u) = \frac{1}{1-u}\int_u^1 M^2(p)dp = \frac{\sigma^2\alpha}{(\alpha-1)^2(\alpha-2)}(1-u)^{-\frac{2}{\alpha}}.$$

Also, the functions $\Lambda(u)$, $R(u)$ and $D(u)$ can also be similarly found.

Chandra and Singpurwalla [134] obtained the following relationships between $T(u)$, $L(u)$ and the sample analogs corresponding to them:

(a)

$$T(u) = (1-u)Q(u) + \mu L(u). \tag{5.19}$$

Equation (5.19) is obtained by integrating by parts the right-hand side of (5.6) and then using (5.15). Since $Q(u) = \mu L'(u)$, (5.19) has the alternative form

$$T(u) = \mu[(1-u)L'(u) + L(u)],$$

or equivalently

$$\phi(u) = (1-u)L'(u) + L(u).$$

Now, upon treating the last relationship as a linear differential equation in u and solving it, we obtain an integral expression for $L(u)$ as

$$L(u) = (1-u)\int_0^u \frac{\phi(p)}{(1-p)^2}dp.$$

(b) We also have

$$C\phi(u) = 2CL(u),$$

where $C\phi(u) = \int_0^1 \phi(p)dp = \frac{1}{\mu}\int_0^1 T(p)\,dp$ is called the cumulative total time on test transform. To establish the above assertion, we note that

$$\int_0^1 \int_0^u Q(p)dp = -\int_0^1 \left(\frac{d}{dp}(1-p)\int_0^u Q(p)dp\right) du$$
$$= \int_0^1 (1-p)Q(p)dp \text{ (by partial integration)},$$

$$\int_0^u (1-p)q(p)dp = (1-u)Q(u) + \int_0^u Q(p)dp.$$

Thus, we get

$$\begin{aligned} C\phi(u) &= \frac{1}{\mu}\int_0^1 \int_0^u (1-p)q(p)dpdu \\ &= \frac{1}{\mu}\int_0^1 \left\{(1-u)Q(u) + \int_0^u Q(p)dp\right\} du \\ &= \frac{1}{\mu}\int_0^1 \left\{\int_0^u Q(p)dp\right\} du + \frac{1}{\mu}\int_0^1 \left\{\int_0^u Q(p)dp\right\} du \\ &= 2CL(u), \end{aligned}$$

as required.

(c) $G = 1 - C\phi(u)$, which is seen as follows:

$$\begin{aligned} G &= 1 - 2\mu^{-1}\int_0^1 \left\{\int_0^u Q(p)dp\right\} du \\ &= 1 - 2CL(u) = 1 - C\phi(u) \text{ (by using (b))}. \end{aligned}$$

If we denote the sample Lorenz curve and the sample Gini index by

$$L_n(u) = \frac{\sum_{r=1}^{[nu]} X_{r:n}}{\sum_{r=1}^{n} X_{r:n}},$$

and

$$G_n = \frac{\sum_{r=1}^{n-1} r(n-r)(X_{r+1:n} - X_{r:n})}{(n-1)\sum_{r=1}^{n} X_{r:n}},$$

respectively, and the cumulative total time on test statistic by

$$V_n = \frac{1}{n-1}\sum_{r=1}^{n-1} \phi_{r:n},$$

then we have

$$\phi_{r:n} = L_n\left(\frac{r}{n}\right) + \frac{(n-r)X_{r:n}}{\sum_{j=1}^{n} X_{r:n}}$$

and

$$V_n = 1 - G_n.$$

Chandra and Singpurwalla [134] also pointed out the potential of the Lorenz curve in comparing the heterogeneity in survival data and also in characterizing the extremes of life distributions. The latter aspect is illustrated by the following theorem.

Theorem 5.1. *If X is IHR with mean μ, then*

$$L_G(u) \le L_F(u) \le L_D(u), \quad 0 \le u \le 1,$$

and if X is DHR with mean μ, then

$$L_F(u) \begin{cases} \le L_G(u), & 0 < u \le 1 \\ \ge 0, & 0 \le u < 1 \\ = 1, & u = 1. \end{cases}$$

Here, F and G are the distribution functions of X and exponential variable with same mean μ, respectively, and D is the distribution degenerate at μ.

The distribution which is degenerate at μ has $h(x) = \infty$ at μ and so $L_D(u) = u$ characterizes distributions which are most IHR. Likewise, distributions with $L(u) = 0$ for $u < 1$ and $L(u) = 1$ for $\mu = 1$ are the most DHR.

Pham and Turkkan [493] established more results in this direction. They pointed out that $\phi(u)$ strictly increases in the unit square with $\phi(0) = 0$ and $\phi(1) = 1$. Moreover,

(a) $\phi(F(\mu)) = 1 - E(|X - \mu|)$;
(b) $\phi(\text{Med}\,X) = \frac{1}{2} + \frac{(\text{Med}\,X - E|X - \text{Med}\,X|)}{2\mu}$;
(c) In the unit square, the area between $\phi(u)$ and $L(u)$ equals the area below $L(u)$. The area above $\phi(u)$ is G;
(d) $L(u) = (1-u)\int_0^u \frac{\phi(p)}{(1-p)^2}dp$;
(e) If X is NBUE, then the Pietra index is less than the reliability at μ and $E(|X - \text{Med}\,X|) < \text{Med}\,X$;
(f) When $\frac{1}{2} < G \le 1$ $(0 \le G < \frac{1}{2})$ and $F(x)$ is a family of IHR (DHR) distributions with common mean μ, $F(x)$ becomes more IHR (DHR) when $L(u)$ gets closer to the diagonal and $\phi(u)$ get closer to the upper (lower) side. Further, when $G = \frac{1}{2}$,

$F(x)$ is exponential. When $0 \leq P < e^{-1}$, X is IHR and the closer P is to zero, the more IHR X becomes. X is exponential when $P = e^{-1}$. Also, $e^{-1} < P < 1$ provides DHR and $P \to 1$ corresponding to the most DHR.

Another curve that has been used in the context of income inequality is the Bonferroni curve. For a non-negative random variable X, the first moment distribution of X is defined by the distribution function

$$F_1(x) = \frac{\int_0^x t f(t) dt}{\mu}.$$

The Bonferroni curve is defined in the orthogonal plane as $(F(x), B_1(x))$ within the unit square, where

$$B_1(x) = \frac{F_1(x)}{F(x)}.$$

In terms of quantile functions, we have

$$B(u) = B_1(Q(u)) = \frac{\int_0^u Q(p) dp}{\mu u}. \tag{5.20}$$

One may refer to Giorgi [218], Giorgi and Crescenzi [219] and Pundir et al. [498] and the references therein for a study of (5.20) and its properties. As $u \to 0$, $B(u)$ has the indeterminate form $\frac{0}{0}$ and hence the curve does not begin from the origin. It is strictly increasing but can be convex or concave in parts of the plane. Several results concerning $B_1(x)$ have been given by Pundir et al. [498]. We now make a comparative study of $B(u)$ with $L(u)$ and $\phi(u)$. First, we note that $B(u)$ characterizes the distribution of X through

$$Q(u) = \mu(B(u) + uB'(u)). \tag{5.21}$$

Also,

$$B(u) = u^{-1} L(u)$$

and

$$\phi(u) = \frac{(1-u)Q(u)}{\mu} + \frac{1}{\mu}\int_0^u Q(p) dp,$$

or equivalently

$$\phi(u) = B(u) + u(1-u)B'(u). \tag{5.22}$$

Solving (5.22) as a linear differential equation, we get

$$B(u) = \frac{1-u}{u}\int_0^u \frac{\phi(p)}{(1-p)^2} dp,$$

relating scaled TTT and the Bonferroni curve. Equation (5.20) verifies

$$\mu u B(u) = \int_0^u Q(p)dp = \mu - \int_u^1 Q(p)dp,$$

and hence

$$M(u) = \frac{\mu(1-uB(u))}{1-u} - Q(u) = \mu\left\{\frac{1-B(u)}{1-u} - uB'(u)\right\}$$

by virtue of (5.21). Rewriting the above equation as

$$B'(u) + \frac{B(u)}{u(1-u)} = \frac{1}{u(1-u)} - \frac{M(u)}{u\mu}$$

and solving it, we see that $B(u)$ is uniquely determined by $M(u)$ as

$$B(u) = \frac{1-u}{u}\int_0^u \frac{1}{p}\left\{\frac{1}{1-p} - \frac{M(p)}{\mu}\right\}dp.$$

A more concise relationship exists between $B(u)$ and the reversed mean residual quantile function $R(u)$ in the form

$$R(u) = \mu u B'(u).$$

As in the case of $L(u)$, all other reliability functions can be derived using the relations they have with $M(u)$ and $R(u)$. Pundir et al. [498] showed that the Bonferroni index

$$B = 1 - \int_0^1 B(u)du$$

is such that

$$B \leq \frac{1}{2}(1+G) \quad \text{and} \quad B \leq 1 - \frac{V}{2},\ V = 1-G.$$

The Leimkuhler curve, which is closely related to the Lorenz curve, is also discussed recently for its relationships with the reliability functions. It is used in economics as a plot of cumulative proportion of productivity against cumulative proportion of sources and is also used in studying concentration of bibliometric distributions in information sciences. A general definition of the curve is given in Sarabia [518] and methods of generating such curves have been detailed in Sarabia et al. [519]. Balakrishnan et al. [60] have pointed out the relationships between reliability functions and the Leimkuhler curve. The Leimkuhler curve is defined in terms of quantile function as

$$
\begin{aligned}
K(u) &= \frac{1}{\mu}\int_{1-u}^{1} Q(p)dp \\
&= \frac{1}{\mu}\left\{\int_0^1 Q(p)dp - \int_0^{1-u} Q(p)dp\right\} \\
&= 1 - \frac{1}{\mu}\int_0^{1-u} Q(p)dp.
\end{aligned} \tag{5.23}
$$

Evidently,

$$K(u) = 1 - L(1-u) \quad \text{or} \quad K(1-u) = 1 - L(u)$$

and so $K(u)$ characterizes the distribution of X. The relation in (5.23) gives

$$
\begin{aligned}
M(u) &= \frac{\mu\{1-K(1-u)\}}{1-u} - Q(u) \\
&= \mu\left\{\frac{1-K(1-u)}{1-u} - K'(1-u)\right\}.
\end{aligned}
$$

Similarly, from

$$\mu(1-K(u)) = \int_0^{1-u} Q(p)dp$$

and the definition of $R(u)$, we obtain

$$\mu(1-K(u)) = (1-u)\{Q(1-u) - R(1-u)\}.$$

Since

$$Q(1-u) = \mu K^{-1}(u),$$

upon combining the expressions, we obtain

$$R(u) = \mu u^{-1}[K'(1-u) + K(1-u) - 1].$$

Regarding the geometric properties, it is seen from the definition that $K(u)$ is continuous, concave and increasing with $K(0) = 0$ and $K(1) = 1$. The main difference between the Lorenz curve and the Leimkuhler curve $K(u)$ is that in the Lorenz curve the sources are arranged in increasing order of productivity, while in the Leimkuhler curve the sources are arranged in decreasing order. The expressions of $B(u)$, $L(u)$ and $K(u)$ for some distributions are presented in Table 5.2.

Table 5.2 Expressions of $L(u)$, $B(u)$ and $K(u)$ for some distributions

Distribution	$L(u)$	$B(u)$	$K(u)$
Power	$u^{\frac{1}{\beta}+1}$	$u^{\frac{1}{\beta}}$	$1-(1-u)^{\frac{1}{\beta}+1}$
Exponential	$u+(1-u)\log(1-u)$	$1+\frac{1-u}{u}\log(1-u)$	$u(1-\log u)$
Pareto II	$c(1-(1-u)^{1-\frac{1}{c}})-u(c-1)$	$\frac{c(1-(1-u))^{1-\frac{1}{c}}}{u}-(c-1)$	$u[1-c+cu^{-\frac{1}{c}}]$
Rescaled beta	$c(1-(1-u)^{1+\frac{1}{c}-1})+u(1+c)$	$1+c+\frac{c}{u}((1-u)^{1+\frac{1}{c}}-1)$	$u[c+1-cu^{\frac{1}{c}}]$
Pareto I	$\alpha[1-(1-u)^{-\frac{1}{\alpha}+1}]$	$\frac{\alpha}{u}[1-(1-u)^{-\frac{1}{\alpha}+1}]$	$\alpha u^{-\frac{1}{\alpha}+1}$

5.4 Characterizations of Ageing Concepts

In this section, we discuss the role of TTT in detecting different ageing properties. In this regard, the new definitions offered below in terms of TTT provide alternative ways of interpreting and analysing lifetime data. The proofs given here assume that F is continuous and strictly increasing.

Theorem 5.2 (Barlow and Campo [66]). *A lifetime random variable X is IHR (DHR) if and only if the scaled transform $\phi(u)$ is concave (convex) for* $0 \le u \le 1$.

From (5.8), we have

$$T'(u) = \frac{1}{H(u)}$$

and so

$$\frac{1}{H^2(u)}H'(u) = -T''(u).$$

Thus, $H'(u)$ is positive (negative) or X is IHR (DHR) if and only if $T''(u)$ is negative (positive). This is equivalent to the concavity (convexity) of $T(u)$ or $\phi(u)$. It now follows that if $\phi(u)$ has an inflexion point u_0 such that $0 < u_0 < 1$ and $\phi(u)$ is convex (concave) on $[0, u_0]$, and concave (convex) on $[u_0, 1]$, then X has a bathtub (upside-down bathtub)-shaped hazard quantile function. This can be used for constructing life distributions with BT (UBT) hazard quantile functions.

Barlow and Campo [66] have also shown that if X is IHRA (DHRA), then $\frac{\phi(u)}{u}$ is decreasing (increasing) in $0 < u < 1$. This condition is not sufficient as seen from the following life distribution (Barlow [64]) which is not IHRA, but at the same time $\frac{\phi(u)}{u}$ is decreasing:

$$F(x) = \begin{cases} 0, & 0 \le x < \frac{1}{2} \\ 1-\exp[-(c+x)], & x \ge \frac{1}{2}. \end{cases}$$

In this regard, we have the following results.

Theorem 5.3 (Asha and Nair [39]). *A necessary and sufficient condition for X to be DMTTF (IMTTF) is that* $\frac{\phi(u)}{u}$ *is decreasing (increasing).*

Theorem 5.4. *A necessary and sufficient condition for X to be IHRA (DHRA) is that*

$$\frac{1}{t(u)}\int_0^u \frac{t(p)}{1-p}dp \geq (\leq) -\log(1-u), \tag{5.24}$$

where $t(u) = T'(u)$.

The proof follows from (5.7), (5.8) and the definition of IHRA distributions.

Remark 5.1. Since $T(u)$ is the quantile function of the transformed distribution, $t(u)$ is the corresponding quantile density function. From (5.7), $t(u) = (1-u)q(u)$ and so (5.24) is equivalent to

$$t(u) \leq (\geq) -\frac{Q(u)}{\log(1-u)}. \tag{5.25}$$

Bergman [89] has proved that X is NBUE (NWUE) if and only if $\phi(u) \geq u$ ($\phi(u) \leq u$). This follows from

$$\begin{aligned}\phi(u) \geq u &\Leftrightarrow \frac{1}{\mu}\int_0^u (1-p)q(p)dp \geq u \\ &\Leftrightarrow \frac{1}{\mu}[\mu(1-u)M(u)] \geq u \\ &\Leftrightarrow M(u) \leq \mu.\end{aligned}$$

The proof in the case of NWUE involves simply reversing the inequalities.

Theorem 5.5 (Klefsjö [333]). *A lifetime random variable X is*

(a) DMRL (IMRL) if and only if $\frac{1-\phi(u)}{1-u}$ *is decreasing (increasing) in u;*
(b) HNBUE (HNWUE) if and only if

$$\phi(u) \leq (\geq) 1 - \exp[-\frac{Q(u)}{\mu}], \quad 0 \leq u \leq 1.$$

These results are direct consequences of (5.9) and the definition of HNBUE (HNWUE).

In view of the definitions of ageing concepts in the quantile set-up in Chap. 4 and the identities between $T(u)$, $Q(u)$, $H(u)$ and $M(u)$, more ageing classes can be characterized in terms of $T(u)$ or $\phi(u)$ as follows.

Theorem 5.6. *We say that X is*

(a) NBUHR (NWUHR) if and only if $t(u) \leq (\geq) t(0)$;
(b) NBUFHA (NWUHRA) if and only if $-\frac{\log(1-u)}{Q(u)} \leq (\geq) t(0)$;
(c) IHRAt_0 if and only if*

$$\int_0^u \frac{t(p)}{(1-p)} dp \geq \frac{Q(u_0)}{\log(1-u_0)} \log(1-u) \text{ for all } u \geq u_0;$$

(d) UBAE (UWAE) if and only if $T(u) \leq (\geq) \mu - (1-u)M(1)$, *where* $T(1) = \lim_{u \to 1-} T(u)$ *is finite;*
(e) DMRLHA (IMRLHA) if and only if

$$-\frac{1}{Q(u)} \log(1-\phi(u))$$

is increasing (decreasing) in u;
(f) DVRL (IVRL) if and only if

$$\int_u^1 \left(\frac{1-\phi(p)}{1-p} \right)^2 dp \leq (\geq) \frac{(1-\phi(u))^2}{1-u};$$

(g) NBU (NWU) if and only if

$$\int_0^{u+v-uv} \frac{t(p)dp}{1-p} \leq (\geq) Q(u) + Q(v),\ 0 < v < 1,\ u+v-w < 1;$$

(h) NBU-t_0 (NWU-u_0) if and only if

$$\int_0^{u+u_0-uv} \frac{t(p)dp}{1-p} \leq (\geq) Q(u) + Q(u_0)$$

for some $0 < u_0 < 1$ *and all u;*
(i) NBUu_0 (NWU*u_0) if and only if*

$$\int_0^{u+v-uv} \frac{t(p)}{1-p} dp \leq (\geq) Q(u + Q(v))$$

for some $v \geq u_0$ *and all u.*

Note that in (g)–(i), Q(s) is evaluated as $\int_0^s \frac{t(p)dp}{1-p}$.

Ahmad et al. [25] defined a new ageing class of life distributions called the new better than used in total time on test transform order (NBUT). They defined the class as distributions for which the inequality

$$\int_0^{F_t^{-1}(u)} \bar{F}(x+t)dt \leq \bar{F}(t) \int_0^{F^{-1}(u)} \bar{F}(x)dx$$

is satisfied. It was proved that the NBUT class has the following preservation properties:

(i) Let $X_1, X_2, \ldots, X_N$ be a sequence of independent and identically distributed random variables and N be independent of the X_i's. If X_i's are NBUT, so is $\min(X_1, X_2, \ldots, X_N)$;
(ii) The NBUT class is preserved under the formation of series systems provided that the constituent lifetime variables are independent and identically distributed;
(iii) If X_1, X_2 and X_3 are independent and identically distributed, then

$$E\min(X_1,X_2,X_3) \geq \frac{2}{3} E\min(X_1,X_2).$$

This result is used to test exponentiality against non-exponential NBUT alternatives.

5.5 Some Generalizations

Several generalizations of the TTT have been proposed in the literature. The earliest one is that of Barlow and Doksum [67]. If F and G are absolutely continuous distribution functions with positive right continuous densities f and g, respectively, then the generalized total time on test transform is defined as

$$H_F^{-1}(x) = \int_{F^{-1}(0)}^{F^{-1}(x)} g[G^{-1}F(t)]dt, \quad 0 \leq x \leq 1.$$

As before, $H_F(\cdot)$ is a distribution function and $H_G^{-1}(u) = u, 0 \leq u \leq 1$.

The generalized version can also be shown to possess properties similar to $T(u)$. For instance, the density C_F of H_F is such that

$$C_F(H_F^{-1}(u)) = \frac{f(Q_F(u))}{g(Q_G(u))} = h_F(Q(u)), \quad 0 \leq u \leq 1,$$

where

$$h_F(x) = \frac{f(x)}{g[G^{-1}(F(x))]}$$

is referred to as the generalized failure rate function. Further, if $S_n(\cdot)$ is the empirical distribution function based on a sample of size n from life distribution F, then H_F^{-1} is estimated as

$$H_{S_n}^{-1}(u) = \int_0^{S_n^{-1}} g[G^{-1}S_n(t)]dt$$

and so

$$H_{S_n}^{-1}\left(\frac{r}{n}\right) = \int_0^{X_{r:n}} g[G^{-1}F_n(u)]du = \sum_{j=1}^{r} gG^{-1}\left(\frac{j-1}{n}\right)(X_{j:n} - X_{j-1:n})$$

for $r = 1,2,\ldots,n$. Neath and Samaniego [468] proved that if G is exponential and F is IFRA, then $\frac{H_F^{-1}}{u}$ is decreasing in u. Many reliability properties of the generalized transform like those of $T(u)$ are still open problems. For a study of the order relations of the general form, we refer to Bartoszewicz [73]. Yet another extension due to Li and Shaked [388] is of the form

$$T_2(u) = \int_0^u h(p)q(p)dp,$$

where $h(u)$ is positive on $(0,1)$ and zero elsewhere. The usual TTT results when $h(p) = 1-p$. While the main focus of Li and Shaked [388] is on stochastic orders, they also point out some applications of the order considered by them in reliability context. Various results regarding orderings can be seen in Bartoszewicz [74, 75] and Bartoszewicz and Benduch [76].

In a slightly different direction, Nair et al. [447] studied higher order TTT by applying Definition 5.3, recursively, to the transformed distributions.

Definition 5.6. The TTT transform of order n (TTT-n) of the random variable X is defined recursively as

$$T_n(u) = \int_0^u (1-p)t_{n-1}(p)dp, \qquad n = 1,2,\ldots, \tag{5.26}$$

where $T_0(u) = Q(u)$ and $t_n(u) = \frac{dT_n(u)}{du}$, provided that $\mu_{n-1} = \int_0^1 T_{n-1}(p)dp < \infty$.

The primary reasons for defining the above generalization are (i) the hierarchy of distributions generated by the iterative process reveals more clearly the reliability characteristics of the transformed models than that of $T(u)$ and (ii) the results obtained from (3.27) subsume those for $T(u) = T_1(u)$ and will generate new models and properties. We denote by Y_n the random variable with quantile function $T_n(u)$, mean μ_n, hazard quantile function $H_n(u)$, and mean residual quantile function $M_n(u)$. Recall that $T(u)$, the transform of order one, is a quantile function and consequently the successive transforms T_n, $n = 2,3,\ldots$, are also quantile functions with support $(0,\mu_n)$. Differentiating (5.26), we obtain the quantile density function of Y_n as

$$t_n(u) = (1-u)t_{n-1}(u) = (1-u)^n t_0(u) = (1-u)^n q(u), \tag{5.27}$$

and hence

$$t_n(u) = [H_{n-1}(u)]^{-1} = (1-u)^{n-1}(H(u))^{-1},$$

where $H(u)$ is the hazard quantile function of $X = Y_0$. Thus, we have an identity connecting the hazard quantile function of the baseline distribution $F(x)$ of X and that of Y_n in the form

$$H(u) = (1-u)^n H_n(u),\ n = 0,1,2,\ldots. \tag{5.28}$$

Using (5.9), we have

$$T_{n+1}(u) = \mu_n - (1-u)M_n(u),$$

or equivalently

$$t_{n+1}(u) = M_n(u) - (1-u)M_n'(u).$$

This, along with $t_{n+1}(u) = (1-u)^n t_1(u)$ and

$$t_1(u) = t(u) = M(u) - (1-u)M'(u),$$

yields a relationship between the mean residual quantile functions of X and Y_n as

$$M_n(u) - (1-u)M_n'(u) = (1-u)^n\{M(u) - (1-u)M'(u)\}. \tag{5.29}$$

Incidentially, the definition in (5.26) is also true for negative integers, since $Q(u)$ can be thought of as a transform of $T_{-1}(u)$ and so on. Thus,

$$t_{-n}(u) = (1-u)^{-n}q(u)$$

and

$$H(u) = (1-u)^{-n}H_{-n}(u), \quad n = 1,2,\ldots$$

A remarkable feature of the recurrent transform $T_n(u)$ is that the sequence $\langle H_n(u)\rangle$ increases for positive n and decreases for negative n. Thus, Y_n provides a life distribution whose failure rate is larger (smaller) than that of Y_{n-1} when n is positive (negative). It is therefore of interest to know and compare the ageing patterns of Y_n and Y_{n-1}.

Theorem 5.7. *(i) If X is IHR, then Y_n is IHR for all n;*
(ii) If X is DHR, then Y_n is DHR (IHR) if $Q(u) \geq (\leq) Q_L(k, \frac{1}{n})$ and is bathtub shaped if there exists a u_0 for which $Q(u) \geq Q_L(k, \frac{1}{n})$ in $[0, u_0]$ and $Q(u) \leq Q_L(k, \frac{1}{n})$ in $[u_0, 1]$, where $Q_L(\alpha, C)$ is the quantile function of the Lomax distribution (see Table 1.1).

Proof. Since $t_{n+1}(u) = (1-u)^n t_1(u)$, we have

$$t'_{n+1}(u) = (1-u)^{n-1}\{(1-u)t'_1(u) - nt_1(u)\}.$$

Thus,

$$\begin{aligned} X \text{ is IHR } &\Rightarrow t_1(u) \text{ is decreasing} \\ &\Rightarrow t'_{n+1}(u) < 0 \\ &\Rightarrow T_{n+1}(u) \text{ is concave} \\ &\Rightarrow Y_n \text{ is IHR.} \end{aligned}$$

Similarly, when X is DHR, $T_1(u)$ is convex and accordingly

$$\begin{aligned} Y_n \text{ is DHR (IHR) } &\Rightarrow (1-u)t'_1(u) \geq (\leq) nt_1(u) \\ &\Rightarrow t_1(u) \geq (\leq) k(1-u)^{-n} \\ &\Rightarrow Q(u) \geq (\leq) Q_L\left(k, \frac{1}{n}\right). \end{aligned}$$

The last part follows from the definition of bathtub-shaped hazard quantile function in Chap. 4.

In a similar manner, by backward iteration of a $Q(u) = T_0(u)$ and using

$$t'_1(u) = (1-u)^{-n}(n(1-u)^{-1}t_{n+1}(u) + t'_{n+1}(u)),$$

we get the following result.

Theorem 5.8. *(i) If Y_n is DHR, then X is DHR;*
(ii) If Y_n is IHR, then X is IHR (DHR) if $T_n(u) \leq (\geq) Q_B(k(n+1)^{-1}, (n+1)^{-1})$, and is upside-down bathtub shaped if there exists a u_0 for which $T_n(u) \leq Q_B(k(n+1)^{-1}, (n+1)^{-1})$ in $[0, u_0]$ and $T_n(u) \geq Q_B(k(n+1)^{-1}, (n+1)^{-1})$ in $[u_0, 1]$. Here, $Q_B(R, C)$ denotes the quantile function of the rescaled beta distribution.

Using Theorems 5.7 and 5.8, it is possible to construct BT and UBT distributions with finite range. Generation of BT distributions is facilitated by the choice of DHR distributions for which $t_{n+1}(u)$ has a point of inflexion. On the other hand, IHR distributions can provide UBT models provided $t_{n+1}(u)$ has an inflexion point for negative integers n. The following examples illustrate the procedure.

Example 5.3. Consider the Weibull distribution with

$$Q(u) = \sigma(-\log(1-u))^{\frac{1}{\lambda}}.$$

In this case, we have

$$q(u) = \frac{\sigma}{\lambda(1-u)}(-\log(1-u))^{\frac{1}{\lambda}-1}$$

and

$$t_n(u) = \frac{\sigma}{\lambda}(1-u)^{n-1}(-\log(1-u))^{\frac{1}{\lambda}-1}.$$

Hence,

$$t_n'(u) = \frac{\sigma}{\lambda}(1-u)^{n-2}(-\log(1-u))^{\frac{1}{\lambda}-2}\left[\frac{1}{\lambda} - 1 + (n-1)\log(1-u)\right].$$

Thus, when $0 < \lambda \leq 1$, $T_{n+1}(u)$ is convex in $[0, u_0]$ and concave in $[u_0, 1]$, where

$$u_0 = 1 - \exp\left\{\frac{\lambda - 1}{(n-1)\lambda}\right\}.$$

It follows that Y_n has BT hazard quantile function for $n \geq 1$. Notice that with increasing values of n, the change point u_0 becomes larger. For $\lambda \geq 1$ and every n, Y_n is IHR.

Example 5.4. The Burr distribution with $k = 1$ (see Table 1.1) has

$$Q(u) = u^{1/\lambda}(1-u)^{-\frac{1}{\lambda}}$$

and

$$t_{n+1}'(u) = \frac{1}{\lambda}u^{\frac{1}{\lambda}-2}(1-u)^{n-\frac{1}{\lambda}-1}\left\{\frac{1}{\lambda} - 1 - u(n-1)\right\}.$$

Therefore, $u_0 = \frac{\frac{1}{\lambda}-1}{n-1}$ is a point of inflexion when $n\lambda > 1$. Thus, Y_n is BT in this case.

Theorem 5.9. *(i) X is DMRL implies that Y_n is DMRL;*
(ii) Y_n is IMRL implies that X is IMRL.

Proof. Theorem 5.3 gives the necessary and sufficient condition for X to be DMRL as $(1-u)^{-1}(\mu - T_1(u))$ is decreasing in u. This condition is equivalent to

$$\mu - T_1(u) - (1-u)t_1(u) \leq 0. \tag{5.30}$$

Further,

$$T_{n+1}(u) = \int_0^u (1-p)^n t_1(p)dp = (1-u)^n T_1(u) + A(u),$$

where

$$A(u) = n\int_0^u (1-p)^{n-1}T_1(p)dp > 0 \text{ for all } u.$$

This gives

$$\begin{aligned}\mu_n - T_{n+1}(u) - (1-u)t_{n+1}(u) &= \mu_n - (1-u)^n T_1(u) - (1-u)^{n+1}t_{n+1}(u) - A(u)\\ &\leq \mu_n - (1-u)^n T_1(u) - (1-u)^{n+1}t_{n+1}(u)\\ &\leq \mu_1 - T_1(u) - (1-u)t_1(u) \leq 0.\end{aligned}$$

Hence, X is DMRL according to (5.29). This proves (i) and the proof of (ii) follows similarly by taking n as a negative integer.

Theorem 5.10. *(i) X is IHRA implies that X_n is IHRA;*
(iii) X_n is DHRA implies that X is DHRA.

Proof. We prove only (i) since the proof of (ii) follows on the same lines. In view of Theorem 5.2, X is IHRA if and only if $u^{-1}T_1(u)$ is decreasing, or equivalently

$$t_1(u) \leq u^{-1}T(u). \tag{5.31}$$

Considering $T_n(u)$, we can write

$$\begin{aligned}t_{n+1}(u) - u^{-1}T_{n+1}(u) &= (1-u)^n t_1(u) - u^{-1}(1-u)^n T_1(u) - u^{-1}A(u)\\ &\leq (1-u)^n(t_1(u) - u^{-1}T_1(u))\\ &\leq t_1 - u^{-1}T_1(u) \leq 0.\end{aligned}$$

Result in (i) now follows by using (5.31).

Theorem 5.11. *(i) X is NBUE implies that Y_n is NBUE;*
(ii) Y_n is NWUE implies that X_n is NWUE.

Proof. Recall that X is NBUE if and only if $\mu^{-1}T_1(u) > \mu$ for all u. Hence,

$$\begin{aligned}u^{-1}T_n(u) - \mu_n &= u^{-1}\{(1-u)^n T_1(u) + A(u)\} - \mu_n\\ &\geq u^{-1}(1-u)^n T_1(u) - \mu_1\\ &\geq (1-u)^n\{u^{-1}T_1(u) - \mu_1\} \geq 0\end{aligned}$$

which implies that Y_n is NBUE. Part (ii) follows similarly.

From the above theorems, it is evident that when X is ageing positively, the successive transforms are also ageing positively. Similar results can also be established in the case of other ageing concepts discussed in Chap. 4. It is important to mention that the converses of the above theorems need not be true (see next section).

5.6 Characterizations of Distributions

Various identities between the hazard quantile function, mean residual quantile function and the density quantile function of X and Y_n enable us to mutually characterize the distributions of X and Y_n. A preliminary result is that $T_n(u)$ characterizes the distribution of X. This follows from

$$t_n(u) = (1-u)^n q(u)$$

and

$$Q(u) = \int_0^u (1-p)^{-n} t_n(p)dp.$$

The following theorems have been proved by Nair et al. [447].

Theorem 5.12. *The random variable Y_n, $n = 1, 2, \ldots$, has rescaled beta distribution*

$$Q(u) = R(1-(1-u)^{\frac{1}{c}})$$

if and only if X is distributed as either exponential, Lomax or rescaled beta.

Proof. To prove the if part, we observe that in the exponential case

$$t_n(u) = (1-u)^n q(u) = \lambda^{-1}(1-u)^{n-1}$$

and

$$T_n(u) = \int_0^u t_n(p)dp = (\lambda n)^{-1}\{1-(1-u)^n\}$$

which is the quantile function of the rescaled beta distribution with parameters $((\lambda n)^{-1}, n^{-1})$ in the support $(0, \frac{1}{n\lambda})$. Similar calculations show that when X is Lomax, Y_n is rescaled beta $(\alpha(nC-1)^{-1}, C(nC-1)^{-1})$ with support $(0, \alpha(nC-1)^{-1})$, and when X is rescaled beta (R, C), Y_n has the same distribution with parameters $(R(1+nC)^{-1}, C(1+nC)^{-1})$. Conversely, if we now assume that Y_n is rescaled beta, its quantile function has the form

$$T_n(u) = R_n(1-(1-u)^{\frac{1}{C_n}})$$

for some constants R_n and $C_n > 0$. This gives

$$t_n(u) = \frac{R_n}{C_n}(1-u)^{\frac{1}{C_n}-1} = (1-u)^n q(u).$$

The last equation means that $(1-u)^n$ is a factor of the left-hand side and so

$$\frac{1}{C_n} = k_n + n$$

for some real k_n. Thus,

$$q(u) = (k_n + n)(1-u)^{k_n - 1} R_n.$$

Since $q(u)$ is independent of n, taking $n = 1$, we have

$$Q(u) = k_1^{-1} R_1 (k_1 + 1)\{1 - (1-u)^{k_1}\}.$$

Hence, for $k_1 > 0$, X follows rescaled beta distribution $(0, R_1 k_1^{-1}(k_1 + 1))$, Lomax law for $-1 < k_1 < 0$, and exponential distribution as $k_1 \to 0$. Hence, the theorem.

Theorem 5.13. *The random variable X follows the generalized Pareto distribution with quantile function (see Table 1.1)*

$$Q(u) = \frac{b}{a}\left\{(1-u)^{-\frac{a}{a+1}} - 1\right\} \quad a > -1, b > 0, \tag{5.32}$$

if and only if, for all $n = 0, 1, 2, \ldots$ and $0 < u < 1$,

$$M_n(u) = (na + n + 1)^{-1}(1-u)^n M(u). \tag{5.33}$$

Proof. Assuming (5.33) to hold, we have

$$M_n(u) - (1-u)M_n'(u) = \frac{1}{na+n+1}\{M(u) - (1-u)M'(u) + nM(u)\},$$

and then using the identity (5.29), we get

$$\frac{1}{na+n+1}[M(u) - (1-u)M'(u) + nM(u)] = M(u) - (1-u)M'(u).$$

The above equation simplifies to

$$aM(u) = (a+1)(1-u)M'(u)$$

solving which we get

$$M(u) = K(1-u)^{-\frac{a}{a+1}}.$$

Noting that $M(0) = \mu = b$, we have $K = b$. Since the mean residual quantile function determines the distribution uniquely, we see from (2.48) that X has a generalized

Pareto distribution with parameters (a,b). Next, we assume that X has the specified generalized Pareto distribution. Then,

$$q(u) = \frac{b}{a+1}(1-u)^{-\frac{a}{a+1}-1}$$

and

$$M_n(u) = \int_u^1 (1-p)^n q(p)dp,$$

and so

$$M_n(u) = \frac{b}{na+n+1}(1-u)^{n-\frac{a}{a+1}}.$$

Using the expression (see Table 2.5)

$$M(u) = b(1-u)^{-\frac{a}{a+1}},$$

the relationship in (5.33) is easily verified. Hence, the theorem.

There are other directions in which characterizations can be established. For instance, the relationship $T(u)$ has with any reliability function is a characteristic property. It is easy to see that the simple identity

$$T(u) = A + B\log H(u)$$

holds true if and only if X follows the linear hazard quantile distribution. Recall that $T(u)$ is also a quantile function representing some distribution. Thus, when X has a life distribution, the corresponding $T(u)$ may also be a known life distribution. As an example, X follows power distribution if and only if the associated $T(u)$ corresponds to the Govindarajulu distribution.

5.7 Some Applications

A direct approach to see the application of TTT in data analysis is through the model selection for an observed data. One can either derive a model based on physical conditions or postulate one that gives a reasonable fit. The TTT can then be derived and the data is analysed therefrom. An alternative approach is to start with a functional form of TTT and then choose the parameter values that give a satisfactory fit for the observations. The main point here is that the functional form should be flexible enough to represent different data situations. Since many of the quantile functions discussed in Chap. 3 provide great flexibility, their TTTs can provide candidates for this purpose. In such cases, to compute the descriptive measures of the distribution, one need not revert the TTT to the corresponding quantile function. We show that the descriptors can be obtained directly from $T(u)$ and its derivative $t(u)$.

For this purpose, we recall (1.38)–(1.41) and the identity $t(u) = (1-u)q(u)$. Then, the first four L-moments are as follows:

$$L_1 = \int_0^1 (1-p)q(p)dp = \int_0^1 t(p)dp,$$

$$L_2 = \int_0^1 (p-p^2)q(p)dp = \int_0^1 pt(p)dp,$$

$$L_3 = \int_0^1 (3p^2-2p^3-p)q(p)dp = \int_0^1 p(2p-1)t(p)dp,$$

$$L_4 = \int_0^1 (p-6p^2+10p^3-5p^4)q(p)dp = \int_0^1 p(1-5p+5p^2)t(p)dp.$$

Example 5.5. The quantile function of the generalized Pareto distribution (see Table 1.1) yields

$$t(u) = \frac{b}{a+1}(1-u)^{-\frac{a}{a+1}}.$$

Then, direct calculations using the above formulas result in

$$L_1 = b, \quad L_2 = \frac{b(a+1)}{a+2},$$

$$L_3 = \frac{b(a+1)(2a+1)}{(a+2)(2a+3)}, \quad L_4 = \frac{b(a+1)(2a+1)(3a+2)}{(a+2)(2a+3)(3a+4)}.$$

With these L-moments, descriptive measures like L-skewness and L-kurtosis can be readily derived from the formulas presented in Chap. 1.

In preventive maintenance policies, TTT has an effective role to play. At time $x=0$, a unit starts functioning and is replaced upon age T or its failure which ever occurs first, with respective costs C_1 and C_2, with $C_1 < C_2$. If the unit lifetime is X, the first renewal occurs at $Z = \min(X,T)$ and

$$E(Z) = \int_0^T \bar{F}(x)dx.$$

The mean cost for one renewal period is

$$\bar{F}(T)C_1 + (1-\bar{F}(T))C_2$$

and so the cost per unit time under age replacement model is

$$C(T) = \frac{\bar{F}(T)C_1 + (1-\bar{F}(T))C_2}{\int_0^T \bar{F}(x)dx}.$$

This is equivalent to

$$C(T) = \frac{C_1 + KF(T)}{\int_0^T \bar{F}(x)dx}, \tag{5.34}$$

where $K = C_2 - C_1$. The simple replacement problem is to find an optimal interval $T = T^*$ such that it minimizes (5.34). In practice, one may not know the life distribution but only some observations, and so the optimal age replacement interval has to be estimated from the data. Assuming $K = 1$, without loss of generality, a value u^* determined by $u^* = F(T^*)$ maximizes

$$\frac{1}{C(Q(u))} = \frac{T(u)}{u + C_1}, \quad 0 \le u \le 1,$$

or one that maximizes

$$\frac{\phi(u)}{u + C_1}.$$

Bergman [89] and Bergman and Klefsjö [95] provide a nonparametric estimation concerning age replacement policies. Let $(X_{1:n}, X_{2:n}, \ldots, X_{n:n})$ be an ordered sample from an absolutely continuous distribution. For estimating $\phi(u)$, we use

$$u_r = \frac{H_n^{-1}(\frac{r}{n})}{H_n^{-1}(1)}$$

and determine

$$\hat{T}_n = x_{v:n},$$

where v is such that

$$\frac{u_v}{\frac{v}{n} + C_1} = \max_{1 \le r \le n} \frac{u_r}{(\frac{r}{n}) + C_1}.$$

Then,

(i) $C(\hat{T}_n)$ tends with probability one to $C(T^*)$ as $n \to \infty$;
(ii) the optimal cost $C(T^*)$ may be estimated by $C_n(\hat{T}_n)$, where

$$C_n(X_{r:n}) = \frac{C_1 + F_n(X_{r:n})}{\int_0^{X_{r:n}} \bar{F}_n(t)dt}$$

which is strongly consistent. If a unique optimal age replacement interval exists, then $\hat{T}_n$ is strongly consistent. Bergman [89] explains a graphical method of determining T^*. Draw the line passing through $(-\frac{C}{K},0)$ which touches the scaled transform $\phi(u)$ and has the largest slope. The abscissa of the point of contact is u^*. One important advantage of the graphical method is that it is convenient for performing sensitivity analysis. For example, T^* may be compared for different combinations of K and C_1. Suppose that instead of age replacement at T^*, replacement can be thought of at T_1 and T_2 satisfying $T_1 < T^* < T_2$. Which of these ages give the minimum cost per unit time can also be addressed with the help of TTT (Bergman [91]).

The term availability refers to the probability that a system is performing satisfactorily at a given time and is equal to the reliability if no repair takes place. A second optimality criterion is to replace the unit at age T for which the asymptotic availability is maximized. This is equivalent to minimizing

$$A(T) = \frac{m_1 + (m_1 - m_2)F(T)}{\int_0^T \bar{F}(t)dt},$$

where m_1 is the mean time of preventive maintenance and m_2 is the mean time of repair (Chan and Downs [132]). Since this expression is similar to (5.34), the same method of analysis can be adopted here as well.

Klefsjö [338] discusses the age replacement problem with discounted costs, minimal repair and replacements to extend system life. When costs have to be discounted at a constant rate α, the problem ends up to minimizing

$$C(\alpha,T) = \frac{C_1 + K(1 - e^{-\alpha T}\bar{F}(T))}{\alpha \int_0^T e^{-\alpha t}\bar{F}(t)dt} - \alpha(C_1 + K)\int_0^T e^{-\alpha T}\bar{F}(t)dt;$$

see Bergman and Klefsjö [92] for details. The above expression has a minimum at the same value of T as

$$\frac{C_1 + K(1 - e^{-\alpha T}\bar{F}(T))}{\int_0^T e^{-\alpha t}\bar{F}(t)dt},$$

which is of the same form as (5.34) in which $\bar{F}(t)$ is replaced by $\bar{G}(t) = e^{-\alpha T}\bar{F}(t)$. Consequently, the optimization problem permits the usual analysis with $\phi(u)$ for $\bar{G}$. The estimation problem is also dealt with likewise by minimizing

$$\frac{C + KG_n(T)}{\int_0^T \bar{G}_n(t)dt},$$

where

$$\bar{G}_n(t) = e^{-\alpha t}\left(1 - \frac{r}{n}\right),$$

$X_{r:n} \leq t \leq X_{r+1:n}$, for $r = 0, 1, \ldots, n-1$.

The condition of replacement that the unit replacing the older one is as good as new is not always tenable. We assume a milder condition that the replacement is done by a new unit with probability p and a minimal repair is accomplished with probability $(1-p)$. In other words, the unit is repaired to the same state with the same hazard rate as just before failure.

If C^* denotes the average repair cost, the long run average cost per unit is (Cleroux et al. [151])

$$C_p(t) = \frac{C_1 + (K + \frac{C^*}{P})F^p(T)}{\int_0^T F^{-p}(t)dt}.$$

Using the transform of F^p, the above expression can also be brought to the standard form in (5.34). When the costs are discounted, the same kind of analysis is available in this case also.

Assume that the main objective is to extend system life, where the system has a vital component for which n spares are available. When the vital component fails, the system fails. Derman et al. [171] and Bergman and Klefsjö [93] then discussed the schedule of replacements of the vital component such that the system life is as long as possible. If v_n is the expected life when an optimal schedule is used, they showed that $v_0 = \mu$ and

$$v_n = v_{n-1} + \mu \max_{0 \le u \le 1} \left\{ \phi(u) - \frac{v_{n-1}}{\mu} u \right\}.$$

Draw a line touching the $\phi(u)$ curve which is parallel to the line $y = \frac{v_{n-1}}{\mu}$. If the touching point is $(u_n, \phi(u_0))$, then the optimal replacement age is x_n obtained by solving $F(x_n) = u_n$.

It is customary to test certain devices, which have high initial hazard rates under conditions of field operation, to eliminate or reduce such early failures before sending them to the customers. Such an operation of screening equipments for the above purpose is called burn-in. If the burn-in is excessive, it will result in a loss to the manufacturer in terms of several kinds of costs. On the other hand, if burn-in is on a reduced scale, the problem of early failures may still persist among a percentage of products thus resulting in a return cost. So, an important problem in conducting the test is the determination of the optimal time point up to which the test has to be carried out. Test procedures based on hazard rate, mean residual life, coefficient of variation of residual life and so on have been proposed in the literature. Consider the case when a non-repairable component is scrapped if it fails during the burn-in period. Our problem is to determine the length T_0 of the burn-in period for which $C(T)$, the expected long run cost per unit time of useful operation is minimized. Let b be the fixed cost per unit and d be the cost per unit time of burn-in. A unit which fails in useful operation after the burn-in results in a cost C_1. Then, Bergman and Klefsjö [94] have shown that

$$C(T) = \frac{1 + b + C_1\bar{F}(T) + d\int_0^T \bar{F}(t)dt}{\mu - \int_0^T \bar{F}(t)dt}$$

which is minimized for the same value of T as

$$\frac{\alpha - F(T)}{1 - \int_0^T \bar{F}(t)dt},$$

where $\alpha = (1 + b + d\mu + C_1)C_1^{-1}$. Hence, T_0 is obtained by first graphically determining the value of u, say u_0, for which

$$\frac{\alpha - u}{1 - \phi(u)}$$

is minimized and then solving $F(T_0) = u_0$; see Klefsjö [339]. Klefsjö and Westberg [340] point out that if the life distribution $\bar{F}(T)$ is not known, it has to be estimated from the data. For complete samples, the empirical distribution function is the estimate of F. If the data is censored, i.e., in a set of n observations, k parts are observed to fail and $n - k$ are withdrawn from observation, then the Kaplan–Meier estimator

$$F_n^K(t) = 1 - \prod_r \frac{n-r}{n-r+1},$$

where r runs through integer values for which $t_{j:n} \leq t$ and $t_{j:n}$ are observed failure times, could be used. The optimal replacement age is found by (1) drawing the TTT plot based on times to failure, (2) drawing a line from $(-\frac{C_1}{K}, 0)$ which touches TTT plot and has largest possible slope, and (3) taking the optimum replacement age as the failure time corresponding to the optimal point of contact. If the point of contact is (1,1), no preventive maintenance is necessary. Another major aspect of analysis of failure data for repairable systems is the possible trend in inter-failure times. Kvaloy and Lindqvist [365] used some tests based on TTT for this purpose. Some test statistics have also been proposed for testing exponentiality against IFRA alternative (Bergman [89]), for testing whether one distribution is more IFR than another (Wie [580]) and for testing exponentiality against IFR (DFR) alternative (Wie [579]).

Chapter 6
L-Moments of Residual Life and Partial Moments

Abstract The residual life distribution and various descriptive measures derived from it form the basis of modelling, characterization and ageing concepts in reliability theory. Of these, the moment-based descriptive measures such as mean, variance and coefficient of variation of residual life and their quantile forms were all discussed earlier in Chaps. 2 and 4. The role of L-moments as alternatives to conventional moments in all forms of statistical analysis was also highlighted in Chap. 1. L-moments generally outperform the usual moments in providing smaller sampling variance, robustness against outliers and easier characterization of distributional characteristics, especially in the case of models with explicit quantile functions but no tractable distribution functions. For this reason, we discuss in this chapter the properties of the first two L-moments of residual life. After introducing the definitions, several identities that connect L-moments of residual life with the hazard quantile function, and mean and variance of residual quantile function, are derived. A comparison between the second L-moment and variance of residual life points out the situations in which the former is better. Expressions for the L-moments of residual life of quantile function models of Chap. 3 are derived and their behaviour is discussed in relation to the mean residual quantile function. Characterization of lifetime models based on the functional form of the second L-moment as well as in terms of its relationship with the hazard and mean residual quantile functions are also presented. The upper and lower partial moments have been found to be of use in reliability analysis, economics, insurance and risk theory. Quantile-based definitions of these moments and their relationships with various reliability functions are presented in this chapter. Many of the results on L-moments of residual life have potential applications in economics. For example, income distributions can be characterized by means of some simple properties of concepts like income gap ratio, truncated Gini index and poverty measures. Quantile forms of all these measures are defined and their usefulness in establishing characterizations are explored.

N.U. Nair et al., *Quantile-Based Reliability Analysis*, Statistics for Industry and Technology, DOI 10.1007/978-0-8176-8361-0_6,

6.1 Introduction

The notion of residual life, based on the information that a unit has functioned satisfactorily for a specified period of time, has been fundamental in reliability theory and practice. As seen already in Chaps. 2 and 4, the residual life distribution and various descriptive measures derived from it form the basis for the definition of various ageing concepts. Of these measures, the moment-based descriptive measures such as mean, variance and coefficient of variation of residual life are used commonly in modelling lifetime data, characterizing life distributions, defining classes of life distributions, and in evolving strategies for maintenance and repair of equipments. The Lorenz curve and Bonferroni curve used in measuring income inequality in economics and the Leimkuhler curve in informatics are all characterized by the mean residual life and variance residual life along with other reliability functions; see Chap. 5 for details. Upper and lower partial moments of X are closely related to the moments of residual life. If X has finite moment of order r, the rth upper partial moment (also called the stop-loss moment) about x is defined as

$$p_r(x) = E[(X-x)^+]^r = \int_x^\infty (t-x)^r dF(t),$$

where $(X-x)^+ = \max(X-x,0)$. The quantity $(X-x)^+$ is interpreted as a residual life in the context of lifelength studies (Lin [401]) and the moments $p_r(x)$ are used in actuarial studies in the analysis of risks (Denuit [170]). In the assessment of income tax, x can be taken as the tax exemption level, so that $(X-x)^+$ then becomes the taxable income. Obviously, from the expression

$$m(x) = \bar{F}(x)p_1(x),$$

various identities connecting $p_1(x)$ and the different reliability functions follow. For characterizations of distributions using $p_r(x)$ for $r=1$ and in the general case, we refer to Chong [147], Nair [438], Lin [401], Sunoj [554] and Abraham et al. [14]. If we consider

$$(X-x)^- = \begin{cases} x-X & \text{if } X \le x \\ 0 & \text{if } X > x \end{cases},$$

we have similarly the lower partial moments as $E[(X-x)^-]^r$. Sunoj and Maya [555] have discussed characterizations of distributions and various applications of lower partial moments in the context of risk analysis and income analysis for the poor.

The use of L-moments as an alternative to the conventional moments, for all purposes in which the latter is prescribed, is well known. Our discussions and the references earlier in Chap. 1 do emphasize this aspect. L-moments generally outperform the usual moments in providing smaller sampling variance, robustness against outliers and easier characterization of distributional characteristics, especially for models with explicit quantile functions but no tractable distribution functions. All these considerations apply to lifetime data analysis as well as in the discussion of properties of residual life distributions. Heavy-tailed distributions occur as models of reliability data in which case the usual sample moments lack efficiency. Nair and Vineshkumar [452] pointed out the usefulness of L-moments of residual life in reliability analysis and then studied their properties in comparison with the mean and variance of residual life.

6.2 Definition and Properties of L-Moments of Residual Life

Recall from Sect. 1.6 that the L-moment of order r is given by

$$
\begin{aligned}
L_r &= \frac{1}{r}\sum_{k=0}^{r-1}(-1)^k\binom{r-1}{k}E(X_{r-k:r}),\ r=1,2,\ldots \\
&= \sum_{k=0}^{r-1}(-1)^k\binom{r-1}{k}^2\int_0^\infty x(F(x))^{r-k-1}(1-F(x))^k f(x)dx. \qquad (6.1)
\end{aligned}
$$

The truncated variable $X(t)=X|(X>t)$ has its survival function as

$$\bar{F}_{(t)}(x)=\frac{\bar{F}(x)}{\bar{F}(t)},\quad x>t,$$

so that the L-moment of $X(t)$ is given by

$$L_r(t)=\sum_{k=0}^{r-1}(-1)^k\binom{r-1}{k}^2\int_t^\infty x\left(\frac{\bar{F}(t)-\bar{F}(x)}{\bar{F}(t)}\right)^{r-k-1}\left(\frac{\bar{F}(x)}{\bar{F}(t)}\right)^k\frac{f(x)}{\bar{F}(t)}dx. \qquad (6.2)$$

In particular, setting $r=1$ in (6.2), we obtain

$$L_1(t)=\frac{1}{\bar{F}(t)}\int_t^\infty xf(x)dx=E[X|(X>t)]$$

which is the conditional mean function studied in Chap. 3. Further, $r=2$ in (6.2) leads to

$$
\begin{aligned}
L_2(t) &= \sum_{k=0}^{1}(-1)^k\binom{1}{k}^2\int_t^\infty x\left(\frac{\bar{F}(t)-\bar{F}(x)}{\bar{F}(t)}\right)^{1-k}\left(\frac{\bar{F}(x)}{\bar{F}(t)}\right)^k\frac{f(x)}{\bar{F}(t)}dx \\
&= \frac{1}{\bar{F}^2(t)}\int_t^\infty x[\bar{F}(t)-2\bar{F}(x)]f(x)dx
\end{aligned}
$$

$$= \frac{1}{\bar{F}(t)}\int_t^\infty xf(x)dx - \frac{2}{\bar{F}^2(t)}\int_t^\infty x\bar{F}(x)f(x)dx$$
$$= L_1(t) - t - \frac{1}{\bar{F}^2(t)}\int_t^\infty \bar{F}^2(x)dx$$
$$= m(t) - \frac{1}{\bar{F}^2(t)}\int_t^\infty \bar{F}^2(x)dx, \tag{6.3}$$

where $m(t)$ is the usual mean residual life function. It thus follows that $L_2(t) \leq m(t)$, where the equality sign does not hold for any non-degenerate distribution. Thus, $L_2(t)$ is strictly less than the mean residual life function for all non-degenerate distributions. Differentiating (6.2) and simplifying the resulting expression, we get

$$L_2'(t) = h(t)(2L_2(t) - m(t)). \tag{6.4}$$

Now, setting $F(x) = p$ and $F(t) = u$ in (6.2), we get

$$l_r(u) = L_r(Q(u))$$
$$= \sum_{k=0}^{r-1}(-1)^k\binom{r-1}{k}^2\int_u^1\left(\frac{p-u}{1-u}\right)^{r-k-1}\left(\frac{1-p}{1-u}\right)^k\frac{Q(p)}{1-u}dp.$$

In particular, we have

$$l_1(u) = (1-u)^{-1}\int_u^1 Q(p)dp \tag{6.5}$$

and

$$l_2(u) = (1-u)^{-2}\int_u^1 (2p-u-1)Q(p)dp. \tag{6.6}$$

The properties of $l_1(u)$, equivalent to $E[X|(X > t)]$, have been studied rather extensively and so we concentrate here more on $l_2(u)$. However, notice that $l_1(u)$ uniquely determines the distribution through the formula

$$Q(u) = l_1(u) - (1-u)l_1'(u) \tag{6.7}$$

which is evident from (6.5). From (6.6), as $u \to 0$, we have

$$l_2(0) = \int_0^1 (2p-1)Q(p)dp = \int_0^1 p(1-p)q(p)dp = 2\Delta,$$

where Δ is the mean difference of X as defined in (1.12).

Theorem 6.1. *The functions $l_1(u)$, $l_2(u)$ and $M(u)$ determine each other and $Q(u)$ uniquely.*

Proof. We begin with $M(u)$, and the identity

$$\begin{aligned} M(u) &= l_1(u) - Q(u) \\ &= l_1(u) - \{l_1(u) - (1-u)l_1'(u)\} \\ &= (1-u)l_1'(u). \end{aligned} \tag{6.8}$$

Differentiating (6.6), we have

$$\begin{aligned} (1-u)^2 l_2'(u) - 2(1-u)l_2(u) &= -2uQ(u) + (u+1)Q(u) - \int_u^1 Q(p)dp \\ &= Q(u) - uQ(u) - \int_u^1 Q(p)dp \\ &= Q(u) - uQ(u) - (1-u)(M(u) + Q(u)) \\ &= -(1-u)M(u), \end{aligned}$$

or equivalently

$$M(u) = 2l_2(u) - (1-u)l_2'(u). \tag{6.9}$$

Finally, from (2.38), we have

$$Q(u) = \mu - M(u) + \int_0^u \frac{M(p)}{1-p}dp, \tag{6.10}$$

and thus $M(u)$ determines $Q(u)$, and $l_1(u)$ and $l_2(u)$ determine $M(u)$. In the case of $l_1(u)$, we have

$$l_1(u) = \int_0^u \frac{M(p)}{1-p}dp \tag{6.11}$$

$$= \int_0^u \frac{2l_2(p) - (1-u)l_2'(p)}{1-p}dp. \tag{6.12}$$

Equations (6.11) and (6.12) express $l_1(u)$ in terms of $M(u)$, while $l_2(u)$ and (6.7) recover $Q(u)$ from $l_1(u)$. We also have

$$\begin{aligned} l_2(u) &= (1-u)^{-2}\int_u^1 (1-p)M(p)dp \\ &= (1-u)^{-2}\int_u^1 (1-p)^2 l_1'(p)dp \end{aligned}$$

determining $l_2(u)$ from $M(u)$ and $l_1(u)$. Given $l_2(u)$, $M(u)$ is derived from (6.9) and so $Q(u)$ from (6.10). This completes the proof of the theorem.

Gini's mean difference of $X(t)$ is

$$G(t) = 2\int_t^{\infty} F_{(t)}(x)\bar{F}_{(t)}(x)dx.$$

In terms of quantile functions, we have

$$\Delta(u) = G(Q(u)) = 2\int_u^1 \frac{(1-p)(p-u)}{(1-u)^2} q(p)dp. \tag{6.13}$$

Integrating the RHS of (6.13) by parts, we obtain

$$\Delta(u) = 2l_2(u).$$

Thus, the second L-moment of the conditional distribution of $X|(X>t)$ is half the mean difference of $X|(X>t)$. Since the mean difference is location invariant, the second L-moment of $X(t)$ is the same as that of $X_t = X - t|(X>t)$, and so we refer to $l_2(u)$ as the second L-moment of residual life. Mean difference is a measure of dispersion and so $l_2(u)$ will be treated as a measure of variation in the residual life. Thus, $l_2(u)$ can be viewed as an alternative to the variance residual life in future discussions.

In addition to the mean residual quantile function, other quantile-based reliability functions are also connected with $l_2(u)$. Some typical examples are worked out below. The others can be obtained by exploiting various identities presented earlier in Chap. 2. Invoking (2.36), we have

$$l_2(u) = (1-u)^{-2}\int_u^1 \left(\int_p^1 H^{-1}(s)ds\right) dp.$$

Similarly, from (2.46), we have

$$\begin{aligned} V(u) &= \frac{1}{1-u}\int_u^1 M^2(p)dp \\ &= \frac{1}{1-u}\int_u^1 \left\{2l_2(p) - (1-u)l_2'(p)\right\}^2 dp. \end{aligned}$$

Using the relation (2.36) once again, the total time on test transform satisfies

$$l_2(u) = \frac{1}{(1-u)^2}\int_u^1 (\mu - T(p))dp.$$

Table 6.1 Second L-moment of residual life for some distributions

Distribution	$l_2(u)$
Exponential	$(2\lambda)^{-1}$
Pareto II	$\frac{\alpha c}{(c-1)(2c-1)}(1-u)^{-\frac{1}{c}}$
Rescaled beta	$\frac{Rc}{(c+1)(2c+1)}(1-u)^{-\frac{1}{c}}$
Half-logistic	$\frac{2\sigma}{(1-u)^2}\{1-u-(1+u)\log\left(\frac{2}{1+u}\right)\}$
Power	$\frac{\alpha}{(1+\beta)(1-u)^2}\left\{\beta+(\beta+1-u)u^{\frac{1}{\beta}}\right\}$
Exponential geometric	$\frac{1-p}{p(1-u)^2}\left\{\frac{1-pu}{p}\log\left(\frac{1-pu}{1-p}\right)-(1-u)\right\}$

Example 6.1. The linear hazard quantile distribution is specified by

$$q(u) = [(1-u)(a+bu)]^{-1}$$

and so

$$\begin{aligned} l_2(u) &= \frac{1}{(1-u)^2}\int_u^1 (1-p)(p-u)q(p)dp \\ &= \frac{1}{(1-u)^2}\int_u^1 \frac{p-u}{a+bp}dp \\ &= \frac{1}{b(1-u)^2}\left\{1-u+\frac{a+bu}{b}\log\left(\frac{a+bu}{a+b}\right)\right\}. \end{aligned}$$

The expressions of $l_2(u)$ of some life distributions are presented in Table 6.1.

Since both variance residual life quantile function and $l_2(u)$ are measures of variability, it is appropriate to compare the two. The functional form of $l_2(u)$ characterizes the life distribution and hence it can be used to identify the distribution. Although $V(u)$ also characterizes the distribution, unlike $l_2(u)$, there is no simple expression that relates $Q(u)$ in terms of $V(u)$ or between $\bar{F}(x)$ and $\sigma^2(x)$. See the corresponding discussion in Sect. 2.1.3. Yitzhaki [596] has pointed out that the mean difference is a better measure than variance in deriving properties of distributions which are non-normal. Nair and Vineshkumar [452] have provided empirical evidence that supports this observation. They simulated random samples from the exponential population with varying parameter values. Using $V(u) = \lambda^{-2}$ and $l_2(u) = (2\lambda)^{-1}$, the parameter λ was estimated by equating the sample and population values. They then noted that $l_2(u)$ gave a better approximation to the model as well as estimates with less bias in at least 75 % of the samples.

Another important advantage of the L-moments is that, if the mean exists, all higher-order L-moments exist, which may not be the case with the usual moments. The data on annual flood discharge rates of Floyd river at James, Iowa, considered by Mudholkar and Hutson [423], was reanalysed using the power-Pareto law which gave a good fit at the parameter values $\hat{c} = 3{,}495.2$, $\hat{\lambda}_1 = 0.6226$ and $\hat{\lambda}_2 = 0.5946$. Note that, since $\lambda_2 > 0.5$, the function $V(u)$ does not exist for the distribution, while $l_2(u)$ can be used for further analysis.

The variance residual quantile function has an important role in analysing the ageing aspects. Some additional life distributions were identified based on their monotone behaviour, such as DVRL and IVRL classes (Chap. 4). An important implication observed earlier was that decreasing (increasing) mean residual quantile function implied decreasing (increasing) $V(u)$. By comparison, $V(u)$ and $l_2(u)$ may not show the same type of monotonicity. Even when $V(u)$ increases for larger u, $l_2(u)$ may show a decreasing trend. For example, for the distribution

$$Q(u) = 4u^3 - 3u^4, \quad 0 \le u \le 1,$$

after performing some algebra, we obtain

$$V(u) = \frac{1}{175}\left\{22 - 6u - 34u^2 - 62u^3 + 50u^4 + 78u^5 + 106u^6 + 9u^7 - 38u^8\right\}.$$

Thus,

$$\frac{dV(u)}{du} = \frac{1}{175}\left\{-6 - 68u - 186u^2 + 200u^3 + 390u^4 + 636u^6 + 63u^6 - 304u^7\right\},$$

from which we find that $V(u)$ decreases in $(0, u_0)$ and then increases in $(u_0, 1)$ with the change point u_0 which is approximately 0.554449. At the same time, we have

$$l_2(u) = \frac{(1-u^2)^2}{5}$$

and

$$\frac{dl_2(u)}{du} = -\frac{4}{5}u(1-u^2) < 0$$

showing that $l_2(u)$ is decreasing for all u in $(0,1)$. Neither the implications between the mean residual quantile function $M(u)$ and $V(u)$ hold good when $V(u)$ is replaced by $l_2(u)$. This is well established in the following illustrations that involve some quantile function models discussed earlier in Chap. 3.

Example 6.2. The generalized Tukey-lambda distribution of Freimer et al. [203] with

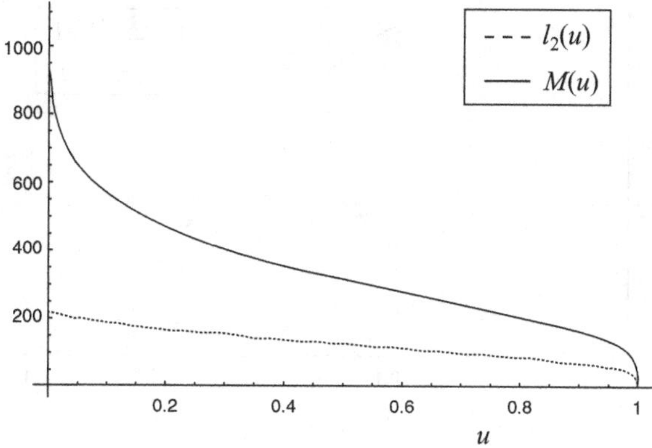

Fig. 6.1 Plot of $M(u)$ and $l_2(u)$ for the data on lifetimes of aluminum coupons

$$Q(u) = \lambda_1 + \lambda_2^{-1}\left\{\frac{u^{\lambda_3}-1}{\lambda_3} - \frac{(1-u)^{\lambda_4}-1}{\lambda_4}\right\}$$

has

$$M(u) = \frac{(1-u)^{\lambda_4}}{\lambda_2(\lambda_4+1)} - \frac{u^{\lambda_3}}{\lambda_2\lambda_3} + \frac{1-u^{\lambda_3+1}}{\lambda_2(1+\lambda_3)(1-u)}$$

and

$$l_2(u) = \frac{1-u}{\lambda_2\lambda_4} - \frac{2(1-u^{\lambda_3+2})}{\lambda_2\lambda_3(\lambda_3+1)(\lambda_3+2)(1-u)^2} + \frac{1-u^{\lambda_4}}{\lambda_2(1+\lambda_4)(2+\lambda_4)} + \frac{(1-u)(1+u^{\lambda_3+1})}{\lambda_2\lambda_3(\lambda_3+1)(1-u)^2}.$$

The distribution provides satisfactory fit to the aluminum coupon data discussed earlier (first 100 observations) with parameter values

$$\hat{\lambda}_1 = 1382.18, \quad \hat{\lambda}_2 = 0.0033, \quad \hat{\lambda}_3 = 0.2706 \text{ and } \hat{\lambda}_4 = 0.2211.$$

The graphs of $M(u)$ and $l_2(u)$ given in Fig. 6.1 show that both are decreasing functions of u.

Example 6.3. Govindarajulu [224] fitted the distribution

$$Q(u) = ((\beta+1)u^{\beta} - \beta u^{\beta+1})$$

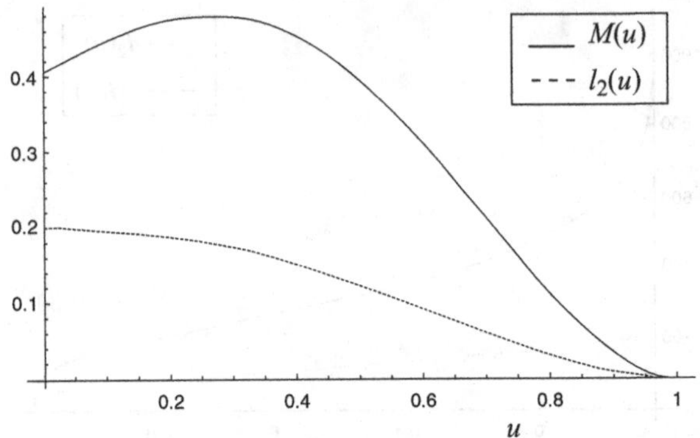

Fig. 6.2 Plot of $M(u)$ and $l_2(u)$ for the data on failure times of a set of refrigerators

to the data on failure times of a set of refrigerators, with the estimate of β being $\hat{\beta} = 2.94$. The mean residual quantile function and the second L-moment function for the distribution are

$$M(u) = \frac{2 - (\beta+1)(\beta+2)u^{\beta} + 2\beta(\beta+2)u^{\beta+1} - \beta(\beta+1)u^{\beta+2}}{(\beta+2)(1-u)}$$

and

$$l_2(u) = \frac{2\beta - 2(\beta+3)u + (\beta+2)(\beta+3)u^{\beta+1} - 2\beta(\beta+3)u^{\beta+2} + \beta(\beta+1)u^{\beta+3}}{(\beta+2)(\beta+3)(1-u)^2}.$$

In the case of the data mentioned above, $M(u)$ initially increases and then decreases with approximate change point $u = 0.2673$, but $l_2(u)$ decreases for all u, as displayed in Fig. 6.2.

Example 6.4. Consider the power-Pareto distribution with

$$Q(u) = Cu^{\lambda_1}(1-u)^{-\lambda_2}, \quad C, \lambda_1, \lambda_2 > 0,$$

for which

$$M(u) = c(1-u)^{-1}\{B_{1-u}(\lambda_1+1, 1-\lambda_2) - u^{\lambda_1}(1-u)^{1-\lambda_2}\}$$

and

$$l_2(u) = c(1-u)^{-2}\{2B_{1-u}(\lambda_1+2, 1-\lambda_2) - (u+1)B_{1-u}(\lambda_1+1, 1-\lambda_2)\},$$

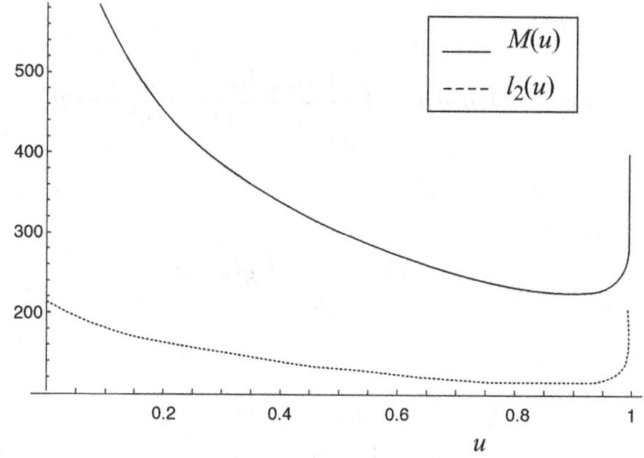

Fig. 6.3 Plot of $M(u)$ and $l_2(u)$ for the data on failure times of electric carts

where

$$B_u(p,q) = \int_0^u t^{p-1}(1-t)^{q-1}dt$$

is the incomplete beta integral. Applying the model to the times of failure of 20 electric carts reported in Zimmer et al. [604], the fit by the method of L-moments with

$$\hat{\lambda}_1 = 0.234612, \quad \hat{\lambda}_2 = 0.09669912, \quad \hat{C} = 1530.53,$$

is observed to be satisfactory. Both $M(u)$ and $l_2(u)$ are seen to possess the similar behaviour, decreasing first and then increasing, as displayed in Fig. 6.3.

Arising from the mean and variance of residual life, the coefficient of variation of residual life is also of importance in reliability. We refer to Sects. 2.1.3 and 2.5 for pertinent definitions and other details. Just as the coefficient of variation of residual life uniquely determines a distribution, it is possible to show that the L-coefficient of variation $c(u) = \frac{l_2(u)}{l_1(u)}$ also possesses a similar property. Nair and Vineshkumar [452] have shown that if $C(u)$ is differentiable, from the definitions of $l_2(u)$ and $l_1(u)$, we can write

$$\int_0^1 (2p-u-1)Q(p)dp = (1-u)c(u)\int_u^1 Q(p)dp.$$

Differentiating this expression and simplifying, we obtain

$$\frac{Q(u)}{\int_u^1 Q(p)dp} = \frac{(1-u)c'(u)-c(u)+1}{(1-u)(1+c(u))}.$$

Upon integrating, we get

$$-\log \int_u^1 Q(p)dp = \int \frac{(1-u)c'(u) - c(u) + 1}{(1-u)(1+c(u))} du$$

so that

$$Q(u) = g(u) \exp\left\{-\int g(u)du\right\}, \tag{6.14}$$

where

$$g(u) = \frac{(1-u)c'(u) - c(u) + 1}{(1-u)(1+c(u))}.$$

Equation (6.14) retrieves $Q(u)$ from $c(u)$ only up to a change of scale. We illustrate this result in the next theorem.

Theorem 6.2. *X has L-coefficient of variation of the form*

$$c(u) = \frac{1-u}{3(1+u)} \tag{6.15}$$

if and only if it has uniform distribution.

Proof. When X has uniform distribution over (α, β), $0 < \alpha < \beta$, we have $Q(u) = u(\beta - \alpha)$,

$$l_1(u) = \frac{(\beta-\alpha)}{2}(1+u)$$

and

$$l_2(u) = \frac{(\beta-\alpha)}{6}(1-u)$$

giving (6.15). Conversely, applying (6.14) with $c(u)$ as in (6.15), we get

$$g(u) = \frac{2u}{(1-u)(1+u)}.$$

Upon substituting this in (6.14), we obtain

$$Q(u) = 2u$$

which is uniform (with a change of scale).

6.3 L-Moments of Reversed Residual Life

On lines similar to those in the preceding section, we can look at the L-moments of ${}_tX = X|(X \leq t)$ whose distribution function is $\frac{F(x)}{F(t)}$, $0 < x \leq t$. Using (6.1), the rth L-moment of ${}_tX$ has the expression

$$B_r(t) = \sum_{k=0}^{r-1}(-1)^k \binom{r-1}{k}^2 \int_0^t x\left(\frac{F(x)}{F(t)}\right)^{r-k-1}\left(1-\frac{F(x)}{F(t)}\right)^k \frac{f(x)}{F(t)}dx.$$

In particular, we have

$$B_1(t) = \int_0^t \frac{xf(x)}{F(t)}dx = E[X|X \leq x], \tag{6.16}$$

$$B_2(t) = \frac{1}{F^2(t)}\int_0^t (2F(x) - F(t))xf(x)dx. \tag{6.17}$$

Setting $u = F(t)$ and $p = F(x)$, we have

$$\theta_1(u) = B_1(Q(u)) = u^{-1}\int_0^u Q(p)dp \tag{6.18}$$

and

$$\theta_2(u) = B_2(Q(u)) = u^{-2}\int_0^u (2p-u)Q(p)dp. \tag{6.19}$$

By differentiating (6.16) and using the definitions of the reversed hazard rate in (2.22) and the reversed mean residual life in (2.24), we get

$$\lambda(t) = \frac{B_1'(t)}{t - B_1(t)}$$

and

$$r(t) = t - B_1(t)$$

so that

$$\lambda(t) = B_1'(t)r(t).$$

Also, by differentiating (6.17) and simplifying, we get

$$B_2'(t) = \lambda(t)[r(t) - 2B_2(t)].$$

Table 6.2 Expressions of $\theta_2(u)$ for some quantile models

Distribution	$\theta_2(u)$
Power	$\frac{\alpha\beta}{(\beta+1)(2\beta+1)}u^{\frac{1}{\beta}}$
Govindarajulu	$\frac{\sigma\beta u^{\beta}}{(\beta+2)(\beta+3)}\{\beta+3-(\beta+1)u\}$
Generalized lambda	$\frac{1}{u^2}\Big[\frac{\lambda_3 u^{\lambda_3+2}}{\lambda_2(\lambda_3+1)}$ $+\frac{1}{\lambda_2(\lambda_4+1)}\Big(u\{(1-u)^{\lambda_3+1}-1\}+\frac{2}{\lambda_4+2}\{(1-u)^{\lambda_4+2}-1\}\Big)\Big]$
Power Pareto	$Cu^{-2}\{2B_u(\lambda_1+2,1-\lambda_2)-uB(\lambda_1+1,1-\lambda_2)\}$

Likewise, we have the following relationships connecting $\theta_1(u)$ and $\theta_2(u)$ with the reliability functions

$$\begin{aligned}R(u) &= Q(u)-\theta_1(u)\\ &= \theta_1(u)+u\theta_1'(u)-\theta_1(u)=u\theta_1'(u),\\ \theta_2(u) &= \frac{1}{u^2}\int_0^u pR(p)dp,\\ Q(u) &= R(u)+\int_0^u p^{-1}R(p)dp,\end{aligned} \tag{6.20}$$

and

$$R(u)=u\theta_2'(u)+2\theta_2(u).$$

As in Sect. 6.2, each of $Q(u)$, $\theta_1(u)$, $R(u)$ and $\theta_2(u)$ determine others uniquely. We further have

$$D(u)=u^{-1}\int_0^u p\{\theta_2'(p)+2\theta_2(p)\}^2dp.$$

Examples of $\theta_2(u)$ for some quantile models are presented in Table 6.2.

The L-coefficient of variation of ${}_tX$, i.e., $\theta(u)=\frac{\theta_2(u)}{\theta_1(u)}$, determines the distribution of X up to a change of scale through the formula

$$Q(u)=\frac{u\theta'(u)+\theta(u)+1}{u(1-\theta(u))}\exp\left\{\int\frac{u\theta'(u)+\theta(u)+1}{u(1-\theta(u))}du\right\}.$$

Theorem 6.3. *X follows the power distribution if and only if $\theta(u)$ is a constant.*

Proof. For the power distribution with

$$Q(u) = \alpha u^{\frac{1}{\beta}}, \quad 0 \le u \le 1,\ \beta \ne 0, \beta > 0,$$

we have

$$\theta_1(u) = \frac{\alpha\beta}{1+\beta} u^{\frac{1}{\beta}} \quad \text{and } \theta_2(u) = \frac{\alpha\beta}{(\beta+1)(2\beta+1)} u^{\frac{1}{\beta}}$$

so that

$$\theta(u) = \frac{1}{2\beta+1},$$

a constant. Conversely when $\theta(u) = c$, a constant, the expression given above for $Q(u)$ in terms of $\theta(u)$ yields

$$\begin{aligned} Q(u) &= \frac{c+1}{u(1-c)} \exp\left\{ \int \frac{c+1}{u(1-c)} du \right\} \\ &= \frac{c+1}{1-c} u^{\frac{2c}{1-c}}, \quad c \ne 1, \end{aligned}$$

which corresponds to a power distribution. Hence, the theorem.

The relevance of this characterization in economics is explained later in Sect. 6.5.

6.4 Characterizations

Like other reliability functions, the second L-moments $l_2(u)$ and $\theta_2(u)$ also characterize life distributions through special relationships. We now present several such results. The first result is the characterization of the generalized Pareto distribution by simple relationships between $l_2(u)$, $M(u)$ and $l_1(u)$.

Theorem 6.4. *Let X be a continuous non-negative random variable with $E(X) < \infty$. Then, X follows the generalized Pareto distribution with*

$$Q(u) = \frac{b}{a}\left\{ (1-u)^{-\frac{a}{a+1}} - 1 \right\}, \quad a > -1, b > 0, \tag{6.21}$$

if and only if the following conditions are satisfied:

(i) $l_2(u) = CM(u)$, $0 < C < 1$;
(ii) $l_2(u) = a_1 l_1(u) + a_2$, $a_1 > -1$, $a_2 > 0$;
(iii) $l_1(u) = AM(u) + B$.

Proof. First, we calculate $l_1(u)$, $l_2(u)$ and $M(u)$, using (6.21), to be

$$l_1(u) = ba^{-1}(a+1)\left\{(1-u)^{-\frac{a}{a+1}} - 1\right\},$$
$$l_2(u) = b(a+1)(a+2)^{-1}(1-u)^{-\frac{a}{a+1}},$$
$$M(u) = b(1-u)^{-\frac{a}{a+1}},$$

so that

$$l_2(u) = \frac{a+1}{a+2}M(u),$$
$$l_2(u) = \frac{a}{a+2}l_1(u) + \frac{b(a+1)}{a+2},$$
$$l_1(u) = \frac{a}{a+1}M(u) - \frac{ba}{a+1}.$$

Thus, the conditions (i), (ii) and (iii) are satisfied for the generalized Pareto distribution. Conversely, condition (i) is equivalent to

$$C(1-u)^2M(u) = \int_0^1 (1-p)M(p)dp$$

or

$$\frac{(1-u)M(u)}{\int_u^1(1-p)M(p)dp} = \frac{1}{C(1-u)}.$$

Upon solving the last equation, we get

$$M(u) = K(1-u)^{\frac{1-2C}{C}}, \tag{6.22}$$

where K is found to be $K = M(0) = \mu$. Since $0 < C < 1$, we can write it as $C = \frac{a+1}{a+2}$ for $a > -1$ and obtain (6.21). To prove the sufficiency of (ii), we note that it implies

$$(1-u)^{-2}\int_u^1 (2p-u-1)Q(p)dp = (1-u)^{-1}a_1\int_u^1 Q(p)dp + a_2.$$

Differentiating the above equation twice, we get

$$Q'(u) - \frac{2a_1Q(u)}{(1+a_1)(1-u)} = \frac{2a_2}{(1+a_1)(1-u)}. \tag{6.23}$$

Noticing that (6.23) is a first-order linear differential equation with integrating factor $(1-u)^{\frac{2a_1}{1+a_1}}$, we have its unique solution as

$$(1-u)^{\frac{2a_1}{1+a_1}}Q(u) = K - \frac{a_2}{a_1}(1-u)^{\frac{2a_1}{1+a_1}}.$$

Evaluating K at $u=0$, we obtain $K=\frac{a_2}{a_1}$, and thus

$$Q(u)=\frac{a_2}{a_1}\left\{(1-u)^{-\frac{2a_1}{1+a_1}}-1\right\}$$

which corresponds to a generalized Pareto $\left(\text{reduces to (6.21) when } a_1=\frac{a}{a+2}, a_2=\frac{b}{a+2}\right)$. The result in (iii) is a consequence of (i) and (ii), and this completes the proof of the theorem.

Remark 6.1. Conditions (i), (ii) and (iii) show that each of $l_1(u)$, $l_2(u)$ and $M(u)$ is a linear function of the other.

Remark 6.2. The generalized Pareto law reduces to the exponential distribution as $a\to 0$, rescaled beta for $-1<a<0$, and Pareto II for $a>0$. Thus, the exponential (rescaled beta; Pareto II) is characterized by $l_2(u)=\frac{1}{2}M(u)$ $\left(<\frac{1}{2}M(u);\ >\frac{1}{2}M(u)\right)$ corresponding to the values $C=\frac{1}{2}$ $\left(<\frac{1}{2},\ >\frac{1}{2}\right)$ in result (i).

Remark 6.3. It is seen from direct calculations that

$$V(u)=\frac{1+a}{1-a}b^2(1-u)^{-\frac{2a}{a+1}}=Kl_2^2(u), \quad \text{with } K=\frac{(a+2)^2}{1-a^2}.$$

But, Nair and Vineshkumar [452] have shown that this is not a characteristic property of the generalized Pareto.

Life distributions characterized by simple forms of various reliability functions have been of interest in reliability theory. They are quite useful in modelling lifetime data. The linear and quadratic hazard rate distributions belong to this category. One may refer to Bain [45, 46], Sen and Bhattacharya [525] and Gore et al. [223] for details. A second example is the generalized Pareto distribution which is uniquely determined by a linear mean residual life function (also by a reciprocal linear hazard rate function). It has been seen that $L_2(t)=c$ $(l_2(u)=c)$, where c is a constant characterizing the exponential law. In the same manner, let us consider the linearity

$$L_2(t)=A+Bt,$$

or equivalently

$$l_2(u)=A+BQ(u) \tag{6.24}$$

and identify the corresponding life distribution. Using (6.9), we then have

$$M(u)=2A+2BQ(u)-B(1-u)q(u.)$$

To express the RHS also in terms of $M(u)$, we make use of (2.38) and (2.39) to arrive at

$$M(u) = 2A + 2B\left\{\int_0^u \frac{M(p)dp}{1-p} - M(u) + \mu\right\} - B[M(u) - (1-u)M'(u)].$$

Differentiating with respect to u and simplifying the resulting expression, we obtain the homogeneous linear differential equation of order two with variable coefficients

$$B(1-u)^2 M''(u) - (4B+1)(1-u)M'(u) + 2BM(u) = 0. \tag{6.25}$$

To solve (6.25), we set $M(u) = (1-u)^m$ to get the auxiliary equation

$$Bm(m-1) + (4B+1)m + 2B = 0,$$

or the quadratic equation (in m)

$$Bm^2 + (3B+1)m + 2B = 0. \tag{6.26}$$

Suppose (6.26) has two distinct roots m_1 and m_2. Then, the general solution of (6.25) is of the form

$$M(u) = C_1(1-u)^{m_1} + C_2(1-u)^{m_2}. \tag{6.27}$$

As u tends to zero, we get

$$\mu = C_1 + C_2. \tag{6.28}$$

Thus, from (6.27), the distribution satisfying (2.26) is recovered as

$$\begin{aligned} Q(u) &= \int_0^u \frac{M(p)}{1-p}dp - M(u) + \mu \\ &= \mu + C_1\left\{\frac{1}{m_1} - \frac{1}{m_1+1}(1-u)^{m_1}\right\} + C_2\left(\frac{1}{m_2} - \frac{1}{m_2+1}\right)(1-u)^{m_2}. \end{aligned}$$

Upon substituting for μ from (6.28), we obtain the final expression

$$Q(u) = C_1\frac{1+m_1}{m_1}\{1-(1-u)^{m_1}\} + C_2\frac{1+m_2}{m_2}\{1-(1-u)^{m_2}\}. \tag{6.29}$$

When the roots are equal, we must have

$$(3B+1)^2 - 8B^2 = 0,$$

or equivalently

$$B^2 + 6B + 1 = 0. \tag{6.30}$$

Also, in this situation, we have

$$m_1 = -\frac{(1+3B)}{2B}.$$

Since the product of the roots is 2, $m_1 = \pm\sqrt{2}$, and therefore from (6.30), we have

$$B = -\frac{1}{-3-2\sqrt{2}} = -3+2\sqrt{2}$$

or

$$B = -\frac{1}{-3+2\sqrt{2}} = -3-2\sqrt{2}.$$

Both values satisfy (6.30). We then use the method of variation of parameters to extract the solution of (6.25). Assume that the solution in (6.27), of the form

$$M(u) = C_1M_1(u) + C_2M_2(u),$$

where $M_i(u) = (1-u)^{m_i}$, $i = 1,2$, is such that $M_2(u) = yM_1(u)$ is a solution with y being some function of u. Then,

$$M_2'(u) = yM_1'(u) + y'M_1(u)$$

and

$$M_2''(u) = yM_1''(u) + 2y'M_1'(u) + M_1(u)y''.$$

Substituting these in (6.25), we get

$$\begin{aligned}[B(1-u^2)M_1''(u) - 4(B+1)(1-u)M_1'(u) + 2BM_1(u)]y \\ + (1-u)^2M_1(y)y'' - (4B+1)(1-u)y'M_1(u) + 2(1-u)^2y'M_1'(u) = 0.\end{aligned}$$

Since $M_1(u)$ is a particular solution of (6.25), the first term vanishes and so we get

$$(1-u)^2M_1(u)y'' - (4B+1)(1-u)y'M_1(u) + 2(1-u)B^2y'M_1'(u) = 0. \tag{6.31}$$

The transformation

$$M_1(u) = (1-u)^{-\frac{3B+1}{2B}}$$

in (6.31) shows that

$$(1-u)y'' - By' = 0,$$

which has its solution as

$$y' = (1-u)^{-B}$$

or

$$y = \frac{(1-u)^{1-B}}{B-1}.$$

Thus, we have

$$M_2(u) = \frac{(1-u)^{1-B}}{B-1} M_1(u)$$

which gives the second solution corresponding to $m_1 = m_2$ as

$$\begin{aligned} M(u) &= C_1 M_1(u) + C_2 \frac{(1-u)^{1-B}}{B-1} M_1(u) \\ &= \left\{ C_1 + \frac{C_2}{B-1}(1-u)^{1-B} \right\} (1-u)^{m_1}. \end{aligned} \tag{6.32}$$

The quantile function corresponding to (6.32) is calculated from (2.38) as

$$Q(u) = C_1 \frac{m_1+1}{m_1}\{1-(1-u)^{m_1}\} + \frac{c_2(m_1-B+2)}{(B-1)(m_1-B+1)}\{1-(1-u)^{m_1-B+1}\}. \tag{6.33}$$

To complete the required characterization, it remains to be shown that the identity in (6.24) holds for the quantile functions in (6.28) and (6.33). By direct calculation from (6.28), we see that

$$\begin{aligned} l_2(u) &= \frac{c_1(1-u)^{m_1}}{m_1+2} + \frac{c_2(1-u)^{m_2}}{m_2+2} \\ &= \frac{c_1(1-u)^{m_1}}{m_1+2} + \frac{c_2 m_1 (1-u)^{\frac{2}{m_1}}}{2(1+m_1)}, \end{aligned}$$

where we have used the fact that $m_1 m_2 = 2$. Then, (6.24) holds with

$$A = \frac{C_1}{2+m_1} + \frac{C_2 m_1}{2(1+m_1)},$$

$$B = -\frac{m_1}{(1+m_1)(2+m_2)}.$$

In the second case, $A = C_1 \frac{m_1+1}{m_1} + \frac{C_2(m_1-B+2)}{(B-1)(m_1-B+1)}$, where m_1 and B have the values determined earlier.

Thus, we have established the following theorem.

Theorem 6.5. *A continuous non-negative random variable X with finite mean satisfies*

$$l_2(u) = A + BQ(u) \quad (L_2(t) = A + Bt)$$

if and only if its distribution is specified by the quantile functions (6.29) *or* (6.33).

Remark 6.4. The conditions on the parameters are determined such that $Q(u)$ is a quantile function. Notice also that the quantile functions in this case cannot be inverted into analytically tractable distribution functions.

Remark 6.5. The generalized Pareto distribution arises as a particular solution when $C_2 = 0$ and $m_1 = -\frac{a}{a+1}$.

The next result is based on a simple relationship between $l_2(u)$ and the hazard quantile function $H(u)$.

Theorem 6.6. *A continuous non-negative random variable with finite mean satisfies*

$$l_2(u) = K[H(u)]^{-1}, \quad K > 0, \tag{6.34}$$

for all $0 < u < 1$, *if and only if*

$$Q(u) = C_1 \frac{1+m_1}{m_1}\{1-(1-u)^{m_1}\} + C_2 \frac{1+m_2}{m_2}\{1-(1-u)^{m_2}\}, \tag{6.35}$$

where m_1 *and* m_2 *are the roots of the quadratic equation*

$$Km^2 + 3Km + (2K-1) = 0.$$

Proof. The condition (6.34) is equivalent to

$$\frac{1}{(1-u)^2}\int_u^1 (1-p)M(p)dp = K\{M(u) - (1-u)M'(u)\},$$

or

$$\int_u^1 (1-p)M(p)dp = K(1-u)^2 M(u) - K(1-u)^3 M'(u).$$

Differentiating and simplifying the expression, we get

$$K(1-u)^2 M''(u) - 4K(1-u)M'(u) + (1-2K)M(u) = 0. \tag{6.36}$$

Now by setting $M(u) = (1-u)^m$ and proceeding as in the previous theorem, we have the auxiliary equation

$$Km^2 + 3Km + 2K - 1 = 0.$$

Let m_1 and m_2 be the roots of this quadratic equation. Then, a general solution to (6.36) is

$$M(u) = C_1(1-u)^{m_1} + C_2(1-u)^{m_2},$$

where $m_1 + m_2 = -3$. Then,

$$Q(u) = \frac{C_1(1+m_1)}{m_1}\{1-(1-u)^{m_1}\} + C_2\frac{1+m_2}{m_2}\{1-(1-u)^{m_2}\}.$$

If the roots are the same, the condition for this is

$$K^2 + 4K = 0.$$

However, the roots $K = 0$ and $K = -4$ are both inadmissible. Hence, (6.35) represents the unique distribution satisfying (6.34).

Now, for the distribution (6.35), we have

$$l_2(u) = \frac{C_1(1-u)^{m_1}}{m_1+2} + \frac{C_2(1-u)^{m_2}}{m_2+2}$$

and

$$q(u) = C_1(1+m_1)(1-u)^{m_1-1} + C_2(1+m_2)(1-u)^{m_2-1}.$$

Hence,

$$\begin{aligned} H(u) &= \{(1-u)q(u)\}^{-1} \\ &= \{C_1(1+m_1)(1-u)^{m_1} + C_2(1+m_2)(1-u)^{m_2}\}^{-1}. \end{aligned}$$

Since $(1+m_1)(2+m_1) = (1+m_2)(2+m_2)$ by virtue of $m_1 + m_2 = -3$, we have

$$l_2(u) = K(H(u))^{-1} \quad \text{with } K = (1+m_1)(2+m_2).$$

The proof of the theorem is thus completed.

There exist similar results for the reversed hazard quantile functions. Since the proof proceeds along the same lines, we just briefly outline the proofs.

Theorem 6.7. *If X is a continuous non-negative random variable with finite mean, then*

$$\theta_2(u)\Lambda(u) = C, \tag{6.37}$$

a positive constant, if and only if

$$Q(u) = \frac{1+m_1}{m_1} C_1 u^{m_1} + \frac{1+m_2}{m_2} C_2 u^{m_2}, \tag{6.38}$$

where C_1 *and* C_2 *are the roots of the quadratic equation*

$$Cm^2 + 3Cm + 2C - 1 = 0.$$

Proof. Condition (6.37) is same as

$$\frac{1}{u^2} \int_0^u pR(p)dp = C\{R(u) + R'(u)\}$$

leading to

$$Cu^2 R''(u) + 4CuR'(u) + (2C-1)R(u) = 0.$$

Assuming $R(u) = u^m$, the auxiliary equation becomes

$$Cm(m-1) + 4m + (2C-1) = 0,$$

and so

$$R(u) = C_1 u^{m_1} + C_2 u^{m_2}$$

which gives $Q(u)$ in (6.38) on applying (2.51). The condition for equal roots is $C = -4$ or 0, which are both inadmissible.

Conversely, when (6.38) holds, we have

$$\begin{aligned}\theta_2(u) &= \frac{C_1 u^{m_1}}{m_1+2} + \frac{C_2 u^{m_2}}{m_2+2} \\ &= C \wedge (u),\end{aligned}$$

where $C^{-1} = (1+m_1)(m_1+2) = (1+m_2)(m_2+2)$, on using $m_1 + m_2 = -3$.

Theorem 6.8. *If X is a non-negative random variable with finite mean, the identity*

$$\theta_2(u) = CR(u) \tag{6.39}$$

holds if and only if X has power distribution

$$Q(u) = \alpha u^{1/\theta}, \quad i.e., F(x) = \left(\frac{x}{\alpha}\right)^{\beta}, 0 \le x \le \alpha. \tag{6.40}$$

Proof. For the power distribution, we have

$$\theta_2(u) = \frac{\alpha\beta}{(\beta+1)(2\beta+1)} u^{\frac{1}{\beta}}$$

and

$$R(u) = \frac{\alpha}{\beta+1} u^{\frac{1}{\beta}}$$

so that (6.39) is satisfied with $C = \frac{\beta}{(2\beta+1)}$. Conversely, (6.39) means that

$$\frac{1}{u^2}\int_0^u pR(p)dp = cR(u),$$

or equivalently

$$CuR'(u) = (1-2c)R(u).$$

The last equation yields the solution as

$$R(u) = Ku^{\frac{1-2c}{c}} \quad \text{and} \quad Q(u) = \frac{K(1-c)}{1-2c} u^{\frac{1-2c}{c}}$$

which is of the from (6.40) with $C = \frac{\beta}{1+2\beta}$ and $\alpha = \frac{K(1-C)}{1-2C}$.

Theorem 6.9. *If X is a non-negative random variable with finite mean, the identity*

$$\theta_2(u) = AQ(u) \tag{6.41}$$

holds if and only if X has a distribution with

$$Q(u) = C_1\left(\frac{1+m_1}{m_1}\right)u^{m_1} + C_2\left(\frac{1+m_2}{m_2}\right)u^{m_2}, \tag{6.42}$$

where m_1 and m_2 are the distinct roots of

$$Am^2 + (3A-1)m + 2A = 0. \tag{6.43}$$

If (6.43) *has equal roots, then*

$$Q(u) = \left\{C_1\left(\frac{1+m_1}{m_1}\right) + C_2\frac{1+m_1}{m_1}\log u - \frac{C_2}{m_1^2}\right\}u^{m_1} \tag{6.44}$$

with $m_1 = \frac{3\sqrt{2}-4}{3-2\sqrt{2}}$ and $A = 3-2\sqrt{2}$.

Proof. Let us assume the identity in (6.41). Then, from (2.51) and (6.20), we have

$$\frac{1}{u^2}\int_0^u pR(p)dp = AR(u) + A\int_0^u \frac{R(p)}{p}dp.$$

Differentiation of this equation yields

$$Au^2R''(u) + (4A-1)uR'(u) + 2AR(u) = 0.$$

Substitution of $R(u) = u^m$ gives the auxiliary equation

$$Am^2 + (3A-1)m + 2A = 0. \tag{6.45}$$

When the roots of the quadratic equation in (6.45) are distinct, we get

$$R(u) = C_1 u^{m_1} + C_2 u^{m_2} \tag{6.46}$$

and then for (2.51)

$$Q(u) = C_1\frac{1+m_1}{m_1}u^{m_1} + C_2\frac{1+m_2}{m_2}u^{m_2},$$

where m_1 and m_2 are such that

$$m_1 m_2 = 2 \quad \text{and} \quad m_1 + m_2 = \frac{1-3A}{A}. \tag{6.47}$$

Using (6.46), we have

$$\theta_2(u) = \frac{1}{u_2}\int_0^u pR(p)dp = \frac{C_1 u^{m_1}}{m_1+2} + \frac{C_2 u^{m_2}}{m_2+2}.$$

One can verify that

$$\theta_2(u) = AQ(u)$$

with

$$A = \frac{m_1}{(1+m_1)(m_1+2)} = \frac{m_2}{(1+m_2)(m_2+2)},$$

where the last equality holds since $m_1 m_2 = 2$. When the roots of (6.45) are equal, say m_1, we see that

$$A^2 - 6A + 1 = 0$$

holds whenever $A = 3 \pm 2\sqrt{2}$ both of which are admissible values. Taking

$$R(u) = C_1 u^{m_1} + C_2 u^{m_2} = C_1 R_1(u) + C_2 R_2(u)$$

from (6.46) and setting $R_2(u) = yR_1(u)$, we get, by the method of variation of parameters,

$$Au^2 R_1(u) y'' + 2Au^2 R_1'(u) y' + u(4A-1) R_1(u) y' = 0$$

when $R_1(u) = u^{\frac{1-3A}{2A}}$,

$$Au^2 y'' + Auy' = 0$$

or

$$uy'' + y' = 0.$$

The solution is $y = \log u$, and so the quantile function simplifies to

$$Q(u) = C_1 \frac{1+m_1}{m_1} u^{m_1} + C_2 \frac{1+m_1}{m_1} u^{m_1} \log u - \frac{C_2 u^{m_1}}{m_1^2},$$

as in (6.44). Notice that $Q(u)$ becomes a quantile function only when $m_1 > 0$. In this case, $m_1 = \frac{3\sqrt{2}-4}{3-2\sqrt{2}}$ and

$$\begin{aligned}\theta_2(u) &= \left\{ \frac{C_1}{m_1+2} + \frac{C_2}{m_1+2} \log u - \frac{C_2}{(m_1+2)^2} \right\} u^{m_1} \\ &= 2(3-2\sqrt{2}) Q(u).\end{aligned}$$

This completes the proof of the theorem.

6.5 Ageing Properties

When conceived as a reliability function, the L-moment $l_2(u)$ can also be employed in distinguishing life distributions based on its monotone behaviour. Since $l_2(u)$ is twice the mean difference residual quantile function, we have the following definitions.

Definition 6.1. A lifetime random variable X is said to have increasing (decreasing) mean difference residual quantile function, IMDR (DMDR), according to whether $l_2(u)$ is an increasing (decreasing) function of u.

Example 6.5. From the expressions in Table 6.1, the Pareto II distribution has increasing mean difference residual quantile function, while the rescaled beta has decreasing mean difference residual quantile function.

The mean difference residual quantile function is known to be

$$\Delta(u) = 2l_2(u),$$

and accordingly, from (6.9), we have

$$\Delta'(u) = \frac{2}{1-u}(\Delta(u) - M(u)).$$

Thus, a necessary and sufficient condition that $\Delta(u)$ is increasing (decreasing) is $\Delta(u) \geq (\leq) M(u)$. It is evident that the graph of $\Delta(u)$ lies above (below) that of $M(u)$ when the former is increasing (decreasing). Also, when $\Delta(u)$ crosses $M(u)$ at some point u_0 from below (above), then it is a change point of $\Delta(u)$ that indicates $\Delta(u)$ is increasing (decreasing) first and then decreasing (increasing). Since $\Delta(u)$ is directly related to $M(u)$, it is also clear that

$$X \text{ is DMRL (IMRL) } \Leftrightarrow 3\Delta'(u) \leq (\geq)(1-u)\Delta''(u).$$

The comparison of the implications of monotonicities of $V(u)$, $\Delta(u)$ and $M(u)$ have all been addressed earlier in Sect. 6.2.

6.6 Partial Moments

The partial moments, whose definitions were given earlier in Sect. 6.1, can also be viewed as reliability functions. Since the first two moments are of interest to the concepts discussed earlier, we recall their definitions as

$$p_1(x) = E[(X-x)^+] = \int_x^\infty (t-x)f(t)dt \tag{6.48}$$

and

$$p_2(x) = E[(X-x)^{+2}] = \int_x^\infty (t-x)^2 f(t)dt. \tag{6.49}$$

Gupta and Gupta [231] have discussed the general properties of the rth partial moment. They proved that the rth moment determines the underlying distribution for any positive real r. Also, when r is a positive integer, there exists a recurrence relation between two consecutive partial moments. Earlier, Chong [147] characterized the exponential distribution by the property

$$E(X-t-s)^+ E(X) = E(X-t)^+ E(X-s)^+.$$

The expression for the survival function in terms of $p_r(x)$ is

$$\bar{F}(x) = \frac{(-1)^r}{r!}\frac{d^r p_r(x)}{dx^r};$$

see Navarro et al. [465] and Sunoj [554]. Sunoj [554] also obtained the partial moments of the length-biased distribution, equilibrium distribution and characterizations thereof. Gupta [236] extended this result to show that the kth order equilibrium distribution has survival function

$$S_K(x) = \frac{E[(X-x)^+]^k}{E(X^k)}.$$

Lin [401] and Abraham et al. [14] characterized the exponential, beta and Lomax distributions by relationships between the partial moments. The quantile forms of (6.48) and (6.49) are

$$\begin{aligned} P_1(u) = p_1(Q(u)) &= \int_u^1 (Q(p)-Q(u))dp \\ &= \int_u^1 (1-p)q(p)dp \\ &= (1-u)M(u) \end{aligned} \tag{6.50}$$

and

$$P_2(u) = p_2(Q(u)) = \int_u^1 [Q(p)-Q(u)]^2 dp.$$

Accordingly, the variance of $(X-x)^+$ has the form

$$V_+(u) = \int_u^1 [Q(p)-Q(u)]^2 dp - P_1^2(u). \tag{6.51}$$

We then have

$$P_1'(u) = -(1-u)q(u) \tag{6.52}$$

and

$$\begin{aligned} V_+(u) &= \int_u^1 Q^2(p)dp - 2Q(u)\int_u^1 Q(p)dp + (1-u)Q^2(u) - P_1^2(u) \\ &= \int_u^1 Q^2(p)dp - 2Q(u)[P_1(u)+(1-u)Q(u)] + (1-u)Q^2(u) - P_1^2(u) \\ &= \int_u^1 Q^2(p)dp - [P_1(u)+Q(u)]^2 + uQ^2(u). \end{aligned}$$

Differentiating the above expression, we get

$$V_+'(u) = -2[P_1(u) + Q(u)][P_1'(u) + q(u)] + 2uQ(u)q(u).$$

Eliminating $Q(u)$ and $q(u)$ by using (6.50) and (6.52), we obtain

$$V_+'(u) = \frac{2uP_1(u)P_1'(u)}{1-u}. \tag{6.53}$$

Equation (6.53) shows that both $P_1(u)$ and $V_+(u)$ determine each other as

$$V_+(u) = -\int_u^1 \frac{2pP_1(p)P_1'(p)}{1-p}dp$$

and

$$P_1^2(u) = -\int_u^1 \frac{(1-p)}{p}V_+'(p)dp.$$

Thus, for all practical purposes, it is enough if the first partial moment (stop loss transform) is available. The relationships that the partial moments have with the reliability functions is immediate from the above discussions. We notice that

$$H(u) = -\frac{1}{P_1'(u)}, \tag{6.54}$$

$$M(u) = (1-u)^{-1}P_1(u),$$

$$V(u) = \frac{1}{(1-u)}\int_u^1 (1-p)^{-2}P_1(p)dp,$$

$$T(u) = \mu - P_1(u,) \tag{6.55}$$

and

$$(1-u)P_1(u) = 2l_2(u) - (1-u)l_2'(u).$$

The ageing properties of X can also be characterized in terms of $P_1(u)$. These can be expressed with the use of Theorems in Sect. 5.4 and (6.55). Some examples are

(i) X is IHR (DHR) if and only if $P_1(u)$ is convex (concave). This result follows from Theorem 5.2 and (6.54);
(ii) A necessary and sufficient condition that X is DMTTF (IMTTF) is that $\frac{\mu - P_1(u)}{\mu u}$ is decreasing (increasing), which simplifies to

$$P_1(u) - uP_1'(u) < \mu.$$

The other ageing properties result from Theorems 5.4–5.6.

Table 6.3 Stop-loss transforms for some distributions

Distribution	$P_1(u)$
Exponential	$\lambda^{-1}(1-u)$
Pareto II	$\frac{\alpha}{C-1}(1-u)^{-\frac{1}{c}+1}$
Rescaled beta	$\frac{C}{R+1}(1-u)^{\frac{1}{c}+1}$
Half logistic	$2\sigma \log \frac{2}{1+u}$
Exponential geometric	$\frac{1-p}{\lambda p} \log \frac{1-pu}{1-p}$
Power	$\frac{\alpha}{1+\beta}\{\beta-(1+\beta)u^{\frac{1}{\beta}}+u^{\frac{1}{\beta}+1}\}$
Linear hazard quantile	$\frac{1}{b}\log \frac{a+b}{a+bu}$
Generalized lambda	$\frac{1}{\lambda_2}\left\{\frac{\lambda_4}{1+\lambda_4}(1-u)^{\lambda_4+1}+\frac{(1-u^{\lambda_3+1})}{1+\lambda_3}-(1-u)u^{\lambda_3}\right\}$
Generalized Tukey lambda	$(1-u)\left\{\frac{(1-u)^{\lambda_4}}{\lambda_2(\lambda_4+1)}+\frac{1-u^{\lambda_3+1}}{\lambda_2\lambda_3(1+\lambda_3)(1-u)}-\frac{u^{\lambda_3}}{\lambda_2\lambda_3}\right\}$
van Staden and Loots	$\lambda_2\left[\frac{(1-\lambda_3)}{\lambda_4}\left\{\frac{1-u^{\lambda_4+1}}{\lambda_4+1}-(1-u)u^{\lambda_4}\right\}+\frac{\lambda_3}{\lambda_4+1}(1-u)^{\lambda_4+1}\right]$
Generalized Weibull	$\frac{\sigma^\alpha}{\lambda^\alpha}(1-u)B_{(1-u)^\lambda}(\frac{1}{\lambda}+1,\alpha)$
Power-Pareto	$c(1-u)\{\lambda_1 B_{1-u}(2-\lambda_2,\lambda_1)+\lambda_2 B_{1-u}(1-\lambda_2,\lambda_1)\}$
Govindarajulu	$\frac{\sigma}{\beta+2}\{2-(\beta+1)(\beta+2)u^\beta+2\beta(\beta+2)u^{\beta+1}-\beta(\beta+1)u^{\beta+2}\}$

The stop-loss transforms of several distributions are presented in Table 6.3.

The lower partial moments of order r in the case of a non-negative random variable is defined as

$$p_r^*(t) = E[(X-t)^-]^r,$$

where

$$(X-t)^- = \begin{cases} t-X, & X \le t \\ 0, & X \ge t \end{cases}.$$

The first two moments, in terms of quantile functions, become

$$\begin{aligned} P_1^*(u) = p_1^*(Q(u)) &= \int_0^u [Q(u)-Q(p)]dp \\ &= \int_0^u pq(p)dp, \end{aligned} \tag{6.56}$$

$$P_2^*(u) = p_2^*(Q(u) = \int_0^u [Q(u)-Q(p)]^2 dp. \tag{6.57}$$

From (6.56) and (6.57), the variance of $(X-t)^-$ is obtained as

$$v_-(u) = \int_0^u [Q(u)-Q(p)]^2 - [P_1^*(u)]^2 dp.$$

Using now the relations

$$\int_0^u Q(p)dp = P_1^*(u) - uQ(u),$$
$$\frac{dP_1^*(u)}{du} = uq(u),$$

we can eliminate $Q(u)$ and $q(u)$ from

$$v'_-(u) = 2q(u)P_1^*(u) - 2uQ(u)q(u) - 2uQ(u)q(u)$$

to arrive at the identity

$$v'_-(u) = \frac{2(1-u)}{u} P_1^*(u) \frac{dP_1^*(u)}{du}. \tag{6.58}$$

Thus, $P_1^*(u)$ determines $v_-(u)$ uniquely as

$$v_-(u) = \int_0^u \frac{2(1-p)}{p} P_1^*(p) \frac{dP_1^*}{dp} dp,$$

and conversely

$$[P_1^*(u)]^2 = \int_0^u \frac{pv'_-(p)}{1-p} dp.$$

From the reliability theory perspective, the partial mean is useful in defining the reversed quantile functions. The basic relationships are as follows:

$$\Lambda(u) = \frac{dP_1^*(u)}{du},$$
$$R(u) = u^{-1}P_1^*(u),$$
$$D(u) = \frac{1}{u}\int_0^u R^2(p)dp = \frac{1}{u}\int_0^u \left\{\frac{P^*(p)}{p}\right\}^2 dp.$$

6.7 Some Applications

The L-moments and the two kinds of partial moments discussed so far are known in some other disciplines than reliability for their applications. We now give a brief account of the important ones, partly because the models discussed have relevance in reliability theory as well. One major application is related to income analysis in economics. Let X denote the non-negative continuous random variable representing incomes of individuals in a population. Income is often conceived as an indicator to differentiate between the strata of the society, notably the poor and the affluent, with generally more attention to the former. A poverty line $X = t$ is set such that those having income below t is considered poor. Then, $\alpha = F(t)$ represents the proportion of poor in the population, and their income has the distribution

$$ {}_tF(x) = \begin{cases} \frac{F(x)}{F(t)}, & x \le t \\ 1, & x > t \end{cases}. $$

The extent to which poverty exists among the poor is measured by the income gap ratio defined as

$$ \begin{aligned} G(t) &= 1 - E\left[\frac{X}{t}|(X \le t)\right] \\ &= 1 - \frac{1}{t}E[X|(X \le t)] = 1 - \frac{B_1(t)}{t}. \end{aligned} \tag{6.59} $$

In terms of quantile functions, we have

$$ g(u) = G(Q(u)) = 1 - \frac{\theta_1(u)}{Q(u)}. \tag{6.60} $$

Traditionally, the income gap ratio is computed from the income distribution; but, the reverse process is also valid. Nair et al. [440] have shown that there exists a one-to-one relationship between income gap ratio and the income distribution and the latter can be retrieved from the former as explained in the following theorems. Empirically, it is possible to draw some ideas about the approximate form of $G(t)$ from the data.

Theorem 6.10. *If X has a finite mean and income gap ratio $G(t)$, then the distribution of X is*

$$ F(x) = \exp\left\{-\int_x^\infty \frac{1 - G(t) - tG'(t)}{tG(t)}dt\right\}, \quad x > 0. $$

Remark 6.6. Using the above theorem, it follows that the only continuous distribution for which $G(t) =$ a constant is the power distribution.

Remark 6.7. The analogue of Theorem 6.10 is

$$Q(u) = \frac{\mu}{u(1-g(u))} \exp\left\{-\int_u^1 \frac{dp}{p(1-g(p))}\right\}.$$

A popular measure for the income inequality in a population is the Gini index defined as

$$I = 1 - \frac{2}{\mu}\int_0^\infty x\bar{F}(x)f(x)dx.$$

In the case of the poor (below the poverty line or $X \leq t$), the index has the form

$$I(t) = 1 - \frac{2}{E[X|(X \leq t)]}\int_0^t x\left(1 - \frac{F(x)}{F(t)}\right)\frac{f(x)}{F(t)}dx. \tag{6.61}$$

Using the transformation $x = Q(u)$, we have the quantile version as

$$i(u) = I(Q(u)) = 1 - \frac{2}{\theta_1(u)}\int_0^u Q(p)\left(\frac{u-p}{u^2}\right)dp. \tag{6.62}$$

From (6.18) and (6.19), we then have

$$\int_0^u Q(p)dp = u\theta_1(u) \tag{6.63}$$

and

$$\int_0^u pQ(p)dp = \frac{u^2}{2}(\theta_1(u) + \theta_2(u)). \tag{6.64}$$

Eliminating the integral on the right-hand side of (6.62) with the use of (6.63) and (6.64), we obtain

$$i(u) = \frac{\theta_2(u)}{\theta_1(u)},$$

which is the L-coefficient of variation $\theta(u)$ considered earlier in Sect. 6.3. By virtue of Theorem 6.10, we conclude that $i(u) =$ a constant if and only if X has power distribution. Theorem 6.10 leaves scope for characterizing income distributions by the form of their truncated Gini index. A further example is that the form

$$i(u) = \frac{(\beta+3)-(\beta+1)u}{(\beta+2)-\beta u}$$

determines the Govindarajulu distribution.

The income gap ratio and truncated Gini index play a crucial role in defining index of poverty. For example, Sen [524] suggested the index

$$s(t) = F(t)[G(t) + (1 - G(t))I(t)]$$

for a measure of poverty. This turns out to be equivalent to

$$\begin{aligned} S(u) &= u[g(u) + (1 - g(u))i(u)] \\ &= u\left\{1 - \frac{\theta_1(u)}{\theta(u)} + \frac{\theta_2(u)}{\theta_1(u)}\left(\frac{\theta_1(u)}{Q(u)}\right)\right\}. \end{aligned}$$

Since $Q(u) = u\theta_1'(u) + \theta_1(u)$, we have on simplification

$$S(u) = u\left[\frac{u\theta_1'(u) + \theta_2(u)}{u\theta_1'(u) + \theta_1(u)}\right]. \tag{6.65}$$

Instead of distribution functions as models of income, Tarsitano [562] used the generalized lambda distribution and Haritha et al. [260] employed the generalized Tukey lambda distribution. Since both these distributions do not have closed-form expressions for their distribution functions, the expressions in (6.57), (6.62) and (6.65) become important.

Theorem 6.11. *Let X be a non-negative random variable with finite mean. Then, $S(u) = cu$ if and only if X has power distribution of the form*

$$Q(u) = \alpha u^{\frac{1}{\beta}}, \quad \alpha, \beta > 0, 0 \le u \le 1.$$

Proof. In the case of the power distribution, we have

$$\theta_1(u) = \frac{\alpha\beta}{\beta+1}u^{\frac{1}{\beta}},$$

$$\theta_2(u) = \frac{\alpha\beta}{(\beta+1)(1+2\beta)}u^{\frac{1}{\beta}},$$

and so from (6.65), we obtain

$$S(u) = \frac{1+3\beta}{(1+\beta)(1+2\beta)}u$$

which proves the 'if' part. Conversely, when $S(u) = cu$, (6.1) provides

$$c[u\theta_1'(u) + \theta_1(u)] = u\theta_1'(u) + u\theta_2(u)$$

or equivalently

$$cu^2Q(u) = u^2Q(u) - 2u\int_0^u Q(p)dp + 2\int_0^u pQ(p)dp$$

upon using the expressions of $\theta_1(u)$ and $\theta_2(u)$. Differentiating and simplifying the resulting expression, we get

$$(c-1)u^2Q''(u) + 4(c-1)uQ'(u) + 2cQ(u) = 0. \tag{6.66}$$

Now, by setting $Q(u) = u^m$, the auxiliary equation for the solution of (6.66) is

$$(c-1)m^2 + 3(c-1)m + 2c = 0,$$

which has its roots as

$$m = -\frac{3(c-1) \pm \sqrt{a(c-1)^2 - 8c(c-1)}}{2(c-1)},$$

that simplify to

$$m = -\frac{3}{2} \pm \frac{1}{2}\sqrt{\frac{c-9}{c-1}}.$$

Hence, the solution of (6.66) becomes

$$Q(u) = C_1 u^{-\frac{3}{2}+\frac{1}{2}(\frac{c-9}{c-1})^{\frac{1}{2}}} + C_2 u^{-\frac{3}{2}-\frac{1}{2}(\frac{c-9}{c-1})^{\frac{1}{2}}}.$$

Since $Q(u)$ has to be increasing for all u, $C_2 = 0$ and so

$$Q(u) = \alpha u^{\frac{1}{\beta}},$$

which corresponds to the power distribution with $\beta = \frac{1}{2}[(\frac{c-9}{c-1})^{\frac{1}{2}} - 3]$ and $\alpha = C_1$. This completes the proof of the theorem.

The lower partial moments have an important role in the measurement of risk associated with management, industrial and insurance strategies. Sunoj and Maya [555] discussed their role in stochastic modelling that includes characterization of distributions, weighted and equilibrium models. In $p_r^*(t)$ defined earlier, t is a target, that separates gains and losses and the main interest is in 'downside risk' measured by $p_1^*(t)$. Portfolio theory is concerned about maximizing the return for a given risk, where X stands for the random return and t the target return. In this context, lower partial moments provides summary measures of downside risk. The second moment $p_2^*(t)$ is called target semi-variance which fits investors' risk preference better than the traditional variance. Some references in this connection are Bawa [81], Fishburn [199], Harlow [261], Brogan and Stidham [121], Willmot et al. [582] and Hesselager et al. [270].

Chapter 7
Nonmonotone Hazard Quantile Functions

Abstract The existence of nonmonotonic hazard rates was recognized from the study of human mortality three centuries ago. Among such hazard rates, ones with bathtub or upside-down bathtub shape have received considerable attention during the last five decades. Several models have been suggested to represent lifetimes possessing bathtub-shaped hazard rates. In this chapter, we review the existing results and also discuss some new models based on quantile functions. We discuss separately bathtub-shaped distributions with two parameters, three parameters, and then more flexible families. Among the two-parameter models, the Topp-Leone distribution, exponential power, lognormal, inverse Gaussian, Birnbaum and Saunders distributions, Dhillon's model, beta, Haupt-Schäbe models, loglogistic, Avinadev and Raz model, inverse Weibull, Chen's model and a flexible Weibull extension are presented along with their quantile functions. The quadratic failure rate distribution, truncated normal, cubic exponential family, Hjorth model, generalized Weibull model of Mudholkar and Kollia, exponentiated Weibull, Marshall-Olkin family, generalized exponential, modified Weibull extension, modified Weibull, generalized power Weibull, logistic exponential, generalized linear failure rate distribution, generalized exponential power, upper truncated Weibull, geometric-exponential, Weibull-Poisson and transformed model are some of the distributions considered under three-parameter versions. Distributions with more than three parameters introduced by Murthy et al., Jiang et al., Xie and Lai, Phani, Agarwal and Kalla, Kalla, Gupta and Lvin, and Carrasco et al. are presented as more flexible families. We also introduce general methods that enable the construction of distributions with nonmonotone hazard functions. In the case of many of the models so far specified, the hazard quantile functions and their analysis are also presented to facilitate a quantile-based study. Finally, the properties of total time on test transforms and Parzen's score function are utilized to develop some new methods of deriving quantile functions that have bathtub hazard quantile functions.

N.U. Nair et al., *Quantile-Based Reliability Analysis*, Statistics for Industry and Technology, DOI 10.1007/978-0-8176-8361-0_7,

7.1 Introduction

The recognition of the existence of nonmonotonic hazard rates dates back to three centuries in the study of human mortality when researchers found that the force of mortality (alternative name for hazard rate) first decreases, then remains more or less constant and then increases. Since then, the problem of modelling such curves through different distributions has been taken up in many disciplines such as reliability, survival analysis, demography and actuarial science. Among nonmonotonic hazard rates, those with bathtub shape or upside-down bathtub shape have received much attention during the last 5 decades. There is an extensive literature on finding appropriate models for representing them and also on methods of analysing their behaviour, in several practical problems. Earlier in Sect. 4.3, we have introduced the notions of bathtub (BT) and upside-down bathtub (UBT) hazard rates and the corresponding hazard quantile functions. Recall that a random variable X with differentiable $h(x)(H(u))$ possesses a BT hazard rate (hazard quantile function) if and only if $h'(x)(H'(u)) < 0$ for $x(u)$ in $(0,x_0)((0,u_0))$, $h'(x_0) = 0$ $(H'(u_0) = 0)$ and $h'(x)(H'(u)) > 0$ for $x(u)$ in $(x_0,\infty)((u_0,1))$. In the UBT case, $H'(u) > 0$ for u in $(0,u_0)$, $H'(u_0) = 0$ and $H'(u_0) < u_0$ in $(u_0,1)$. Thus, BT distributions are characterized by a hazard rate (hazard quantile function) that is first decreasing and then increasing with a unique change point. A more general definition that considers $H(u)$ as a constant in an interval (see Definition 4.2) is also available, but this extended definition will not be considered in the sequel. The three phases of a BT hazard rate represent an 'infant mortality' period in which $H(u)$ decreases, a 'useful period' in which $H(u)$ is approximately constant, and a 'wear out' stage in which the hazard function increases leading to the ultimate failure of a unit. To avoid infant mortality in large proportions, 'burn-in' procedures are often employed to enhance the reliability of products. On the other hand, replacement policies aim at improving the reliability of units by eliminating those with short lives before the wear out process is at an advanced stage. We make use of the sign of the derivative of $H(u)$ $(h(x))$ to ascertain the nonmonotonicity. In case when the survival function is not tractable, giving complicated expressions for $H(u)$ or $h(x)$, Theorems 4.1 and 4.2 will be employed, as demonstrated in Example 4.4. Several methods of construction of models with BT or UBT have been proposed in the literature. However, in the following discussion, we will distinguish the models by the number of parameters they involve. Quantile functions and corresponding hazard quantile functions are presented whenever the proposed distributions have such functions in tractable forms. A review of bathtub-shaped distributions is given in Rajarshi and Rajarshi [500] and Lai et al. [369],

7.2 Two-Parameter BT and UBT Hazard Functions

The Topp-Leone [566] distribution with density function

$$f(x) = \frac{2\alpha}{\theta}\left(\frac{x}{\theta}\right)^{\alpha-1}\left(1-\frac{x}{\theta}\right)\left(2-\frac{x}{\theta}\right)^{\alpha-1}, \quad 0 \le x \le \theta,\, 0 < \alpha < 1,$$

has its survival function as

$$\overline{F}(x) = 1 - \left(\frac{x}{\theta}\right)^{\alpha}\left(2-\frac{x}{\theta}\right)^{\alpha}, \quad 0 \le x \le \theta. \tag{7.1}$$

Thus, the hazard rate turns out to be

$$h(x) = \frac{2\alpha}{\theta}\frac{(\frac{x}{\theta})^{\alpha-1}(1-\frac{x}{\theta})(2-\frac{x}{\theta})^{\alpha-1}}{1-(\frac{x}{\theta})^{\alpha}(2-\frac{x}{\theta})^{\alpha}}.$$

By differentiating $h(x)$, it can be seen that $h(x)$ has a bathtub shape with change point x_0 for every α, where x_0 satisfies the equation

$$\left(\frac{x_0}{\theta}\right)^{\alpha} + \frac{2\alpha(\theta - x_0)}{2\theta - x_0} - 1 = 0.$$

The distribution in (7.1) admits a convenient quantile function. Applying the transformation $Y = 1 - \frac{X}{\theta}$ to (7.1), we have

$$F_Y(x) = (1-x^2)^{\alpha}$$

and so

$$Q_Y(u) = (1-u^{\frac{1}{\alpha}})^{\frac{1}{2}}.$$

Retransforming this expression to X, we readily obtain

$$\begin{aligned} Q_X(u) &= \theta - \theta Q_Y(u) \\ &= \theta\left\{1-(1-u^{\frac{1}{\alpha}})^{\frac{1}{2}}\right\}. \end{aligned}$$

The corresponding hazard quantile function is

$$H(u) = \frac{1}{(1-u)q(u)} = \frac{2\alpha}{\theta}\frac{(1-u^{\frac{1}{\alpha}})^{\frac{1}{2}}}{u^{\frac{1}{\alpha}-1}(1-u)}.$$

Smith and Bain [544] introduced the exponential power model with survival function

$$\overline{F}(x) = \exp[-e^{(\lambda x)^{\alpha}}+1], \quad 0 < x < \infty. \tag{7.2}$$

Notice that the hazard rate is

$$h(x) = \lambda^{\alpha}\alpha x^{\alpha-1}e^{(\lambda x)^{\alpha}}, \tag{7.3}$$

which is strictly convex in $(0,\infty)$ satisfying $\int_0^\infty h(x)dx = \infty$. Sometimes, the choice of such a function is adopted as a method of deriving a BT distribution. A feature of the function (7.3) is that $h(x) \to \infty$ when $x \to 0$ or ∞. The hazard function is BT for $\alpha < 1$ with a change point at $x_0 = \frac{(1-\alpha)}{(\lambda\alpha)^{\frac{1}{\alpha}}}$. For further detailed study of the distribution including the estimation of parameters and applications to other disciplines, one may refer to Dhillon [175], Paranjpe and Rajarshi [481], Leemis [379] and Chen [141]. A closed-form expression is available for the quantile function of (7.2) as

$$Q(u) = \frac{1}{\lambda}[\log\{1-\log(1-u)\}]^{\frac{1}{\alpha}},$$

which can be used to simulate observations from the distribution from the uniform (0,1) random numbers. Observing that the quantile density function is

$$q(u) = \frac{[\log(1-\log(1-u))]^{\frac{1}{\alpha}-1}}{\lambda\alpha(1-u)(1-\log(1-u))},$$

it becomes clear that

$$H(u) = \frac{\lambda\alpha(1-\log(1-u))}{[\log(1-\log(1-u))]^{\frac{1}{\alpha}-1}}.$$

Two standard distributions possessing nonmonotone hazard rates that were considered reliability analysis are the lognormal and the inverse Gaussian. The lognormal distribution has its density function as

$$f(x) = \frac{1}{x\sqrt{2\pi}\sigma}\exp\left[-\frac{(\log x-\mu)^2}{2\sigma^2}\right], \quad x \geq 0,\ -\infty < \mu < \infty,\ \sigma > 0, \qquad (7.4)$$

and survival function as

$$\bar{F}(x) = 1-\Phi\left(\frac{\log x - x}{\sigma}\right) = 1-\Phi[\log(\alpha x)^{\frac{1}{\sigma}}], \quad \text{with } \alpha = e^{-\mu},$$

where Φ is the distribution function of the standard normal distribution. The hazard rate is

$$h(x) = \frac{1}{\sqrt{2\pi}\sigma x}\frac{\exp[-(\log\alpha x)^2/2\sigma^2]}{1-\Phi(\log(\alpha x)/\sigma)}.$$

A detailed study of the hazard rate has been carried out by Sweet [556]. The book by Crow and Shimizu [159] details all methods and applications of lognormal distribution. Moreover, from Marshall and Olkin [412] and Johnson et al. [303], we note the following properties:

1. $h(x) = \sigma^{-1}h_N(t)\exp[-\sigma t - \mu]$, where $t = \sigma^{-1}(\log x - \mu)$ and $h_N(x)$ is the hazard rate of the normal distribution;
2. For all real θ,

$$\lim_{x\to 0} x^{\theta} h(x) = 0, \quad \lim_{x\to\infty} h(x) = 0;$$

3. $h(x)$ is unimodal with mode at $\exp(\sigma x^* + \mu)$, where x^* is the unique solution of the equation $h_N(x) = x + \sigma$. This solution is less than $\exp[1 + \mu - \sigma^2]$, but greater than $\exp[\mu - \sigma^2]$. As $\sigma \to \infty$, $x^* \to \exp[\mu - \sigma^2]$ and so for large σ, we have

$$\max h(x) \doteq \frac{\exp(\mu - \frac{\sigma^2}{2})}{\sigma\sqrt{2\pi}};$$

as $\sigma \to 0$, $x^* \to \exp[\mu - \sigma^2 + 1]$ and so for small σ, we have

$$\max h(x) \doteq \{\sigma^2 \exp(\mu - \sigma^2 + 1)\}^{-1}.$$

The quantile function corresponding to (7.4) is

$$Q(u) = \exp[\mu + \sigma\Phi^{-1}(u)]$$

and so $H(u)$ does not have a nice algebraic form for manipulations.

Inverse Gaussian distribution, discussed in detail by Chhikara and Folks [146] and Seshadri [528] as a lifetime model, has its density function as

$$f(x) = \frac{\theta\mu}{(2\pi^3 x^3)^{\frac{1}{2}}} \exp\left\{-\frac{(\theta x - \mu)^2}{2\theta x}\right\}, \quad x, \theta, \mu > 0. \tag{7.5}$$

Its survival function is

$$\overline{F}(x) = \frac{1}{2}\left[\overline{G}\left(\frac{(\theta x - \mu)^2}{\theta x}\right) - e^{2\mu}\overline{G}\left(\frac{(\theta x + \mu)^2}{\theta x}\right)\right],$$

where

$$\overline{G}(y) = \int_y^{\infty} (2\pi x)^{-\frac{1}{2}} e^{-\frac{x}{2}} dx$$

is the survival function of a chi-square variable with one degree of freedom. Needless to say, the hazard rate function is of a complicated form to study its behaviour explicitly. The hazard rate is UBT with change point x_0 that is the solution of the equation

$$h(x) = \frac{3}{2x} + \frac{\theta}{2} - \frac{\mu^2}{2\theta x^2}.$$

For various applications in reliability and lifetime data analysis, we refer the readers to Padgett and Tsai [479] and Bhattacharya and Fries [99], and similarly to Hougaard [283] in survival analysis and Feaganes and Suchindran [195] as a distribution of frailty.

A distribution that is related to the inverse Gaussian, but derived independently as a lifetime model based on shocks that arrive at regular intervals of time causing random damages, was derived by Birnbaum and Saunders [105, 106]. It models fatigue life of metals subject to periodic stress. The distribution has density function

$$f(x) = \frac{\lambda}{2\alpha\sqrt{2\pi}} \frac{1}{\sqrt{\lambda x}} \left(1 + \frac{1}{\lambda x}\right) \exp\left\{-\frac{1}{2\alpha^2}\left(\lambda x - 2 + \frac{1}{\lambda x}\right)\right\}. \tag{7.6}$$

Desmond [174] pointed out that (7.6) can be written as a mixture in equal proportions of an inverse Gaussian and a reciprocal inverse Gaussian. The distribution function is given by

$$F(x) = \Phi(\alpha^{-1} g(\lambda x)), \tag{7.7}$$

where $\lambda, \alpha > 0$ and Φ is the standard normal distribution function and

$$g(x) = x^{\frac{1}{2}} - x^{-\frac{1}{2}}.$$

Note the resemblance between (7.7) and the distribution of the lognormal law in which case $g(x) = \log x$. Various properties and inferential procedures of the distribution have been discussed by Chang and Tang [136,137], Johnson et al. [302], Dupuis and Mills [182], Rieck [507], Ng et al. [470, 471], Owen [478], Leomonte et al. [384], Balakrishnan and Zhu [62] and Xie and Wei [591]. Recently, Kundu et al. [359] expressed the hazard rate function of the Birnbaum-Saunders distribution in (7.6) as

$$h(x, \alpha) = \frac{\frac{1}{\sqrt{2\pi}} g'(x) \exp\left\{-\frac{1}{2\alpha^2} g^2(x)\right\}}{\Phi(e^{-\frac{g(x)}{\alpha}})}$$

by taking $\lambda = 1$, without loss of generality, since λ is a scale parameter. They then showed that $h(x)$ is UBT for all $x > 0$ and for all α and λ. The change point x_0 is the solution of the equation

$$\Phi\left(-\frac{1}{\alpha}g(x)\right)\{-(g'(x))^2 g(x) + \alpha^2 g''(x)\} + \alpha\Phi\left(-\frac{1}{\alpha}g(x)\right)(g'(x))^2 = 0,$$

which has to be solved by numerical methods. An approximation has been given as

$$x_0 = (-0.4604 + 1.8417\alpha)^{-2}, \quad \alpha > 0.25;$$

see also Bebbington et al. [84]. A comparison of the hazard rates of (7.7) and the lognormal has been made by Nelson [469].

Some useful generalizations of the Birnbaum-Saunders distribution have been developed in order to provide more flexible models in terms of the range of skewness as well as varying shapes of the hazard function. For example, with the choice of the function $g(x) = x^{\frac{1}{2}} - x^{-\frac{1}{2}}$, instead of basing the distribution in (7.7) on a normal distribution, one could base it on general family of elliptically contoured distributions or scale-mixture distributions; see, for example, Diaz-Garcia and Leiva [176], Leiva et al. [386] and Balakrishnan et al. [54]. Properties of such models and their reliability characteristics have also been studied; for instance, Azevedo et al. [42] recently discussed the shape and change points of the hazard function of the BS-t (Birnbaum-Saunders model based on t-distribution) model.

Dhillon [175] introduced a two-parameter survival function

$$\bar{F}(x) = \exp\left[-\{\log(\lambda x + 1)\}^{\beta+1}\right], \quad x \geq 0, \beta \geq 0, \lambda > 0, \tag{7.8}$$

and density function

$$f(x) = \frac{\lambda(\beta+1)}{\lambda x + 1}\{\log(\lambda x + 1)\}^{\beta} \exp\left[-\{\log(\lambda x + 1)\}^{\beta+1}\right].$$

The corresponding hazard rate is

$$h(x) = \frac{(\beta+1)\lambda\{\log(\lambda x + 1)\}^{\beta}}{\lambda x + 1}.$$

It can be seen that $h(x)$ is UBT with change point $x_0 = \lambda^{-1}(e^{\beta} - 1)$. We see that (7.8) is also expressible as

$$Q(u) = \frac{1}{\lambda}[e^{-\{\log(1-u)\}^{\frac{1}{\beta+1}}} - 1].$$

Its hazard quantile function is

$$H(u) = \frac{(\beta+1)\lambda \exp\left[\{\log(1-u)\}^{\frac{1}{\beta+1}}\right]}{\{\log(1-u)\}^{\frac{1}{\beta+1}-1}},$$

which becomes UBT with change point

$$u_0 = 1 - e^{-\beta(\beta+1)}.$$

Mukherjee and Islam [430] and Lai and Mukherjee [367] considered the power distribution with

$$F(x) = \left(\frac{x}{\alpha}\right)^{\beta}, 0 \le x \le \alpha, \beta < 1, \tag{7.9}$$

and hazard rate

$$h(x) = \frac{\beta x^{\beta-1}}{\alpha^{\beta} - x^{\beta}}$$

which has a BT shape with change point $x_0 = \alpha(1-\beta)^{\frac{1}{\beta}}$. The quantile function of this distributions and its properties has been discussed several times in the preceding chapters. The distribution in (7.9) forms a special case of the beta distribution with density function

$$f(x) = \frac{1}{B(p,q)} x^{p-1}(1-x)^{q-1}, \quad 0 \le x \le 1, \; p,q > 0, \tag{7.10}$$

when $q = 1$ and then rescaled to the interval $(0,\theta)$. Pham and Turkkan [494] have considered standby systems with component lives distributed as beta and Ganter [209] used it in the context of accelerated test of electronic assemblies. However, a detailed analysis of the hazard rate and mean residual life has been carried out much later by Gupta and Gupta [232] and Ghitany [212]. The hazard rate of the beta model is

$$h(x) = \frac{x^{p-1}(1-x)}{B(p,q) - B_x(p,q)},$$

where

$$B_x(p,q) = \int_0^x t^{p-1}(1-t)^{q-1} dt$$

is the incomplete beta integral. Ghitany [212] has shown that Glaser's result mentioned earlier in Sect. 4.3 is valid only when the upper end of the support is ∞ and $f(\infty) = 0$, and that it fails to determine the shape of the hazard rate when the support of a distribution is $(0,b)$ with $b < \infty$. He then modified Glaser's result as follows.

Theorem 7.1. *Let X be a continuous random variable on $(0,b)$, $b < \infty$, with twice differentiable density $f(x)$. Define $\eta(x) = -\frac{f'(x)}{f(x)}$. Then:*

(a) If $\eta(x)$ is decreasing and $f(b) = 0$, then $h(x)$ is decreasing;
(b) If $\eta(x)$ is increasing, then $h(x)$ is increasing;
(c) If $\eta(x)$ is BT and $f(0) = 0$ ($f(0) = \infty$), then $h(x)$ is increasing (BT);

(d) *If* $\eta(x)$ *is UBT,* $f(0) = 0$ $(f(0) = \infty)$*, and* $f(b) = 0$*, then* $h(x)$ *is UBT (decreasing);*

(e) *If* $\eta(x) \leq 0$ *and* $f(b) > 0$*, then* $h(x)$ *is decreasing;*

(f) *If* $\eta(x)$ *is decreasing and* $f(0) = f(b) = \infty$*, then* $h(x)$ *is BT.*

In Theorem 7.1, the monotonicities involved are strict. Using the above results, it has been shown that the hazard rate of the beta distribution is BT (increasing) if $p < 1$ $(p \geq 1)$. Also, the mean residual life is UBT (decreasing) if $p < 1$ $(p \geq 1)$. Notice that the adaptation of Glaser's result in Theorem 4.1 also requires corresponding changes in the cases discussed in Theorem 7.1. For more details on beta distribution and its applications, one may refer to the volume by Gupta and Nadarajah [230].

Haupt and Schäbe [265] proposed the distribution with

$$F(x) = \begin{cases} 1, & x \geq x_0 \\ -\beta + \sqrt{\beta^2 + \frac{(1+2\beta)x}{x_0}}, & 0 \leq x \leq x_0 \end{cases}. \tag{7.11}$$

In this model, β is a shape parameter and it varies over $(-\frac{1}{2}, \infty)$ and x_0 is a scale parameter. The corresponding hazard rate is

$$h(x) = \frac{1+2\beta}{2x_0\left(\beta^2 + \frac{(1+2\beta)x}{x_0}\right)^{\frac{1}{2}}\left\{1+\beta - \left(\beta^2 + \frac{(1+2\beta)x}{x_0}\right)^{\frac{1}{2}}\right\}}$$

which is BT for $\frac{1}{3} < \beta < 1$ and decreasing for $\beta \geq 1$ and $\beta \leq -\frac{1}{3}$. Construction of lifetime distributions with bathtub-shaped hazard rates from DHR distributions was proposed by Schäbe [522]. For $0 < \theta < \infty$, let us define

$$G(x) = \frac{F(x)}{F(\theta)}, \quad x \leq \theta.$$

Then, $G(x)$ has BT hazard rate if

$$h'(x)[\overline{F}(x) - \overline{F}(\theta)] + h^2(x)\overline{F}(\theta)$$

has one and only one zero in the interval $(0, \theta)$ and changes its sign from $-$ to $+$. An illustration of this result has been given with the Pareto II distribution.

Paranjpe and Rajarshi [481] suggested the survival function

$$\overline{F}(x) = \exp[-\exp\{\exp(\beta x^\alpha) - 1\}], \quad \beta > 0, \ \alpha < 1, \tag{7.12}$$

to model BT hazard rates. The hazard rate function has the form

$$h(x) = \beta\alpha x^{\alpha-1}\exp(\beta x^\alpha)\exp\{\exp(\beta x^\alpha) - 1\}.$$

The quantile function of (7.12) becomes

$$Q(u) = \frac{1}{\beta}[\log\{1+\log(-\log(1-u))\}]^{\frac{1}{\alpha}}$$

with hazard quantile function of the form

$$H(u) = -\frac{\beta\alpha\log(1-u)\{1+\log(-\log(1-u))\}}{[\log\{1+\log(-\log(1-u))\}]^{\frac{1}{\alpha}-1}}.$$

Another distribution of interest proposed by Lai et al. [366] has hazard rate

$$h(x) = x^{a-1}(1-x)^{b-1}\{a-(a+b)x\}, \quad 0 < x < 1, a > 0, b < 1. \tag{7.13}$$

In both (7.12) and (7.13), $h(x)$ tends to infinity at both end points of the support thus supporting the BT shape.

A method of constructing BT-shaped hazard rates is given in Haupt and Schäbe [266]. Let $G(u)$ be a twice differentiable function satisfying the following conditions:

(a) $G(0) = 0$, $G(1) = 1$, $0 \le G(u) \le 1$;
(b) the solution $F(x)$ of the differential equation

$$\frac{\theta G(F(x))dF(x)}{\bar{F}(x)} = dx, \quad \theta = T(1) > 0,$$

where $T(u)$ is the TTT;
(c) the scaled TTT $\phi(u)$ (see Chap. 5 for pertinent details) has one inflexion point u_0 such that $0 < u_0 < 1$ and $\phi(u)$ is convex on $[0, u_0]$ and concave on $[u_0, 1]$.

They then illustrated this method for

$$G(u) = -\frac{1}{3}\alpha u^3 + \frac{1}{2}(\alpha - \alpha\beta)u^2 + \alpha\beta u$$

to arrive at the model in (7.11), discussed earlier by Haupt and Schäbe [266]. It appears that (a) and (b) are redundant, since (c) alone can produce a BT curve (see Theorem 5.2).

The loglogistic distribution with density function

$$f(x) = \frac{\alpha\rho^{\alpha}x^{\alpha-1}}{(1+\rho^{\alpha}x^{\alpha})^2} \quad x > 0,\ \alpha,\rho > 0,$$

has its survival function as

$$S(x) = (1+\rho^{\alpha}x^{\alpha})^{-1}, \tag{7.14}$$

and thus the hazard rate as

$$h(x) = \frac{\alpha\rho^{\alpha}x^{\alpha-1}}{1+\rho^{\alpha}x^{\alpha}}$$

which is UBT for $\alpha > 1$ with change point $x_0 = \rho^{-1}(\alpha-1)^{\frac{1}{\alpha}}$. It is easy to convert (7.14) into a quantile function in the simple form

$$Q(u) = \frac{1}{\rho}\left(\frac{u}{1-u}\right)^{\frac{1}{\alpha}}$$

giving the hazard quantile function

$$H(u) = \frac{\rho\alpha(1-u)^{\frac{1}{\alpha}}}{u^{\frac{1}{\alpha}-1}}.$$

A direct differentiation of $H(u)$ shows that it is UBT with change point at $u_0 = \frac{\alpha-1}{\alpha}$. For a detailed discussion of the model in reliability analysis, see Bennet [88] and Gupta et al. [237]. One may also refer to Balakrishnan et al. [55] and Balakrishnan and Saleh [58] for some inferential methods for this model based on censored lifetime data.

Employing what is referred to as the logWeibull time displacement transformation,

$$y = \log(1+\rho x),$$

to the Weibull survival function $\overline{G}(y) = \exp(-y^{\alpha})$, Avinadav and Raz [41] obtained the distribution with survival function

$$\overline{F}(x) = \exp[-\{\log(1+\rho x)\}^{\alpha}]. \tag{7.15}$$

The corresponding density function is

$$f(x) = \frac{\alpha\rho}{(1+\rho x)}\{\log(1+\rho x)\}^{\alpha-1}\exp[-\{\log(1+\rho x)\}^{\alpha}],$$

and so

$$h(x) = \frac{\alpha\rho\{\log(1+\rho x)\}^{\alpha-1}}{1+\rho x}.$$

For $\alpha > 1$, $h(x)$ has upside-down bathtub shape with maximum value at $x_0 = \frac{e^{\alpha-1}-1}{\rho}$. A quantile analysis of the distribution can be made with

$$Q(u) = \frac{1}{\rho}\left[\exp\left[\{-\log(1-u)\}^{\frac{1}{\alpha}}\right]-1\right]$$

and

$$H(u) = \frac{\rho\alpha}{e^{\{-\log(1-u)\}^{\frac{1}{\alpha}}}\{-\log(1-u)\}^{\frac{1}{\alpha}-1}}.$$

An interesting feature of the distribution is that it is closer to the loglogistic distribution until a certain point of time and then becomes closer to the Weibull law.

Remark 7.1. Upon comparing the survival function in (7.15) with that of the Dhillon model in (7.8), we immediately observe that the above distribution is identical to the two-parameter Dhillon model in (7.8) with $\lambda = \rho$ and $\beta = \alpha - 1$.

Applying transformation $X = \frac{\beta^2}{Y}$ when Y is a two-parameter Weibull distribution with survival function

$$\overline{G}(y) = \exp\left\{-\left(\frac{y}{\beta}\right)^{\alpha}\right\}, \quad y > 0; \alpha > 0, \beta > 0,$$

we obtain the inverse Weibull law with distribution function

$$F(x) = \exp\left\{-\left(\frac{\beta}{x}\right)^{\alpha}\right\}, \quad \alpha, \beta > 0; x > 0, \tag{7.16}$$

and density function

$$f(x) = \alpha\beta^{\alpha}x^{-\alpha-1}\exp\left\{-\left(\frac{\beta}{x}\right)^{\alpha}\right\}.$$

Applications of (7.16) in lifetime modelling has been discussed by Erto [188] and Jiang et al. [297]. In this case, we have

$$h(x) = \frac{\alpha\beta^{\alpha}x^{-\alpha-1}\exp[-(\frac{\beta}{x})^{\alpha}]}{1+\exp[-(\frac{\beta}{x})^{\alpha}]}$$

which is UBT shaped with change point x_0 as the solution of the equation

$$\frac{(\frac{\beta}{x})^{\alpha}}{1-e^{-(\frac{\beta}{x})^{\alpha}}} = \frac{\alpha+1}{\alpha}.$$

The quantile function has a simple form

$$Q(u) = \beta(-\log u)^{-\frac{1}{\alpha}}$$

and

$$H(u) = \frac{\alpha u(-\log u)^{\frac{1}{\alpha}+1}}{\beta(1-u)}.$$

Differentiation of $H(u)$ yields the change point as u_0 satisfying the equation

$$\log u_0 + \frac{\alpha+1}{\alpha}(1-u_0) = 0.$$

Cooray [156] and de Gusmao et al. [168] have discussed a generalization of the inverse Weibull distribution.

Chen [142] modified the exponential power distribution in (7.2) by setting $\lambda = 1$ and introducing a new parameter by taking the survival function as

$$\overline{F}(x) = \exp[-\lambda(e^{x^\alpha} - 1)]. \tag{7.17}$$

The hazard rate function has the modified form

$$h(x) = \lambda \alpha x^{\alpha-1} e^{x^\alpha}.$$

Since the parameter λ does not alter the monotonic behaviour of $h(x)$, we have its shape identified to that of (7.3). However, the form (7.17) becomes amenable to developing a three-parameter model as discussed later (see also Tang et al. [560]). Bebbington et al. [83] proposed a flexible Weibull extension by the model

$$\overline{F}(x) = \exp\left[e^{-\alpha x} - \frac{\beta}{x}\right], \quad x > 0; \alpha, \beta > 0, \tag{7.18}$$

with its hazard rate given by

$$h(x) = \left(\alpha + \frac{\beta}{x^2}\right) \exp\left(\alpha x - \frac{\beta}{x}\right).$$

In this case, $\lim_{x \to 0} h(x) = 0$ and so a pure bathtub curve is not envisaged. When $\alpha\beta < \frac{27}{64}$, the hazard rate is strictly increasing in $(0, x_0)$, strictly decreasing in (x_0, x_1), and strictly increasing on (x_1, ∞), where

$$x_0 = \frac{1}{2}\left[-\frac{4\beta}{3\alpha} + A + B\right]^{\frac{1}{2}} - \frac{1}{2}\left[\frac{8\beta}{3\alpha} - A - B + \frac{4\beta}{\alpha^2(-\frac{4\beta}{3\alpha} + A + B)^{\frac{1}{2}}}\right]^{\frac{1}{2}},$$

$$x_1 = \frac{1}{2}\left[-\frac{4\beta}{3\alpha} + A + B\right]^{\frac{1}{2}} + \frac{1}{2}\left[\frac{8\beta}{3\alpha} - A - B + \frac{4\beta}{\alpha^2(-\frac{4\beta}{3\alpha} + A + B)^{\frac{1}{2}}}\right]^{\frac{1}{2}},$$

with

$$A = \frac{2^{\frac{11}{3}}\beta^2}{3[27\alpha^2\beta^2 - 32\alpha^3\beta^3 + 3\sqrt{3}(27\alpha^4\beta^4 - 64\alpha^5\beta^5)^{\frac{1}{2}}]^{\frac{1}{3}}}$$

and

$$B = 2^{\frac{1}{3}}[27\alpha^2\beta^2 - 32\alpha^3\beta^3 + 3\sqrt{3}(27\alpha^4\beta^4 - 64\alpha^5\beta^5)^{\frac{1}{2}}]^{\frac{1}{3}}.$$

7.3 Three-Parameter BT and UBT Models

A majority of models described in the context of nonmonotonic hazard functions contain three parameters, some of them being extensions of two-parameter versions discussed in the preceding section. Some others are postulated in terms of hazard rates, rather than distribution functions. The quadratic hazard rate

$$h(x) = a + bx + cx^2, \quad a \geq 0, -2(ac)^{\frac{1}{2}} \leq b < 0, c > 0,$$

generating the survival function

$$\bar{F}(x) = \exp\left[-\left(ax + \frac{bx^2}{2} + \frac{cx^3}{3}\right)\right], \quad x > 0, \tag{7.19}$$

is one such model discussed at some length in Bain [45, 46] and Gore et al. [223]. The parameters of the model are estimated by the method of maximum likelihood or by regression of the empirical hazard rate on a quadratic polynomial. Hazard rates of the form

$$h(x) = \exp(a_0 + a_1 x + a_2 x^2)$$

were studied by Lewis and Shedler [385, 387] using simulations of homogeneous Poisson process. The truncated normal distribution as a failure time model, with only one failure mechanism, has been studied by Bosch [118]. Generalizing this, Glaser [220], Cobb [152] and Cobb et al. [153] have studied the distribution with density function

$$f(x) = C\exp[-\alpha x - \beta x^2 + r\log x], \quad x > 0,$$

with α real, $\beta > 0$, $\gamma > -1$ or $\alpha > 0$, $\beta = 0$ and $\gamma > -1$ (giving also extended gamma densities) which gives a BT hazard rate for $\gamma < 0$. They also discussed the cubic exponential family with density function

$$f(x) = C\exp[-\alpha x - \beta x^2 - \gamma x^3]$$

with $C < \alpha$ resulting in BT hazard rate.

The lifetime model introduced by Hjorth [272] is an interesting one as it has some physical interpretations. Relying upon the practical interest in mechanical units that are subject to wear, a distribution with minimal number of parameters and with enough flexibility lead to the study of a distribution with survival function

$$\overline{F}(x) = \frac{\exp\left(-\alpha\frac{x^2}{2}\right)}{(1+\beta x)^{\frac{\theta}{\beta}}} \quad x \geq 0,\ \alpha,\beta,\theta > 0,\ \alpha+\theta > 0, \tag{7.20}$$

and density function

$$f(x) = \frac{(1+\beta x)\alpha x + \theta}{(1+\beta x)^{\frac{\theta}{\beta}+1}} e^{-\frac{\alpha x^2}{2}}.$$

So, the hazard rate is given by

$$h(x) = \alpha x + \frac{\theta}{1+\beta x}.$$

As special cases, we have the Rayleigh distribution ($\theta = 0$), exponential ($\alpha = \beta = 0$), decreasing hazard ($\alpha = 0$), increasing hazard ($\alpha \geq \theta\beta$), and the bathtub curve ($0 < \alpha < \theta\beta$). Hjorth [272] has given two physical interpretations for the model in (7.20). Assuming that every produced or maintained unit has linear hazard rate

$$h^*(x) = u + \alpha x,$$

where α is the same for all units, but u is the realization from the gamma distribution with density

$$g(u) = \frac{u^{a-1}e^{-\frac{u}{\beta}}}{\beta^a \Gamma a},$$

we have

$$\overline{F}(x) = \frac{1}{(1+\beta x)^a}\exp\left(-\frac{\alpha x^2}{2}\right)$$

which is of the same form as (7.20). Alternatively, if failures are classified as type A and type B caused by $h_A(x) = \alpha x$ or $h_B(x) = \frac{\theta}{(1+\beta x)}$, then (7.20) is the distribution of $\min(X_A, X_B)$, where X_A and X_B are independent lifetimes with hazard rates h_A

and h_B, respectively. Maximum likelihood method can be used to estimate the parameters. We can also see that (7.20) belongs to the class of additive hazard models discussed in Nair and Sankaran [446].

Let X be a lifetime random variable with hazard rate $h(x)$ and Y be a non-negative random variable representing changes in the conditions so that h has an additive effect on X through the relationship

$$h(x|y) = a(y) + h(x) \tag{7.21}$$

for some positive function $a(y)$. If X^* is the random variable corresponding to X satisfying the relationship in (7.21), then the survival function of X^* is

$$S^*(x) = S(x)S_E(x),$$

where

$$S_E(x) = \int_0^\infty e^{-xa(y)}g(y)dy$$

and $g(y)$ is the density function of Y. Equivalently, we arrive at the additive hazard model

$$h^*(x) = h(x) + h_E(x),$$

where $h^*(\cdot)$ and $h_E(\cdot)$ are the hazard rates of $S^*(x)$ and $S_E(x)$, respectively. Now, when

$$g(y) = [\Gamma(\alpha)]^{-1} c\lambda^{C\alpha} y^{C\alpha-1} \exp[-(\lambda y)^c],$$

and $a(y) = y^c$ for $c > 0$, we get

$$S_E(x) = \lambda^{c\alpha}(x+\lambda^c)^{-\alpha}.$$

Then, the additive model takes on the form

$$h^*(x) = h(x) + \alpha(x+\lambda^c)^{-1}.$$

It is easy to see that the Hjorth model arises from a particular choice of the hazard rate function $h^*(x) = \alpha x$.

Mudholkar and Kollia [426] and Mudholkar et al. [428] introduced a generalization of the Weibull distribution with its survival function as

$$\overline{F}(x) = \left\{1 - \lambda\left(\frac{x}{\alpha}\right)^\beta\right\}^{\frac{1}{\lambda}}, \quad \alpha, \beta > 0, \tag{7.22}$$

where X has support $(0,\infty)$ for $\lambda \le 0$ and $\left(0, \frac{\alpha}{\lambda^{\frac{1}{\beta}}}\right)$ for $\lambda > 0$. It is easy to see that, as $\lambda \to 0$,

$$\overline{F}(x) = \exp\left[-\left(\frac{x}{\alpha}\right)^{\beta}\right],$$

which is the standard two-parameter Weibull model. The hazard rate corresponding to (7.22) is

$$h(x) = \frac{\beta(\frac{x}{\alpha})^{\beta-1}}{\alpha\{1-\lambda(\frac{x}{\alpha})^{\beta}\}},$$

which is BT for $\beta < 1$, $\lambda > 0$; UBT for $\beta > 1$, $\lambda < 0$; IHR for $\beta \ge 1$, $\lambda > 0$; and DHR for $\beta \le 1$, $\lambda \le 0$. For the estimation of parameters, they discussed the maximum likelihood method. Corresponding to (7.22), the quantile function is

$$Q(u) = \begin{cases} \alpha\left\{\frac{1-(1-u)^{\lambda}}{\lambda}\right\}^{\frac{1}{\beta}}, & \lambda \ne 0 \\ \alpha\{-\log(1-u)\}^{\frac{1}{\beta}}, & \lambda = 0 \end{cases},$$

which has been discussed in detail in Chap. 3. Another modification to the Weibull model is the exponentiated Weibull distribution with distribution function

$$F(x) = \left\{1-\exp\left(-\left(\frac{x}{\alpha}\right)^{\beta}\right)\right\}^{\theta}, \quad \alpha,\beta,\theta > 0,\ x \ge 0, \tag{7.23}$$

density function

$$f(x) = \frac{\beta\alpha\theta}{\alpha^{\beta}} e^{-(\frac{x}{\alpha})^{\beta}} \left\{1-\exp\left(-\left(\frac{x}{\alpha}\right)^{\beta}\right)\right\}^{\theta-1},$$

and hazard function

$$h(x) = \frac{\beta\theta e^{-(\frac{x}{\alpha})^{\beta}}\left\{1-\exp(-(\frac{x}{\alpha})^{\beta})\right\}^{\theta-1}}{1-\left\{1-\exp(-(\frac{x}{\alpha})^{\beta})\right\}^{\theta}}.$$

The nature of $h(x)$ within the parameter space, other properties, and estimation of parameters have all been discussed by Mudholkar et al. [428], Mudholkar and Hutson [423], Jiang and Murthy [296], Nassar and Eissa [463, 464], Singh et al. [543] and Shanmukhapriya and Lakshmi [534]. It is seen that $h(x)$ is BT for $\beta > 1$, $\beta\theta < 1$; UBT for $\beta < 1$, $\beta\theta > 1$; IHR for $\beta \ge 1$, $\beta\theta \ge 1$ and DHR for $\beta \le 1$; and $\beta\theta \le 1$. Reverting to quantile function, (7.23) becomes

$$Q(u) = \alpha\left\{-\log(1-u^{\frac{1}{\theta}})\right\}^{\frac{1}{\beta}}.$$

The hazard quantile function is then

$$H(u) = \frac{\beta\theta(1-u^{\frac{1}{\theta}})}{\alpha(1-u)u^{\frac{1}{\theta}-1}\left\{-\log(1-u^{\frac{1}{\theta}})\right\}^{\frac{1}{\beta}-1}}. \tag{7.24}$$

Differentiating (7.24), we find that the change points in (7.24) are the solutions of

$$\left\{1-u-\theta(1-u^{\frac{1}{\theta}})\right\}\log(1-u^{\frac{1}{\theta}}) = \frac{1-\beta}{\beta}(1-u)u^{\frac{1}{\theta}}.$$

Marshall and Olkin [411] devised a new method of introducing more flexibility to a given distribution $G(x)$ by adding a new parameter. Their scheme is to construct distribution $\overline{F}(x)$ from $\overline{G}(x)$ through the formula

$$\overline{F}(x) = \frac{\theta\overline{G}(x)}{1-(1-\theta)\overline{G}(x)}, \quad \theta > 0. \tag{7.25}$$

Assuming $\overline{G}(x)$ to be a two-parameter Weibull, $\overline{G}(x) = \exp\left\{-(\frac{x}{\alpha})^{\beta}\right\}$, for example, we find from (7.25) that

$$\overline{F}(x) = \frac{\theta\exp\left\{-(\frac{x}{\alpha})^{\beta}\right\}}{1-(1-\theta)\exp\left\{-(\frac{x}{\alpha})^{\beta}\right\}}$$

yielding the density function

$$\begin{aligned} f(x) &= \frac{\theta g(x)}{[1-(1-\theta)\overline{G}(x)]^2} \\ &= \frac{\theta\beta}{\alpha}\frac{(\frac{x}{\alpha})^{\beta-1}\exp\left\{-(\frac{x}{\alpha})^{\beta}\right\}}{\left\{1-(1-\theta)\exp(-(\frac{x}{\alpha})^{\beta})\right\}^2} \end{aligned}$$

and hazard rate

$$h(x) = \frac{(\frac{\beta}{\alpha})(\frac{x}{\alpha})^{\beta-1}}{1-(1-\theta)\exp\left\{-(\frac{x}{\alpha}\right\}^{\beta})}.$$

Teiling and Xie [565] have carried out a failure time data analysis by using (7.25).

It is of interest to know the quantile version of (7.25) for studying the properties of the new distribution further. Setting $x = Q(u)$, where Q is the quantile function of G, after writing

$$F(x) = \frac{1-\overline{G}(x)}{1-(1-\theta)\overline{G}(x)},$$

$$F(Q(u)) = \frac{u}{1-(1-\theta)(1-u)} = \frac{u}{u+\theta-\theta u},$$

we have

$$Q(u) = Q_1\left(\frac{u}{u+\theta-\theta u}\right)$$

with $Q_1(u)$ being the quantile function corresponding to $F(x)$. Equivalently, we have

$$Q_1(u) = Q\left(\frac{u\theta}{1-(1-\theta)u}\right). \tag{7.26}$$

Applying (7.26) in the case of the Weibull distribution for which

$$Q(u) = \alpha\{-\log(1-u)\}^{\frac{1}{\beta}},$$

we obtain

$$Q_1(u) = \alpha\left\{-\log\left(\frac{1-u}{1-u+u\theta}\right)\right\}^{\frac{1}{\beta}}.$$

The hazard quantile of $Q_1(u)$ is

$$H(u) = \frac{\beta(1-u+u\theta)}{\alpha\theta\left\{-\log\left(\frac{1-u}{1-u+u\theta}\right)\right\}^{\frac{1}{\beta}-1}}.$$

The sign of $H'(u)$ depends on the expression

$$(\theta-1)\log\frac{1-u+u\theta}{1-u} - \frac{\theta(1-\beta)}{\beta(1-u)}$$

and $F(x)$ is IHR for $\theta > 1$, $\beta > 1$, and DHR for $0 < \theta < 1$ and $0 < \beta < 1$. Change points of $H(u)$ are the solutions of the equation

$$(1-u)\log\frac{1-u+u\theta}{1-u} = \frac{\theta(1-\beta)}{\beta(\theta-1)}.$$

A special case of the exponentiated Weibull distribution is the exponentiated exponential distribution considered in Gupta et al. [239] and Gupta and Kundu [250, 253]. We look at the general form with survival function

$$F(x) = \left\{1 - \exp\left(-\frac{x-\mu}{\sigma}\right)\right\}^{\alpha}, \quad x > \mu;\ \alpha, \sigma > 0. \tag{7.27}$$

We find the expression for the hazard rate as

$$h(x) = \frac{\alpha}{\sigma} \frac{\{1 - \exp(-\frac{x-\mu}{\sigma})\}^{\alpha-1} \exp(-\frac{x-\mu}{\sigma})}{1 - \left[1 - \{1 - \exp(-\frac{x-\mu}{\sigma})\}^{\alpha}\right]}.$$

It could be seen that $h(x) = \frac{1}{\sigma}$ for $\alpha = 1$, increases from 0 to $\frac{1}{\sigma}$ for $\alpha > 1$, and decreases from ∞ to $\frac{1}{\sigma}$ for $\alpha < 1$. Quantile analysis of (7.27) is straightforward with

$$Q(u) = \mu - \sigma \log(1 - u^{\frac{1}{\alpha}})$$

and

$$H(u) = \frac{\sigma}{\alpha} \frac{u^{\frac{1}{\alpha}-1}}{1 - u^{\frac{1}{\alpha}}}.$$

A comparative study of (7.27) with the gamma, Weibull and lognormal distributions has been carried out by Gupta and Kundu [251], Kundu et al. [358] and Gupta and Kundu [252].

The two-parameter Chen's [142] model in (7.17) has been generalized by Xie et al. [595] to provide a new distribution with survival function

$$\bar{F}(x) = \exp\left[-\lambda\alpha\left\{\exp\left(\frac{x}{\alpha}\right)^{\beta} - 1\right\}\right], \quad x \geq 0;\ \alpha, \beta, \lambda > 0, \tag{7.28}$$

called the modified Weibull extension. From the corresponding density function

$$f(x) = \exp\left[-\lambda\alpha\left\{\exp\left(\frac{x}{\alpha}\right)^{\beta} - 1\right\}\right] \lambda\beta e^{(\frac{x}{\alpha})^{\beta}} \left(\frac{x}{\alpha}\right)^{\beta-1},$$

we obtain the hazard function as

$$h(x) = \lambda\beta e^{(\frac{x}{\alpha})^{\beta}} \left(\frac{x}{\alpha}\right)^{\beta-1}.$$

The name Weibull extension comes from the fact that (7.28) reduces to the Weibull distribution when $\lambda \to \alpha$ in such a way that $\alpha^{\beta-1}\lambda^{-1}$ remains constant. Two special cases are the exponential power distribution in (7.2) when $\lambda = 1$ and the Chen's model when $\alpha = 1$. When $\beta \geq 1$, $h(x)$ is IHR and when $0 < \alpha < 1$, $h(x) \to \infty$ as $x \to 0$ or ∞. In this case, we have BT shape with change point $x_0 = \alpha\left(\frac{1-\beta}{\beta}\right)^{\frac{1}{\beta}}$. The change point increases as β decreases from 1 to 0. The quantile function of (7.28) takes on the expression

$$Q(u) = \alpha \left[\log \left\{ 1 - \frac{\log(1-u)}{\lambda\alpha} \right\} \right]^{\frac{1}{\beta}}$$

and therefrom we get

$$H(u) = \frac{\beta(\lambda\alpha - \log(1-u))}{\alpha \left[\log \left\{ 1 - \frac{\log(1-u)}{\lambda\alpha} \right\} \right]^{\frac{1}{\beta}-1}}.$$

Yet another extension of the Weibull law is due to Lai et al. [370], called the modified Weibull distribution. It has its density function as

$$f(x) = \beta(\alpha + \lambda x)x^{\alpha-1}\exp\{\lambda x - \beta x^{\alpha} e^{\lambda x}\} \tag{7.29}$$

and survival function as

$$\overline{F}(x) = \exp(-\beta x^{\alpha} e^{\lambda x}).$$

As $\lambda \to 0$, we have the usual Weibull distribution. Note that (7.29) has a hazard rate of the form

$$h(x) = \beta(\alpha + \lambda x)x^{\alpha-1}e^{\lambda x},$$

so that the shape of $h(x)$ is independent of β and λ. For $\alpha \geq 1$, the distribution is IHR, and for $0 < \alpha < 1$, we have BT shape with change point $x_0 = \frac{\alpha^{\frac{1}{2}} - \alpha}{\lambda}$. There is no simple closed-form expression for the quantile function. Nikulin and Haghighi [472] (see also Dimitrakopoulas et al. [178]) proposed a generalized power Weibull distribution with survival function

$$\overline{F}(x) = \exp\left[1 - \left\{ 1 + \left(\frac{x}{\beta}\right)^{\alpha} \right\}^{\theta} \right], \quad x \geq 0,\, \alpha, \beta > 0,\, \theta > 0, \tag{7.30}$$

which is a general family consisting of Weibull ($\theta = 1$) and exponential ($\theta = 1$, $\alpha = 1$) distributions as particular members.

The transformation $Y = 1 + (\frac{X}{\beta})^{\alpha}$ gives the Weibull distribution with parameters 1 and θ in $(1, \infty)$. Similarly, transforming by $\log\{1 + (\frac{x}{\alpha})^{\beta}\}$ and $[\log\{1 + (\frac{x}{\alpha})^{\beta}\}]^{\frac{1}{\beta}}$, respectively, we obtain the modified extreme value distribution and the power exponential distribution of Smith and Bain [544] in (7.2). From the density function

$$f(x) = \frac{\theta\alpha}{\beta^\alpha} \exp\left[1 - \left\{1 + \left(\frac{x}{\beta}\right)^\alpha\right\}^\theta\right] \left\{1 + \left(\frac{x}{\beta}\right)^\alpha\right\}^{\theta-1} x^{\alpha-1},$$

we obtain the hazard function as

$$h(x) = \frac{\theta\alpha}{\beta^\alpha} \left\{1 + \left(\frac{x}{\beta}\right)^\alpha\right\}^{\theta-1} x^{\alpha-1}.$$

The above expression yields flexible hazard rate shapes, like IHR if either $\alpha > 1$ and $\alpha > \frac{1}{\theta}$ or $\alpha = 1$ and $\theta > 1$, DHR if either $0 < \alpha < 1$ and $\alpha < \frac{1}{\theta}$ or $\alpha\theta = 1$ and $0 < \alpha < 1$, and UBT whenever $\frac{1}{\theta} > \alpha > 1$. For a quantile-based analysis, we can use

$$Q(u) = \beta\left[\{1 - \log(1-u)\}^{\frac{1}{\theta}} - 1\right]^{\frac{1}{\alpha}}$$

and

$$H(u) = \frac{\alpha\theta}{\beta}\left[\{1 - \log(1-u)\}^{\frac{1}{\theta}} - 1\right]^{1-\frac{1}{\alpha}} \{1 - \log(1-u)\}^{1-\frac{1}{\theta}}.$$

Differentiating $H(u)$ and setting $H'(u) = 0$, we find the change point u_0 as

$$u_0 = 1 - \exp\left\{1 - \left(\frac{\alpha\theta - \alpha}{1 - \alpha\theta}\right)^\theta\right\}.$$

Lan and Leemis [372] presented the logistic exponential distribution as a model for lifetimes with flexible hazard rate shapes. Their two-parametric version has its survival function as

$$\overline{F}(x) = \left\{1 + (e^{\lambda x} - 1)^k\right\}^{-1}, \quad x \geq 0. \tag{7.31}$$

Clearly, the distribution reduces to the exponential case when $k = 1$ having constant hazard rate. In general, the hazard function is

$$h(x) = \frac{\lambda k e^{\lambda x}(e^{\lambda x} - 1)^{k-1}}{1 + (e^{\lambda x} - 1)^k}.$$

For $0 < k < 1$, $h(x)$ is BT, while for $k > 1$, it is UBT. The change point in both cases is

$$x_0 = \lambda^{-1} \log(x_k + 1),$$

where x_k is the positive root of the equation

$$kx - x^k = 1 - k.$$

It can be shown that the quantile function of (7.31) is

$$Q(u) = \frac{1}{\lambda}\log\left\{1 + \left(\frac{u}{1-u}\right)^{\frac{1}{k}}\right\}$$

and

$$H(u) = k\lambda\frac{u^{\frac{1}{k}} + (1-u)^{\frac{1}{k}}}{u^{\frac{1}{k}} - 1}.$$

The change point of $H(u)$ is the solution of the equation

$$ku^{\frac{1}{k}} + (1-u)^{\frac{1}{k}-1}(k - ku - 1) = 0.$$

Introducing yet another parameter into (7.31), we have the more general model with survival function

$$\overline{F}(x) = \frac{1 + (e^{\lambda\theta} - 1)^k}{1 + \{e^{\lambda(x+\theta)} - 1\}^k}, \quad \theta \geq 0,\ k > 0, \lambda > 0, \tag{7.32}$$

and corresponding hazard function

$$h(x) = \frac{C\lambda k\{e^{\lambda(x+\theta)} - 1\}^{k-1}e^{\lambda(x+\theta)}}{1 + \{e^{\lambda(x+\theta)} - 1\}^k},$$

where

$$c = 1 + (e^{\lambda\theta} - 1)^k.$$

By proceeding along the same lines as in the reduced model, we see that $\overline{F}(x)$ has highly flexible hazard rate being exponential for $k = 1$; BT for $0 < k < 1$, $\lambda\theta < \log(x_k + 1)$ with minimum at $\frac{\log(1+x_k)}{\lambda} - \theta$; IHR for $0 < k < 1$ and $\lambda\theta > \log(x_k + 1)$; and UBT for $k > 1$ and $\lambda\theta > \log(1 + x_k)$. The quantile function, with a slightly more complicated from than the two-parameter version, given by

$$Q(u) = \frac{1}{\lambda_2}\log\left\{1 + \left(\frac{c+u-1}{1-u}\right)^{\frac{1}{k}} - \theta\right\}$$

can be employed to find $H(u)$ and its change points as done before.

By exponentiating the linear failure rate model, Sarhan and Kundu [520] arrived at the generalized linear failure rate distribution with survival function

$$\overline{F}(x) = \left[1 - \exp\left\{-\left(ax + \frac{b}{2}x^2\right)\right\}\right]^{\theta}, \quad x \geq 0. \tag{7.33}$$

It contains as special cases the linear failure rate model, the generalized exponential distribution in (7.27) and generalized Rayleigh distribution discussed by Kundu and Raqab [361]. The hazard rate becomes

$$h(x) = \frac{\theta(a+bx)\left[1 - \exp\{-(ax + \frac{b}{2}x^2)\}\right]^{\theta-1} \exp\left\{-(ax + \frac{bx^2}{2})\right\}}{1 - \left[1 - \exp\left\{-(ax + \frac{bx^2}{2})\right\}\right]^{\theta}}.$$

Analysing $h(x)$, it is seen that the hazard rate is constant or increasing when $\theta = 1$, increasing when $\theta > 1$, either decreasing ($b = 0$) or bathtub ($b > 0$) when $\theta < 1$. The same approach is made by Barreto-Souza and Cribari-Neto [71] to extend the exponential Poisson distribution of Kus [364] given by

$$F(x) = \frac{1 - \exp(-\lambda + \lambda e^{-\beta x})}{1 - e^{-\lambda}}, \quad x, \lambda, \beta > 0,$$

to the general form

$$F(x) = \left\{\frac{1 - \exp(-\lambda + \lambda e^{-\beta x})}{1 - e^{-\lambda}}\right\}^{\theta}, \quad x, \theta > 0. \tag{7.34}$$

The hazard rate of (7.34) is given by

$$h(x) = \frac{\theta\lambda\beta(1 - e^{-\lambda + \lambda e^{-\beta x}})^{\theta-1} e^{-\lambda - \beta x + \lambda e^{-\beta x}}}{(1 - e^{-\lambda})^{\theta} - \{1 - \exp(-\lambda + \lambda e^{-\beta x})\}^{\theta}}$$

which can be IHR, DHR or UBT. A closed-form quantile function for (7.34) is given by

$$Q(u) = \frac{1}{\beta} \log\left[-\frac{1}{\lambda} \log\left\{1 - (1 - e^{-\lambda}) u^{\frac{1}{\theta}}\right\}\right].$$

When the baseline distribution in (7.33) or (7.34) is changed to the exponential power distribution in (7.2), we have the model proposed by Barriga et al. [72], for which the survival function is

$$\overline{F}(x) = 1 - \left[1 - \exp\left\{1 - \exp\left(\frac{x}{\alpha}\right)^{\beta}\right\}\right]^{\theta}, \quad x > 0. \tag{7.35}$$

The corresponding hazard rate function

$$h(x) = \frac{\beta\theta x^{\beta-1}\exp\{1+(\frac{x}{\alpha})^{\beta}-\exp(\frac{x}{\alpha})^{\beta}\}\exp\left[1-\exp\{1-\exp(\frac{x}{\alpha})^{\beta}\}\right]^{\theta-1}}{\alpha^{\beta}[1-\{1-\exp(1-\exp(\frac{x}{\alpha})^{\beta})\}]^{\theta}}$$

has the following properties:

(i) $h(0)=0$ for $\beta>1$ and $h(0)=\frac{1}{\alpha}$ for $\beta=1$, $\theta=1$;
(ii) $h(x)$ is increasing for $\beta>1$, $\theta\leq 1$;
(iii) $h(x)$ is decreasing for $\beta\theta\leq 1$, $\theta>1$;
(iv) $h(x)$ is UBT for $\beta<1$, $\beta\theta>1$;
(v) $h(x)$ is BT for $\theta\leq 1$ or $\beta>1$ and $\beta\theta<1$.

The parameter estimation is carried out by the maximum likelihood method. Distribution (7.35) is specified by the quantile function

$$Q(u) = \alpha\left[\log\left\{1-\log(1-u)^{\frac{1}{\theta}}\right\}\right]^{\frac{1}{\beta}}.$$

Zhang and Xie [600] considered the upper truncated Weibull distribution given by

$$G(x) = \frac{F(x)-F(a)}{F(T)-F(a)}, \quad a\leq x<T,$$

with

$$F(x) = 1-\exp\left\{-\left(\frac{x}{\alpha}\right)^{\beta}\right\}$$

yielding a hazard rate

$$h(x) = \frac{(\frac{\beta}{\eta})(\frac{x}{\eta})^{\beta-1}\exp\left\{-\left(\frac{x}{\eta}\right)^{\beta}\right\}}{F(T)-F(x)}, \quad a\leq x<T,$$

which is increasing for $\beta\geq 1$ and BT for $\beta<1$.

Silva et al. [539] introduced the generalized geometric exponential distribution with distribution function

$$F(x) = \left(\frac{1-e^{-px}}{1-pe^{-\beta x}}\right)^{\theta}, \quad x>0,\ 0<p<1,\ \theta,\beta>0. \tag{7.36}$$

When $\theta>0$ is an integer, (7.36) is the distribution of $X=\max_{1\leq i\leq\alpha} Y_i$, where $Y_1,Y_2,\ldots,Y_{\alpha}$ is a random sample from the exponential geometric distribution. Note that

$$h(x) = \frac{\theta\beta(1-p)e^{-\beta x}(1-e^{-\beta x})^{\theta-1}}{(1-pe^{-\beta x})\{(1-pe^{-\beta x})^{\alpha}-(1-e^{-\beta x})^{\alpha}\}}$$

which has the following properties:

(a) decreasing for p and α in $(0,1)$;
(b) increasing for p in $(0, \frac{\alpha-1}{\alpha+1})$ and α in $(1,\infty)$;
(c) UBT for p in $(\frac{\alpha-1}{\alpha+1}, 1)$ and α in $(1,\infty)$.

Let $Y_1, Y_2, \ldots, Y_N$ be a random sample from a Weibull distribution with density function

$$f(y) = \beta\alpha^{\beta}y^{\beta-1}e^{-(\alpha y)^{\beta}}, \quad y, \alpha, \beta > 0,$$

where N is a zero-truncated Poisson random variable with probability mass function

$$P(N=n) = \frac{e^{-\lambda}\lambda^{n}}{\Gamma(n+1)(1-e^{-\lambda})}, \quad n = 1, 2, \ldots.$$

Assuming that Y_i and N are independent, the distribution of $X = \min(Y_1, Y_2, \ldots, Y_N)$ is called a Weibull-Poisson distribution by Hemmati et al. [267]. It has density function

$$f(x) = \frac{\lambda\beta\alpha}{1-e^{-\lambda}}(\alpha x)^{\beta-1}\exp\left\{-\lambda-(\alpha x)^{\beta}+\lambda e^{-(\alpha x)^{\beta}}\right\}$$

and survival function

$$\bar{F}(x) = \left\{1-\exp(\lambda e^{-(\alpha x)^{\beta}})\right\}(1-e^{-\lambda})^{-1}.$$

The corresponding hazard rate is

$$h(x) = \frac{\lambda\beta\alpha}{1-e^{-\lambda}} \frac{(\alpha x)^{\beta-1}(1-e^{\lambda})\exp\left\{-\lambda-(\alpha x)^{\beta}+\lambda e^{-(\alpha x)^{\beta}}\right\}}{\left\{1-\exp(\lambda e^{-(\alpha x)^{\beta}})\right\}}$$

which can be either increasing, decreasing or modified bathtub shaped.

A lifetime model for bathtub failure rate data by transforming them to the Weibull was considered in Mudholkar et al. [422]. Consider the data in the form of pairs of independent and identically distributed random variables (X_i, δ_i), where $X_i = \min(T_i, C_i)$ and $\delta_i = 1$ when $T_i \le C_i$ (uncensored case) and $\delta_i = 0$ if $T_i > C_i$ (censored case), with T_i as the lifetime and C_i as the censoring time. Mudholkar et al. [422] assumed that there exists a transformation

$$y = g(x, \theta) = \frac{x}{1-\theta x}$$

that transforms the data to the Weibull form $\overline{F}(y) = e^{-(\frac{y}{\alpha})^{\beta}}$. The range of the transformation is $(0,\alpha)$; $g(x,\theta)$ should be invertible and θ can be zero in which case the original data is retained. We have $x = \frac{y}{y+\theta}$, and so

$$\overline{F}(x) = P\Big(\frac{Y}{1+\theta Y} > x\Big) = \exp\Big\{-\Big(\frac{1}{\alpha}\frac{x}{1+\theta x}\Big)^{\beta}\Big\}, \quad 0 < x < \frac{1}{\theta}.$$

The corresponding hazard rate

$$h(x) = \frac{\beta}{\alpha^{\beta}}\frac{x^{\beta-1}}{(1-\theta x)^{\beta+1}}$$

has

$$h'(x) = (1-\theta x)^{\beta} x^{\beta-2}(\beta - 1 + \theta x)$$

which reveals that $h(x)$ is increasing for $\beta > 1$ and BT for $0 < \beta < 1$. The distribution has its quantile function as

$$Q(u) = \frac{\alpha\{-\log(1-u)\}^{\frac{1}{\beta}}}{1+\theta\alpha\{-\log(1-u)\}^{\frac{1}{\beta}}}.$$

7.4 More Flexible Hazard Rate Functions

When we examine the models in the last two sections chronologically, it is seen that model parsimony was an important concern in earlier works with many two-parametric models. With improvement in computational technology, the number of parameters and complexities in estimating them became less problematic. Consequently, models with more than one shape parameter that provide more richness in the shapes of the hazard rate began to appear. In this section, we deal with such models that have at least four parameters.

Models with hazard rates as a sum were proposed by many. These include

$$h(x) = \frac{\theta}{1+\beta x} + \lambda\alpha x^{\alpha-1}, \quad \theta,\beta,\lambda,\alpha > 0, \tag{7.37}$$

suggested by Murthy et al. [435]. Note that the first term is the hazard rate of a Pareto II distribution, while the second is that of the Weibull. Here, $h(0) = \theta$ and $h(\infty) = \infty$ suggesting a bathtub shape. The Hjorth model in (7.20) is a special case of this family with survival function

$$\overline{F}(x) = \frac{e^{-\lambda x^{\alpha}}}{(1+\beta x)^{\frac{\theta}{\beta}}}.$$

Jaising et al. [291] extended the hazard rate in (7.37) to the form

$$h(x) = \lambda + \frac{\theta}{x+\beta} + \lambda x^{\delta}, \tag{7.38}$$

while Canfield and Borgman [126] considered the representation

$$h(x) = \alpha_1 \beta_1 x^{\beta_1 - 1} + \alpha_2 \beta_2 x^{\beta_2 - 1} + \alpha_3 \tag{7.39}$$

with $\beta_2 > 2$ and $\beta_1 < 1$. Notice that (7.38) is a construction of the hazard rates of exponential Pareto II and Weibull, while (7.39) considers the sum of an exponential and two Weibull hazard rates.

In the case of (7.39), the change point is given by

$$x_0 = \left(\frac{\alpha_1 \beta_1 (1-\beta_1)}{\alpha_2 \beta_2}\right)^{\frac{1}{\beta_2 - \beta_1}}, \quad \beta_1 < 1.$$

Similar representation of $h(x)$ with only two Weibull hazard rates as components was also discussed by Xie and Lai [594], Jiang and Murthy [293] and Usagaonkar and Maniappan [570]. Instead of two Weibull distributions, if we take the hazard rates of Burr distributions with survival functions

$$\overline{F}_i(x) = \frac{1}{(1+(\frac{x}{a_i})^{c_i})^{k_i}}, \quad i = 1,2,$$

we get the hazard rate

$$h(x) = \frac{k_1 c_1 x^{c_1 - 1}}{a_1^{c_1}\{1+(\frac{x}{a_1})^{c_1}\}} + \frac{k_2 c_2 x^{c_2 - 1}}{a_2^{c_2}\{1+(\frac{x}{a_2})^{c_2}\}}, \tag{7.40}$$

for $k_1, k_2, a_1, a_2 \geq 0$, $0 < c_1 < 1$ and $c_2 > 2$. A bathtub model arising from the above hazard rate has been discussed by Wang [577].

Phani [495] considered a new distribution with survival function

$$\overline{F}(x) = \exp\left\{-\lambda \frac{(x-\alpha)^{\theta_1}}{(\beta - x)^{\theta_2}}\right\}, \quad \lambda > 0,\ \beta_1, \beta_2 > 0,\ 0 \leq \theta_1 \leq x \leq \theta_2, \tag{7.41}$$

and corresponding hazard rate as

$$h(x) = \frac{\lambda (x-\alpha)^{\theta_1 - 1} (\beta - x)^{\theta_2 - 1} \{\theta_1 (\beta - x) + \theta_2 (x - \alpha)\}}{(\beta - x)^{2\theta_2}}.$$

The sign of $h(x)$ is determined by a quadratic function in x, and so provides a BT shape. In the reduced case of $\theta_1 = \theta_2$, the condition for BT is $0 < \theta_1 < 1$. Subsequently, Moore and Lai [420] proposed the version

$$h(x) = c(x+p)^{a-1}(q-x)^{b-1}, \quad 0<a<1,\, b<-1,\, c>0,\, p\geq 0,\, 0\leq x<q,$$

which also gives a BT form since $h(0) = cp^{a-1}q^{b-1}$ and $h(x) \to \infty$. The change point is $x_0 = (a+b-2)^{-1}\{(a-1)q-(b-1)p\}$.

Mixtures of distributions form an important aspect in the consideration of bathtub-shaped models. Many authors like Glaser [221], Kunitz and Pamme [362], Pamme and Kunitz [480] and Gupta and Warren [249] have focused on this formulation. If $f_1(x)$ and $f_2(x)$ are density functions with hazard rates $h_1(x)$ and $h_2(x)$, respectively, the two-component mixture

$$f(x) = \alpha f_1(x) + (1-\alpha) f_2(x), \quad 0<\alpha<1,$$

has its hazard rate as

$$h(x) = \frac{\alpha f_1(x) + (1-\alpha) f_2(x)}{\alpha \overline{F}_1(x) + (1-\alpha)\overline{F}_2(x)};$$

see the discussion in Sect. 4.2. Assuming

$$f_i(x) = \frac{1}{\beta_i^{\alpha_i}\Gamma(\alpha_i)} x^{\alpha_i - 1} e^{-\frac{t}{\beta_i}}, \quad i = 1,2$$

Gupta and Warren [249] showed that $h(x)$ is BT in the cases (i) $\beta_1 = \beta_2 = \beta$ and $\alpha_1 > 1$, $\alpha_2 > 1$; (ii) $\alpha_2 = 1$, $\alpha_1 > 2$; (iii) $\alpha_1 > 1$, $\alpha_2 > 1$. With $\alpha_1 - \alpha_2 > 1$ and $(\alpha_1 - \alpha_2 - 1)^2 - 4(\alpha_2 - 1) > 0$, $h(x)$ can be UBT.

The properties of mixtures of Weibull distributions have been studied by Jiang and Murthy [295] and Wondamagegnehu [584]. Assuming $\overline{F}_1(x) = \exp(-\lambda_1 x^{\alpha})$ and $\overline{F}_2(x) = \exp(-\lambda_2 x^{\alpha})$, the mixture hazard rate given by

$$h(x) = \frac{pe^{-\lambda_1 x^{\alpha}}\lambda_1 \alpha x^{\alpha-1} + (1-p)e^{-\lambda_2 x^{\alpha}}\lambda_2 \alpha x^{\alpha-1}}{pe^{-\lambda_1 x^{\alpha}} + (1-p)e^{-\lambda_2 x^{\alpha}}}$$

will be modified BT shaped when $0 < p < \theta$ and IHR when $\theta \leq p < 1$, where $\alpha > 1$ and

$$\theta = \frac{\alpha(\beta-1)+A}{2\alpha(\beta-1)\exp\left[\frac{(\alpha-1)(\beta+1)+A}{\alpha(\beta-1)}\right]},$$

$$A = (\alpha^2(\beta-1)^2 + 4(\alpha-1)^2\beta)^{\frac{1}{2}},$$

$$\beta = \frac{\lambda_2}{\lambda_1}.$$

While the mixtures can have IHR, DHR, UBT, modified bathtub or roller-coaster type, they cannot be BT.

Navarro and Hernandez [466] have discussed the nature of the failure rates of truncated normal mixtures. They considered the truncated normal density function

$$f(x) = \frac{1}{\sqrt{2\pi}\sigma\Phi(\frac{\mu}{\sigma})}\exp\Big(-\frac{(x-\mu)^2}{2\sigma^2}\Big), \quad x > 0, \tag{7.42}$$

where $\Phi(x)$ is the standard normal distribution function, and formed the mixture

$$f(x) = pf_1(x) + (1-p)f_2(x), \quad 0 < p < 1, \tag{7.43}$$

where $f_i(x)$ is distributed as (7.42) with parameters (μ_0, σ_0) and (μ_1, σ_1). When $\sigma_1 = \sigma_0$ and $\delta = \frac{\sigma_0^2}{(\mu_0-\mu_1)^2}$, they proved that if

(i) $\delta > \frac{1}{4}$, $f(x)$ is IHR;
(ii) $\delta \le \frac{1}{4}$, $w(0) \le \frac{1}{2}$ and $w(0)(1-w(0)) < \delta$, $f(x)$ is IHR or BT;
(iii) $\delta \le \frac{1}{4}$, $w(0) \le \frac{1}{2}$ and $w(0)(1-w(0)) \ge \delta$, $f(x)$ is IHR or BT;
(iv) $\delta \le \frac{1}{4}$, $w(0) < \frac{1}{2}$ and $w(0)(1-w(0)) \le \delta$, $f(x)$ is IHR, BT or modified BT,

where

$$w(t) = \Big\{1 + \frac{1-p}{p}\frac{f_1(x)}{f_2(x)}\Big\}^{-1}.$$

Further, the change points of $\eta(x) = -\frac{f'(x)}{f(x)}$ in (7.43) is found from the equation

$$w(x)(1-w(x)) = \delta.$$

In the general case when the variances are unequal, let us assume $\sigma_1 > \sigma_0$,

$$x_0 = \frac{\sigma_1^2\mu_0 - \sigma_0^2\mu_1}{\sigma_1^2 - \sigma_0^2} > 0,$$

$$\theta(x) = \frac{\frac{w(x)}{\sigma_1^2} - \frac{(1-w(x))}{\sigma_0^2}}{w(x)(1-w(x))} - \Big(\frac{x-\mu_1}{\sigma_1^2} - \frac{x-\mu_2}{\sigma_0^2}\Big)^2,$$

and

$$y_1 = w(x_1).$$

Then, the following results hold:

1. If $w(x_0) \ge y_1$, then f is IHR;

2. If $w(x_0) < y_1$ and
   (a) $w(0) < y_1$, $\theta(0) \geq 0$ and $\theta(x_1) \geq 0$, then f is IHR;
   (b) $w(0) < y_1$, $\theta(0) < 0$ and $\theta(x_1) \geq 0$, then f is IHR or BT;
   (c) $w(0) < y_1$, $\theta(0) \geq 0$ and $\theta(x_1) < 0$, then f is IHR, BT or modified BT (MBT);
   (d) $w(0) < y_1$, $\theta(0) < 0$ and $\theta(x_1) < 0$, then f is IHR, BT, MBT or BBT (BT in $(0, x_0)$ and BT in (x_0, ∞));
   (e) $w(0) \geq y_1$, $\theta(0) > 0$ and $\theta(x_1) \geq 0$ and $\theta(x_2) \geq 0$, then f is IHR;
   (f) $w(0) \geq y_1$, $\theta(0) \leq 0$ and $\theta(x_2) \geq 0$, then f is IHR or BT;
   (g) $w(0) \geq y_1$, $\theta(0) > 0$ and $\theta(x_i) \geq 0$ for $i = 1$ or 2, then f is IHR, BT or MBT;
   (h) $w(0) \geq y_1$, $\theta(0) \leq 0$ and $\theta(x_2) < 0$, then f is IHR, BT, MBT or BBT;
   (i) $w(0) \geq y_1$, $\theta(0) > 0$ and $\theta(x_1) < 0$, $\theta(x_2) < 0$, then f is IHR, BT, MBT, BBT or IBT.

Sultan et al. [553] considered a mixture of inverse Weibull distributions with survival function

$$\bar{F}(x) = p\left\{1 - \exp(-\alpha_1 x)^{-\beta_1}\right\} + (1-p)\left\{1 - \exp(-\alpha_2 x)^{-\beta_2}\right\}$$

and the corresponding hazard rate

$$h(x) = \frac{p\beta_1 \alpha_1^{-\beta_1} x^{-(\beta_1+1)} e^{-(\alpha_1 x)^{-\beta_1}} + (1-p)\beta_2 \alpha_2^{-\beta_2} x^{-(\beta_2+1)} e^{-(\alpha_2 x)^{-\beta_1}}}{p\{1 - e^{-(\alpha_1 x)^{-\beta_1}}\} + (1-p)\{1 - e^{-(\alpha_2 x)^{-\beta_2}}\}}$$

which can be unimodal and bimodal.

From the above illustrations, one might have noticed that the analysis of hazard rates of mixtures is quite complicated. Also, the shape of the hazard rates changes with the mixing proportion p and the component distributions. In most cases, the quantile functions are not invertible into explicit forms and so have to be evaluated numerically. There are several discussions on the shapes of hazard rates in the general case as well as for mixtures of distributions with specified components. For more details, one may refer to Gurland and Sethuraman [255], Lynch [405], AL-Hussaini and Sultan [29], Shaked and Spizzichino [532], Block et al. [109, 110, 112], Wondamagegnehu et al. [585], Bebbington et al. [82], Sultan et al. [553] and Ahmed et al. [27].

Agarwal and Kalla [20] studied a generalized gamma model of the form

$$f(x) = \frac{x^m e^{-\delta x}(n+x)^\lambda}{\delta^{\lambda-m}\Gamma_\lambda(m+1, n\delta)}, \quad x > 0,\ \lambda, \delta, n, m > 0,$$

where

$$\Gamma_\lambda(m,n) = \int_0^\infty \frac{e^{-t} t^{m-1}}{(t+n)^\lambda} dt$$

which was further extended by Kalla et al. [309] to the model

$$f(x) = \frac{\beta x^{m+\beta-1} e^{-\delta x^{\beta}} (n+x^{\beta})^{\lambda}}{\delta^{\lambda-\frac{m}{\beta}} \Gamma_{\lambda}(\frac{m}{\beta}+1, n\delta)}. \quad (7.44)$$

The distribution in (7.44) includes the Stacy distribution when $\lambda = 1$ and appropriate reparametrization that gives

$$f(x) = Cx^{\alpha\beta-1} \exp\left\{-\left(\frac{x}{\theta}\right)^{\alpha}\right\}. \quad (7.45)$$

The gamma distribution and Weibull distribution are particular cases of (7.45) when $\alpha = 1$ and $\beta = 1$, respectively. Glaser [220] and McDonald and Richards [414,415] have discussed the shape of the hazard rate and conditions on the parameters that produce IHR, DHR, BT and UBT curves. The most general form in (7.44) has been analysed by Gupta and Lvin [248].

In order to accommodate early failures, Muraleedharan and Lathika [433] proposed mixing a Weibull distribution with a singular distribution at $x = \delta$, where δ is small and specified in advance. Thus, their model has the representation

$$F(x) = (1-\alpha)F_1(x) + \alpha F_2(x),$$

where F_1 is the singular component and F_2 is Weibull. Mitra and Basu [418] considered the life distribution of a device subject to a sequence of shocks occurring randomly in time according to a homogeneous Poisson process:

$$\overline{H}(t) = \sum_{k=0}^{\infty} e^{-\lambda t} \frac{(\lambda t)^k}{k!} \overline{P}_K, \quad 1 = \overline{P}_0 \geq \overline{P}_1 \geq \overline{P}_2 \geq \cdots.$$

They derived conditions under which $\overline{H}(t)$ has a BT hazard rate in terms of certain properties of $\overline{P}_K$.

Mitra and Basu [419] have presented some general properties of BT distributions. Their main results resemble the properties of ageing concepts described earlier in Chap. 4. Suppose F has a BT hazard rate with a change point x_0. Then:

1. $\overline{F}(x) < \overline{G}(x)$, where G is exponential with mean $[h(x_0)]^{-1}$;
2. $\mu'_r \leq \frac{\Gamma(r+1)}{[h(x_0)]^k}$ with equality sign holding true when X is exponential;
3. BT-shaped hazard rate distributions are not preserved under convolution or mixing. They are also not closed under the formation of parallel systems. However, if each component in a series system has a BT hazard rate with change point x_0, then the system also has a BT hazard rate with x_0 as one of the change points.

The modified Weibull distribution was further generalized by Carrasco et al. [129] to a density function of the form

$$f(x) = \frac{\alpha\beta x^{r-1}(r+\lambda t)\exp(\lambda x - \alpha x^r e^{\lambda x})}{\{1-\exp(-\alpha x^r e^{\lambda x})\}^{1-\beta}}. \tag{7.46}$$

Correspondingly, the hazard rate function is

$$h(x) = \frac{\alpha\beta x^{r-1}(r+\lambda x)\exp(\lambda x - \alpha x^r e^{\lambda x})\{1-\exp(\alpha x^r e^{\lambda x})\}}{1-\{1-\exp(-\alpha x^r e^{\lambda x})\}^{\beta}}.$$

Special cases of the distribution are Weibull ($\lambda = 0$, $\beta = 1$), type I extreme value ($r = 0, \beta = 1$), exponentiated Weibull ($\lambda = 0$), exponentiated exponential ($\lambda = 0$, $r = 1$), generalized Rayleigh ($r = 2$, $\lambda = 0$), and modified Weibull ($\beta = 1$). We see that $h(x)$ is increasing for $r \geq 1$, $0 < \beta < 1$, decreasing for $0 < r < 1$, $\beta > 1$, unimodal for $0 < r < 1$, $\beta \to \infty$, and BT for $\lambda = 0$, $r > 1$, $r\beta < 1$.

In the past three sections, we have reviewed only models of a representative nature. Further models, inference procedures, and applications to data analysis can all be seen from the papers cited in the text and the references therein. More references and details are available in Lai and Xie [368], Lai et al. [369], Bebbington et al. [85], Nadarajah [437] and Silva et al. [538].

7.5 Some General Methods of Construction

In this section, we present some general methods that lead to the construction of a model with BT-shaped hazard function.

- *Using Glaser's theorem*

  Let X a non-negative random with positive density function $f(x)$ that is twice differentiable. Define $\eta(x) = -\frac{f'(x)}{f(x)}$ and $g(x) = [h(x)]^{-1}$. If there exists a point x_0 such that $\eta'(x) < 0$ for $x < x_0$, $\eta'(x_0) = 0$ and $\eta'(x) > 0$ for $x > x_0$, and further there exists a y_0 such that

  $$g'(y_0) = \int_{y_0}^{\infty} \frac{f(y)}{f(y_0)}[\eta(y_0) - \eta(y)]dy.$$

  Then, the corresponding distribution has BT-shaped hazard rate. Verification of the BT nature of several distributions discussed earlier like (7.4), (7.5), (7.45) and mixtures of gamma is in fact accomplished in this manner.
- *From convex functions*

  A BT hazard rate distribution emerges from a strictly convex positive function on $(0, \alpha)$ satisfying the condition $\int h(x)dx = \infty$. Also, a strictly increasing function of BT hazard rate is also a BT hazard rate. Models (7.12) and (7.19) are examples of this form.

- *Series systems (Addition of hazard rates)*
  The hazard rate of a series system with n independent components is the sum of the hazard rates of the components. By choosing some hazard rates to be IHR and the rest to be DHR, one may arrive at a BT hazard rate. Models (7.19), (7.20), (7.38), (7.39) and (7.40) all belong to this category. Generalizing this idea, lifetime distributions with hazard functions of the form

$$h(x) = A_1 h_1(x) + A_2 h_2(x), \quad x > 0,$$

  were investigated by Shaked [529]. In the above formulation, A_1 and A_2 are independent of $h_1(x)$ and $h_2(x)$, while both $h_1(x)$ and $h_2(x)$ may be assumed to be of known forms. Shaked [529] chose $h_1(x) = 1$ and $h_2(x) = \sin x$, for example, in modelling hazard rate influenced by periodic fluctuations of temperature. Gaver and Acar [210] discussed models with hazard rates of the form

$$h(x) = h_1(x) + \lambda + h_2(x),$$

  where $h_1(x)$ is positive and decreasing and tends to zero as $x \to \infty$, and $h_2(x)$ is increasing. One can see several hazard functions of the above two forms in our earlier discussions. Closely related to these are distributions with polynomial form for $h(x)$.
- *Stochastic hazard rates*
  Rajarshi and Rajarshi [500] identified a stochastic hazard rate as

$$h^*(x) = u + h_1(x),$$

  where u is the realization of a continuous positive random variable U, and $h_1(x)$ need not be a hazard rate, and $h^*(x)$ is the hazard rate of X given $U = u$. It is obvious that the above representation is a special case of the additive hazard rate model of Nair and Sankaran [446] discussed in connection with the Hjorth [272] model. The BT-shaped hazard functions obtained earlier as the sum of hazard rate models of Murthy et al. [435], Shaked [529] and Davis and Feldstein [167] also belong to this category. The ageing properties and stochastic order relations connecting the random variable X^* (corresponding to $h^*(x)$) and the baseline variable X have been studied in Nair and Sankaran [446]).
- *Mixtures*
  Mixtures of two distributions, with one of the components being IHR and the other being DHR, may yield a BT-shaped hazard rate model. See the mixture distributions discussed in Sect. 7.4 for illustration.
- *Introduction of additional parameters*
  Introducing additional parameters that influence the shape of a baseline distribution has become a standard practice to generate new models with BT hazard rates. One simple method is exponentiation, i.e., to consider $[F(x)]^\theta$, $\theta > 0$, where $F(x)$ is a life distribution. The exponentiated Weibull, generalized exponential, and generalized linear failure rate distributions are all examples of this kind.

Another method is to use the Marshall-Olkin [411] method. A given survival function $\overline{G}(x)$ is modified into the form

$$\overline{F}(x) = \frac{\theta\overline{G}(x)}{1-(1-\theta)\overline{G}(x)}.$$

Several such models are discussed, along with their hazard rates, by Marshall and Olkin [411]. See also various generalizations arising from the Weibull distributions in Sects. 7.4 and 7.5.

- *Upside-down mean residual life models*

  Like the hazard rate, the mean residual life function can also have BT and UBT shapes. The following theorem, from Ghai and Mi [211], is of interest in the pursuit of BT or UBT hazard rates.

Theorem 7.2. *Let x_0 be the unique change point of a UBT (BT) mean residual life function $m(x)$. Suppose there exists a $t_0 \in [x_0,\infty)$ such that $m(x)$ is concave (convex) in $[0,t_0]$ and convex (concave) in $[t_0,\infty)$. If $m'(x)$ is convex (concave) in $[x_0,t_0)$, then either of the following is true:*

(a) $h(x)$ exhibits a BT (UBT) that has two change points $x_1 < x_2$, where $x_0 \leq x_1 < x_2 \leq t_0$;
(b) $h(x)$ exhibits a BT (UBT) that has a unique change point x^*, where $x_0 \leq x^* \leq t_0$.

Hence, a known mean residual life satisfying Theorem 7.2 can generate a BT or UBT hazard rate. Other methods of obtaining BT shapes for the hazard rate function will be discussed in the following section.

7.6 Quantile Function Models

So far, we have discussed in this chapter models based on distribution functions that possess nonmonotone hazard rates. Since many of the models have tractable quantile functions, a quantile-based analysis is possible in all such cases. While analysing the standard quantile functions in Chap. 3, the nonmonotonicity of their hazard quantile functions was witnessed to make use of them in data analysis. The primary objective of the present section is to enrich the domain of applications by finding some more new quantile functions.

7.6.1 Bathtub Hazard Quantile Functions Using Total Time on Test Transforms

Recall from Chap. 5 that the total time on test transform (TTT) of order n of a non-negative random variable with quantile functions $Q(u)$ is defined as

$$T_n(u) = \int_0^u (1-p)t_{n-1}(p)dp, \quad n = 1,2,3,\ldots, \tag{7.47}$$

with $T_0(u) = Q(u)$ and $t_n(u) = T_n'(n)$, provided $\mu_{n-1} = \int_0^1 T_{n-1}(p)dp < \infty$. Since $T_n(u)$ is also a quantile function, let us denote by X_n the corresponding random variable. Let $\mu_n = E(X_n)$ and $H_n(u)$ be the hazard quantile function of X_n. We then have

$$t_n(u) = (1-u)t_{n-1}(u) = [H_{n-1}(u)]^{-1}$$

and

$$t_n(u) = (1-u)^n t_0(u) = (1-u)^n q(u) = \frac{(1-u)^{n-1}}{H(u)}.$$

Finally,

$$H(u) = (1-u)^n H_n(u), \quad n = 0,1,2,3,\ldots \tag{7.48}$$

Definition (7.47) applies to negative integers as well; for example, $Q(u)$ can be thought of as the transform of $T_{-1}(u)$. In that case, the hazard quantile function $H_{-n}(u)$ corresponds to

$$H_{-n}(u) = (1-u)^n H(u), \quad n = 0,1,2,3,\ldots. \tag{7.49}$$

Equation (7.48) reveals that, in successive transforms, the hazard quantile function increases when n is positive and decreases when n is negative. The following results (Nair et al. [448]) are useful in this connection.

Theorem 7.3. *1. The random variable X has BT hazard quantile function if there exists a u_0 for which $Q(u) \geq L(u)$ in $[0,u_0]$ and $Q(u) \leq L(u)$ in $[u_0,1]$, where $L(u)$ is the quantile function of the Pareto II distribution with parameters $(k, \frac{1}{n})$. Then, u_0 will be the change point;*

2. The random variable X_n has UBT hazard quantile function if there exists a u_0 for which $T_n(u) \leq B(u)$ in $[0,y_0]$ and $T_n(u) \geq B(u)$ in $[u_0,1]$, where $B(u)$ is the quantile function of the rescaled beta distribution with parameters $(\frac{k}{n+1}, \frac{1}{n+1})$. Then, we have u_0 as the change point.

From (7.48) and (7.49), we see that for DHR (IHR) distributions the hazard quantile function of X_n has a tendency to increase (decrease). In effect, we look at the successive transforms where a change point occurs in the corresponding hazard quantile function to construct a model with BT- or UBT-shaped hazard quantile function. This technique will be used to develop new quantile functions with the above property from some standard distributions.

- *Weibull distribution*

It has been seen in the previous sections that many of the models with nonmonotone hazard rates were generated by either generalizing or modifying the Weibull distribution. In the same spirit, the present example also considers the Weibull distribution with survival function

$$\overline{F}(x) = \exp\left\{-\left(\frac{x}{\alpha}\right)^{\beta}\right\}, \quad x > 0; \alpha, \beta > 0,$$

and mean $\mu = \alpha\Gamma(1+\frac{1}{\beta})$, as the baseline model. Using the quantile function

$$Q(u) = \alpha\{-\log(1-u)\}^{\frac{1}{\beta}},$$

we have

$$H(u) = \beta\alpha^{-1}\{-\log(1-u)\}^{1-\frac{1}{\beta}}$$

and from (7.48),

$$H_n(u) = \beta\alpha^{-1}(1-u)^{-n}\{-\log(1-u)\}^{1-\frac{1}{\beta}}$$

and

$$H_n'(u) = \beta\alpha^{-1}\{-\log(1-u)\}^{-\frac{1}{\beta}}(1-u)^{-n+1}\left\{1-\frac{1}{\beta}-n\log(1-u)\right\}. \quad (7.50)$$

Since $H_n(u)$ has the tendency to increase with n, the only possibility to get a BT hazard quantile function is to consider DHR distributions. Accordingly, we take the DHR Weibull distribution with $\beta \leq 1$. Equation (7.50) reveals that $H_n(u)$ is concave in $[u_0, 1]$ and convex on $[0, u_0]$, where $u_0 = 1-\exp(\frac{\beta-1}{n\beta})$, $\beta \leq 1$. Hence, X_n has BT distribution for $n \geq 1$. As seen from the expression for u_0, the change point u_0 also increases with n so that X_n becomes IHR for a larger range, along with increasing n.

Take the case when $n = 1$. We have the random variable X_1 in the support of $(0, \mu)$ with quantile function $T_1(u)$ and hazard quantile function as

$$H_1(u) = \beta\alpha^{-1}(1-u)^{-1}\{-\log(1-u)\}^{1-\frac{1}{\beta}}.$$

The quantile density function is

$$t_1(u) = \alpha\beta^{-1}\{-\log(1-u)\}^{1-\frac{1}{\beta}}, \quad 0 \leq u \leq 1, \quad (7.51)$$

which is bathtub-shaped hazard quantile function with change point $u_0 = 1 - \exp(\frac{\beta-1}{\beta})$. We can find the distributional characteristics of X_1 from (7.50). Quantile function corresponding to (7.50) is expressed in terms of the incomplete gamma function as

$$T_1(u) = \frac{\alpha}{\beta}\Gamma_{-\log(1-u)}\left(\frac{1}{\beta}\right), \quad 0<\beta<\infty,$$

where

$$\Gamma_x(p) = \int_0^x e^{-t}t^{p-1}dt.$$

The first four L-moments of the distribution are as follows:

$$L_1 = E(X) = \frac{\alpha\Gamma\left(\frac{1}{\beta}\right)}{\beta 2^{\frac{1}{\beta}}},$$

$$L_2 = \frac{\alpha}{\beta}\left(2^{-\frac{1}{\beta}} - 3^{-\frac{1}{\beta}}\right)\Gamma\left(\frac{1}{\beta}\right),$$

$$L_3 = \frac{\alpha}{\beta}\left\{2^{-\frac{1}{\beta}} - 3(3^{-\frac{1}{\beta}}) + 2(4^{-\frac{1}{\beta}})\right\}\Gamma\left(\frac{1}{\beta}\right),$$

$$L_4 = \frac{\alpha}{\beta}\left\{2^{-\frac{1}{\beta}} - 6(3^{-\frac{1}{\beta}}) + 10(4^{-\frac{1}{\beta}}) - 5(5^{-\frac{1}{\beta}})\right\}\Gamma\left(\frac{1}{\beta}\right).$$

Thus, the L-skewness has the simple expression

$$\begin{aligned}\tau_3 &= \frac{2^{-\theta} - 3^{1-\theta} + 4^{\frac{1}{2}-\theta}}{2^{-\theta} - 3^{-\theta}} \\ &= 1 - \frac{2(3^{-\theta} - 4^{-\theta})}{2^{-\theta} - 3^{-\theta}} \\ &= 1 - \frac{2\{1-(\frac{3}{4})^{\theta}\}}{(\frac{3}{2})^{\theta} - 1}, \quad \text{with } \theta = \beta^{-1}.\end{aligned}$$

As $\theta \to \infty$ or $\beta \to 0$, we see that τ_3 tends to 1, and as $\theta \to 0$, we have τ_3 appropriately -0.53. Hence, the distribution covers skewness in the range $(-0.53, 1)$. On the other hand, the L-kurtosis is

$$\tau_4 = 1 - \frac{5 - (\frac{3}{4})^{\theta} + 5(\frac{3}{5})^{\theta}}{(\frac{3}{2})^{\theta} - 1}$$

which tends to 1 as $\beta \to 0$. The parameters of the distribution allows easy estimation by equating the first two L-moments of the sample with those of the population. Thus, (7.50) gives a two-parameter life distribution with BT-shaped hazard quantile function.

7.6.2 Models Using Properties of Score Function

The results discussed here are mainly based on the work of Nair et al. [448]. Recall from Sect. 4.3 that the definition of the score function is

$$J(u) = \frac{q'(u)}{q^2(u)},$$

where $q(u)$ is the quantile density function of the lifetime X. We see that

$$J(u) = -\frac{d}{du}\frac{1}{q(u)} = -\frac{d}{du}(1-u)H(u),$$

or equivalently

$$(1-u)H'(u) = H(u) - J(u).$$

Thus, X is $I(D)$ according as $H(u) \geq J(u)$ for all u. Further, if $H(u)$ is nonmonotonic, the change points of $H(u)$ are zeros of $H(u) - J(u)$. Geometrically, for increasing (decreasing) $H(u)$, the $H(u)$ curve lies above (below) that of $J(u)$ and for BT (UBT) hazard quantile function $H(u)$ crosses $J(u)$ from below (above). An interesting property of $J(u)$ is that there exists some simple relationships between $J(u)$ and $H(u)$ that characterize many life distributions.

Theorem 7.4. *The random variable X is distributed as generalized Pareto with*

$$Q(u) = ba^{-1}\left\{(1-u)^{-\frac{a}{a+1}} - 1\right\}, \quad a > -1, b > 0, \tag{7.52}$$

if and only if

$$J(u) = cH(u) \tag{7.53}$$

for a positive constant c.

Proof. Assuming (7.50), we find $J(u)$ and $H(u)$ as

$$J(u) = \frac{2a+1}{b}(1-u)^{\frac{a}{a+1}}$$

and

$$H(u) = \frac{a+1}{b}(1-u)^{\frac{a}{a+1}}.$$

This readily verifies (7.53) with $c = \frac{2a+1}{a+1}$. Conversely, if (7.53) applies to a random variable X, then

$$\frac{H'(u)}{H(u)} = \frac{1-c}{1-u}$$

and

$$H(u) = K(1-u)^{c-1},$$
$$q(u) = K(1-u)^{c-1},$$

which is the quantile density function of the generalized Pareto with $c = \frac{2a+1}{a+1}$. Hence, the theorem.

Remark 7.2. When $c = 1$, we have the exponential distribution and $c > (<)1$ leads to the Pareto II (rescaled beta) model. It is apparent that by generalizing the identity in (7.53), we can obtain more flexible models. This fact is illustrated in the following theorems.

Theorem 7.5. *The relationship*

$$J(u) = AH(u) + B$$

is satisfied for all u and real constants A and B if and only if the distribution of X is given by

$$Q(u) = \begin{cases} \log\left\{\left(1+\frac{B}{1-A}\right)^{\frac{1}{B}}\left(c+\frac{B}{1-A}(1-u)^{1-A}\right)^{-\frac{1}{B}}\right\}, & c \le 1,\ A \ne 1 \\ \frac{1}{B}\log\left\{\frac{c}{c+B\log(1-u)}\right\}, & A = 1,\ c > 0. \end{cases} \tag{7.54}$$

Theorems 7.4 and 7.5 do not provide models with nonmonotone hazard quantile functions. The distribution in Theorem 7.5 contains known models like the exponential, Pareto, rescaled beta, half-logistic and Gompertz as special cases. In general, $H(u)$ is increasing for (7.54) when $A < 1, 0 < c \le 1$ or $A > 1, C < 0$ and decreasing when $A > 1, C < 0$ or $A > 1, 0 < c \le 1$. Some other properties of the distribution have been studied by Nair et al. [448].

Returning to the construction of bathtub-shaped $H(u)$, we have the following characterization that generates a distribution with BT-shaped hazard quantile function.

Theorem 7.6. *If the functions $J(u)$ and $H(u)$ are such that*

$$J(u) = \left(A + \frac{\alpha}{u}\right)H(u) \tag{7.55}$$

for all u, then it is necessary and sufficient that the distribution is specified by the quantile density function

$$q(u) = Ku^{\alpha}(1-u)^{-(A+\alpha)}, \tag{7.56}$$

where α, A and K are real constants.

Proof. Equation (7.55) is equivalent to

$$\frac{q'(u)}{q^2(u)} = \frac{(A+\frac{\alpha}{u})}{(1-u)q(u)},$$

or

$$\frac{d\log q(u)}{du} = \frac{A+\frac{\alpha}{u}}{(1-u)}.$$

Integrating the above equation, we obtain (7.56). Conversely, if the distribution is of the form (7.56), then by direct calculations, we have

$$H(u) = K^{-1}u^{-\alpha}(1-u)^{A+\alpha-1}$$

and

$$J(u) = K^{-1}u^{-\alpha-1}(1-u)^{A+\alpha-1}(uA+\alpha),$$

thus verifying (7.55).

The family of distributions in (7.55) includes several well-known distributions as special cases. Of these are

- the exponential $(\alpha = 0, A = 1)$ with constant $H(u)$;
- Pareto II $(\alpha = 0, A < 1)$ with decreasing $H(u)$;
- rescaled beta $(\alpha = 0, A > 1)$ with increasing $H(u)$;
- loglogistic $(A = 2, \alpha = \lambda - 1)$, specified by

  $$\bar{F}(x) = \frac{x^{\frac{1}{\lambda}}}{\alpha^{\frac{1}{\lambda}} + x^{\frac{1}{\lambda}}}, \quad x > 0,\ \lambda, \alpha > 0.$$

  The reliability aspects of this distribution have been studied by Gupta et al. [237]. Since

  $$J(u) = \frac{2u+\alpha-1}{u}H(u)$$

  in this case, X is UBT with change point at $u_0 = 1-\lambda$;
- Govindarajulu's distribution with

  $$Q(u) = \theta + \sigma\left\{(\beta+1)u^{\beta} - \beta u^{\beta+1}\right\}$$

  on setting $\alpha = \beta - 1$, $A = -\beta$ and $K = \sigma\beta(\beta+1)$. See Chap. 3 and Nair et al. [448] for a detailed discussion on the properties and reliability implications. For $A > -1$, $H(u)$ is increasing while for $A < -1$, $H(u)$ has bathtub shape with

Table 7.1 Observed and expected frequencies for the gastric carcinoma data

Class intervals	0–111.8	111.8–197	197–289.4	289.4–401	401–550.3	550.3–782	782–1265.5	> 1265.5
Observed frequencies	13	10	9	15	14	10	14	10
Expected frequencies	12	12	12	12	12	12	12	11

change point $u = \frac{A+1}{A-1}$. The hazard quantile can be differentiated to study its shape for various values of the parameters. We have

$$H'(u) = K^{-1}u^{-\alpha-1}(1-u)^{A+\alpha-2}\{-\alpha+u(1-A)\}.$$

Thus, $H(u)$ is increasing for $\alpha < 0, A < 1$, and decreasing for $\alpha < 0, A > 1$ for all u giving the IHR and DHR cases. The BT and UBT cases also hold, respectively, when $\alpha > 0, A < 1$ and $\alpha \not< 0, A > 1$. Accordingly, the model can cover all the cases.

Example 7.1. The use of the model was tested against the data on survival times in days from a clinical trial on gastric carcinoma on 90 patients, as given by Kleinbaum [342], by considering the survival times alone in a single set. In order to estimate the parameters of the model, the 25th, 50th and 75th percentiles of the sample and the population are matched. This procedure results in the estimates

$$\hat{\alpha} = -0.3128,\ \hat{A} = 1.7693 \text{ and } \hat{K} = 296.267.$$

We then calculated the observed and expected frequencies for various classes and these are reported in Table 7.1. The χ^2 value of 3.14 does not reject the model in (7.56) for the data at 5 % level of significance.

Some distributional aspects of (7.56) will also be interest in further analysis. The first four L-moments, for example, are as follows:

$$\begin{aligned}
L_1 &= KB(\alpha+1, 2-A-\alpha), \quad \text{with } A+\alpha < 2,\\
L_2 &= KB(\alpha+2, 2-A-\alpha),\\
L_3 &= K\{B(\alpha+3, 2-A-\alpha) - B(\alpha+2, 3-A-\alpha)\},\\
L_4 &= K\{B(\alpha+2, 2-A-\alpha) - 5B(\alpha+3, 3-A-\alpha)\}.
\end{aligned}$$

Hence, as a location measure, the mean is

$$\mu = KB(\alpha+1, 2-A-\alpha)$$

and as a dispersion measure, the mean difference is

$$\Delta = 2KB(\alpha+1, 2-A-\alpha).$$

The L-skewness is

$$\tau_3 = \frac{L_3}{L_2} = \frac{(\alpha+2)-(2-A-\alpha)(4-A)}{(4-A)},$$

and the L-Kurtosis is

$$\tau_4 = 1 - \frac{5(\alpha+2)(2-A-\alpha)}{(5-A)(4-A)}.$$

Theorem 7.7. *The relationship*

$$J(u) = [A + M\{\log(1-u)\}^{-1}]H(u) \tag{7.57}$$

is satisfied for all u and real A and M if and only if

$$q(u) = K(1-u)^{-A}\{-\log(1-u)\}^{-M}. \tag{7.58}$$

Proof. Rewriting (7.57) as

$$\frac{q'(u)}{q^2(u)} = \left[A + M\{\log(1-u)\}^{-1}\right]\frac{1}{(1-u)q(u)},$$

we have

$$\frac{q'(u)}{q(u)} = \frac{A}{1-u} + \frac{M}{(1-u)q(u)}.$$

Integrating, we obtain (7.58). Conversely, logarithmic differentiation of (7.58) leads to (7.57). Hence, the theorem.

We can write the quantile function in terms of special function as

$$Q(u) = K(1-A)^{M-1}I(1-M, \log(1-u)^{A+1}),$$

where

$$I(a,x) = \int_0^x e^{-t}t^{a-1}dt$$

is the incomplete gamma function. The density function of X can be written in terms of the survival function as

$$f(x) = C[\overline{F}(x)]^A\{1-\log\overline{F}(x)\}^M, \quad x > 0.$$

Some special cases of (7.58) are

- The Weibull distribution with shape parameter λ and scale parameter $\sigma = K\lambda$, and in particular, exponential and Rayleigh distributions when $\lambda = 1$ and 2, respectively;
- Pareto II ($A > 1$, $M = 0$), rescaled beta ($A < 1$, $M = 0$), and uniform ($A = 0$, $M = 0$).

Thus, (7.58) is a generalized Weibull model belonging to the category of several such models discussed in the preceding sections. The hazard quantile function has the form

$$H(u) = K^{-1}(1-u)^{A-1}\{-\log(1-u)\}^{M}.$$

Upon taking the derivative, we get

$$H'(u) = K^{-1}(1-u)^{A-2}\{-\log(1-u)\}^{M-1}\{M + (A-1)\log(1-u)\}. \tag{7.59}$$

Equation (7.59) shows that $H(u)$ is capable of taking on different shapes. In fact,

$$\begin{aligned}
&X \text{ is IHR when } A \leq 1, M > 0; A < 1, M = 0;\\
&X \text{ is DHR when } A \leq 1, M < 0; A > 1, M = 0;\\
&X \text{ is BT when } A < 1, M < 0;\\
&X \text{ is UBT when } A > 1, M > 0;\\
&X \text{ is exponential when } A = 1, M = 0.
\end{aligned}$$

We now look at some distributional properties of this family. First, we see that the members of the family are either unimodal or monotonic with modal value at $u_0 = 1 - \exp(\frac{M}{A})$. The summary measures can be described in terms of the quantiles or. We have the first four L-moments as follows:

$$\begin{aligned}
L_1 &= \frac{K\Gamma(1-M)}{(2-A)^{1-M}}, \quad M < 1, A < 2,\\
L_2 &= \left\{1 - \left(\frac{2-A}{3-A}\right)^{1-M}\right\} L_1,\\
L_3 &= \left\{1 - 3\left(\frac{2-A}{3-A}\right)^{1-M} + 2\left(\frac{1-A}{4-A}\right)^{1-M}\right\} L_1,\\
L_4 &= \left\{1 - 6\left(\frac{2-A}{3-A}\right)^{1-M} + 10\left(\frac{2-A}{4-A}\right)^{1-M} - 5\left(\frac{2-A}{3-A}\right)^{1-M}\right\} L_1.
\end{aligned}$$

The mean, mean difference, L-skewness and L-kurtosis are all readily obtained from the above expressions.

Table 7.2 Observed and expected frequencies of the failure time data

Class intervals	0–3.184	3.184–13.5	13.5–29.48	29.48–48.5	48.5–67.47	67.47–83.25	>83.25
Observed freq.	9	4	6	6	9	6	8
Expected freq.	6	6	6	6	6	6	14

For an empirical validation of the model, the data on the failure times of 50 devices given in Lai and Xie [368, p. 353] is considered. Matching the 25th, 50th and 75th percentiles of the sample with the corresponding percentiles of the population, the estimates of the model parameters are found to be

$$\hat{A} = -1.8224, \quad \hat{M} = -1.2576 \quad \text{and} \quad \hat{K} = 875.927.$$

A χ^2 value of 4.509 is found from the observed and expected frequencies presented in Table 7.2, which does not lead to rejection of the model.

The methods suggested in this section, using the total time on test transform as well as the relationship between $J(u)$ and $H(u)$, are quite general in nature. The above examples illustrate how we can work with them. It will, of course, be of interest to develop more flexible families of distributions that generalize the existing distributions and present varying shapes and characteristics to become practically useful!

Chapter 8
Stochastic Orders in Reliability

Abstract Stochastic orders enable global comparison of two distributions in terms of their characteristics. Specifically, for a given characteristic A, stochastic order says that the distribution of X has lesser (greater) A than the distribution of Y. For example, one may use hazard rate or mean residual life for such a comparison. In this chapter, we discuss various stochastic orders useful in reliability modelling and analysis.

The stochastic order treated here are the usual stochastic order, hazard rate order, mean residual life order, harmonic mean residual life order, renewal and harmonic renewal mean residual life orders, variance residual life order, percentile residual life order, reversed hazard rate order, mean inactivity time order, variance inactivity time order, the total time on test transform order, the convex transform (IHR) order, star (IHRA) order, DMRL order, superadditive (NBU) order, NBUE order, NBUHR and NBUHRA orders and MTTF order. The interpretation of ageing concepts, preservation properties with reference to convolution, mixing and coherent structures are also discussed in relation to each of these orders. Implications among the different orders are also presented. Examples of the stochastic orders and counter examples where certain implications do not hold are also provided. Some special models used in reliability like proportional hazard and reverse hazard models, mean residual life models and weighted distributions have been discussed in earlier chapters. Some applications of these stochastic models are reviewed as well.

8.1 Introduction

There are many situations in practice wherein we need to compare the characteristics of two distributions. In certain cases, descriptive measures like mean and variance have been used for this purpose. Since these measures are summary measures of the data, they become less informative and so cannot capture all the essential features inherent in the data. An alternative approach to assess the relative behaviour of the properties of distributions is provided by stochastic orders which provide a global

N.U. Nair et al., *Quantile-Based Reliability Analysis*, Statistics for Industry and Technology, DOI 10.1007/978-0-8176-8361-0_8,

comparison by taking into account different features of the underlying models. Specifically, for a given characteristic A, a stochastic order says that the distribution F_X of a random variable X has lesser (greater) A than the distribution F_Y of Y and we express it as $F_X \leq_A F_Y$ ($F_X \geq_A F_Y$), or equivalently in terms of the random variables $X \leq_A Y$ ($X \geq_A Y$). For example, in the context of reliability theory, if two manufacturers produce devices for the same purpose, the natural interest is to know which is more reliable. The reliability functions of the two devices then become natural objects for comparison and the characteristic in question may be their mean lives. But, when both devices were working for a specified time, the characteristic in question may change to the mean residual life and the comparison confirms which one of the two has more remaining life on an average. In all cases of comparison, the characteristic of comparison should have an appropriate measure $\omega(A)$, which should satisfy $\omega_X(A) \leq \omega_Y(A)$. Marshall and Olkin [412] point out that Mann and Whitney [409] used this approach initially and Birnbaum [101] subsequently to study peakedness. There is a phenomenal growth in the study of stochastic orders in recent years in such diverse fields as reliability theory, queueing theory, survival studies, biology, economics, insurance, operations research, actuarial science and management. In this chapter, we take up such stochastic orders and present results relevant to reliability analysis using quantile functions. Details of other orderings, proofs of results using the distribution function approach and so on are well documented; see, e.g., Szekli [557] and Shaked and Shantikumar [531].

Some notation need to be introduced first for the developments in subsequent discussions. Let Ω be a nonempty set. A binary relation $\leq$ on this set is called a preorder if

(i) $x \leq x, x \in \Omega$ (reflexivity),
(ii) $x \leq y, y \leq z \Rightarrow x \leq z$ (transitivity).

If, in addition, we also have

(iii) $x \leq y, y \leq x \Rightarrow x = y$ (anti-symmetry),
then $\leq$ is called a partial order. The term stochastic order considered here include both preorders and partial orders.

Let F and G be distribution functions of random variables X and Y, respectively. Then, the function

$$\psi_{F,G}(x) = G^{-1}(F(x)), \tag{8.1}$$

for all real x, is called the relative inverse function of F and G. If F is continuous and supported by an interval of reals, then $\psi(X)$ and Y are identically distributed. If U is uniformly distributed over $[0,1]$, then $\psi_{F_U,G}(U)$ has the same distribution as Y. On the other hand, if Y is exponential, $\psi_{\text{Exp},F}(Y)$ has the same distribution as X for $X \geq 0$. These are easy to verify from the definition of the ψ function. Further, if F and G are strictly increasing with derivatives f and g, then

$$\frac{d}{dx}\psi(x) = \frac{f(x)}{g(G^{-1}F(x))} \tag{8.2}$$

and

$$\frac{d}{dx}FG^{-1}(x) = \frac{f(G^{-1}(x))}{gG^{-1}(x)}. \tag{8.3}$$

If G is continuous with interval support, then

$$\psi_{F,G}^{-1}(x) = \psi_{G,F}(x). \tag{8.4}$$

8.2 Usual Stochastic Order

The usual stochastic order is basic in the sense that it compares the distribution functions of two random variables.

Definition 8.1. Let X and Y be random variables with quantile functions $Q_X(u)$ and $Q_Y(u)$, respectively. We say that X is smaller than Y in the usual stochastic order, denoted by $X \leq_{\text{st}} Y$, if and only if

$$Q_X(u) \leq Q_Y(u) \text{ for all } u \text{ in } (0,1).$$

The $\leq_{\text{st}}$ ordering is usually employed to compare the distributions of two random variable X and Y or to compare the distribution of X at two chosen parameter values.

Example 8.1. Let X follow Pareto II distribution with

$$Q_X(u) = (1-u)^{-\frac{1}{c}} - 1, \quad c > 0,$$

and Y follow the beta distribution with

$$Q_Y(u) = 1 - (1-u)^{\frac{1}{c}}, \quad c > 0.$$

Then,

$$\begin{aligned} Q_Y(u) - Q_X(u) &= 1 - (1-u)^{\frac{1}{c}} - \frac{1-(1-u)^{\frac{1}{c}}}{(1-u)^{\frac{1}{c}}} \\ &= -(1-u)^{-\frac{1}{c}}\left\{1-(1-u)^{\frac{1}{c}}\right\}^2 \\ &\leq 0 \quad \text{for all } u. \end{aligned}$$

Thus, $X \geq_{\text{st}} Y$.

Example 8.2. Assume that X_λ has exponential distribution with

$$Q(u) = -\frac{1}{\lambda}\log(1-u)$$

for $\lambda > 0$. It is easy to verify that for $\lambda_1 < \lambda_2$, $X_{\lambda_1} \le_{\rm st} X_{\lambda_2}$.

There are several equivalent forms of Definition 8.1 that are useful in establishing stochastic ordering results. We list them in the following theorem.

Theorem 8.1. *The following conditions are equivalent:*

(i) $X \le_{st} Y$;
(ii) $\overline{F}_X(x) \le \overline{F}_Y(x)$ *or* $F_X(x) \ge F_Y(x)$ *for all* x;
(iii) $E\phi(X) \le E\phi(Y)$ *for all increasing functions* ϕ *for which the expectations exist. As a consequence, it is apparent that if* $\phi(x) = x^r$, *then*

$$X \le_{st} Y \Rightarrow \begin{cases} E(X^r) \le E(Y^r), & r \ge 0 \\ E(X^r) \ge E(Y^r), & r \le 0 \end{cases}$$

which connects the moments of the two distributions. Another function of interest is $\phi(x) = e^{tx}$, *with which we have a comparison of moment generating functions as*

$$X \le_{st} Y \Rightarrow \begin{cases} E(e^{tX}) \le E(e^{tY}), & t \ge 0 \\ E(e^{tX}) \ge E(e^{tY}), & t \le 0. \end{cases}$$

Proof of the main result is available in Szekli [557]. If ϕ *is strictly increasing and* $X \le_{st} Y$, *then* X *and* Y *are identically distributed if* $E\phi(X) = E\phi(Y)$;
(iv) $\phi(X) \le_{st} \phi(Y)$ *for all increasing functions* ϕ;
(v) $Q_Y^{-1}(Q_X(u)) \le u$;
(vi) $\phi(X,Y) \le_{st} \phi(Y,X)$ *for all* $\phi(x,y)$, *where* $\phi(x,y)$ *is increasing in* x *and decreasing in* y *and* X *and* Y *are independent.*

One important advantage of studying stochastic orders is that many of the ageing concepts discussed earlier in Chap. 4 can be expressed in terms of some ordering. This in turn assists us in deriving many new properties and bounds based on the properties of the orderings, which are otherwise not explicit. We now present some theorems defining the IHR (DHR), NBU (NWU), NBUE, NBUC, RNBU, DMRL and RNBRU classes discussed in Chap. 4.

Theorem 8.2. *The lifetime variable* X *is IHR (DHR) if and only if* $X_t \le_{st} (\ge) X_{t'}$ *whenever* $t < t'$, *where* $X_t = (X - t|X > t)$ *is the residual life.*

Proof. The quantile function of the residual life at t is given by (1.4) as

$$Q_1(u) = Q(u_0 + (1-u_0)u) - Q(u_0),$$

where $u_0 = F(t)$ and $Q(\cdot)$ is the quantile function of X. Similarly, for $X_{t'}$, we have

$$Q_2(u) = Q(u_1 + (1-u_1)u) - Q(u_1),$$

with $u_1 = F(t') > u_0$. Now assume that $X_t \leq_{\text{st}} X_{t'}$. Then, by Definition 8.1, we have

$$\begin{aligned}
&Q(u_0 + (1-u_0)u) - Q(u_0) \leq Q(u_1 + (1-u_1)u) - Q(u_1) \\
&\quad \Leftrightarrow Q(u_1) - Q(u_0) \leq Q(u_1 + (1-u_1)u) - Q(u_0 + (1-u_0)u) \\
&\quad \Leftrightarrow \frac{Q(u_1) - Q(u_0)}{(1-u)(u_1 - u_0)} \leq \frac{Q(u_1 + (1-u_1)u) - Q(u_0 + (1-u_0)u)}{(u_1 + (1-u_1)u) - (u_0 + (1-u_0)u)} \\
&\quad \Rightarrow \frac{1}{1-u} q(u_0) \leq q(u_0 + (1-u_0)u) \qquad (8.5) \\
&\quad \Rightarrow \frac{1}{(1-u_0)q(u_0)} \leq \frac{1}{(1-u_0 - (1-u_0)u)q(u_0 + (1-u_0)u)} \\
&\quad \Rightarrow H(u_0) \leq H(u_0 + (1-u_0)u) \text{ for every } u_0 \text{ in } (0,1). \\
&\quad \Rightarrow X \text{ is IHR.}
\end{aligned}$$

Conversely, when X is IHR, we can retrace the above steps up to (8.5). However, (8.5) is equivalent to

$$\frac{d}{du_0}\left\{\frac{1}{1-u} Q(u_0) - \frac{1}{1-u} Q(u_0 + (1-u_0))u\right\} \leq 0$$

which means that

$$Q(u_0) - Q(u_0 + (1-u_0)u)$$

is a decreasing function of u_0. Hence,

$$Q(u_0) - Q(u_0 + (1-u_0)u) \geq Q(u_1) - Q(u_1 + (1-u_1)u)$$

for $u_1 > u_0$ or $Q_1(u) \leq Q_2(u)$ as we wished to prove. The proof of the DHR case is obtained by simply reversing the above inequalities.

Theorem 8.3. *A lifetime X is NBU (NWU) if and only if $X \geq_{st} (\leq_{st}) X_t$.*

The result is a straightforward application of Definition 4.22.

Theorem 8.4. *If X is a lifetime random variable with $E(X) < \infty$, then X is NBUE (NWUE) if and only if $X \geq_{st} (\leq_{st}) Z$, where Z is the equilibrium random variable with survival function* (4.7).

Proof. Assume that $X \geq_{\text{st}} Z$. Then, from (4.9), we have

$$Q_X(u) \geq Q_Z(u) = \mu Q_X(T_X^{-1}(u)),$$

where $T_X(x) = \int_0^u (1-p)q(p)dp$ and $\mu = E(X)$. This gives

$$X \geq_{\text{st}} Z \Leftrightarrow Q_X(T_X(u)) \geq Q_X(\mu u) \Leftrightarrow \int_0^u (1-p)q(p)dp \geq \mu u$$

$$\Leftrightarrow \mu - \int_u^1 (1-p)q(p)dp \geq \mu u$$

$$\Leftrightarrow \frac{1}{1-\mu}\int_u^1 (1-p)q(p)dp \leq \mu \Leftrightarrow X \text{ is NBUE.}$$

from Definition 4.33.

Theorem 8.5 (Nair and Sankaran [446]).

(a) $X \geq_{st} Z_t$ for all $t \geq 0 \Leftrightarrow X$ is NBUC, where $Z_t = Z - t|(Z > t)$ is the residual life of Z;
(b) $Z \geq_{st} X_t \Leftrightarrow X$ is RNBU;
(c) $X_t \geq_{st} Z_t \Leftrightarrow X$ is DMRL;
(d) $Z \geq_{st} Z_t \Leftrightarrow X$ is RNBRU.

As with ageing criteria, it is customary to study the preservation properties of stochastic orders. With regard to the usual stochastic order, the following properties hold:

1. Let $(X_1, X_2, \ldots, X_n)$ and $(Y_1, Y_2, \ldots, Y_n)$ be two sets of independent random variables. For every increasing function ϕ, we have

$$\phi(X_1, X_2, \ldots, X_n) \leq_{\text{st}} \phi(Y_1, Y_2, \ldots, Y_n)$$

whenever $X_i \leq_{\text{st}} Y_i$. Hence, if $X_i \leq_{\text{st}} Y_i$, then

$$\sum_{i=1}^{n} X_i \leq_{\text{st}} \sum_{i=1}^{n} Y_i.$$

Thus the usual stochastic order preserves convolution property or is closed under the formation of additional lifelengths.
2. The ordering $\leq_{\text{st}}$ is preserved under convergence in distribution. That is, if (X_n) and (Y_n) are sequences such that $X_n \to X$ and $Y_n \to Y$ as $n \to \infty$ in distribution and if $X_n \leq_{\text{st}} Y_n$, $n = 1, 2, \ldots$, then $X \leq_{\text{st}} Y$.
3. Under the formulation of mixture distributions, $\leq_{\text{st}}$ is closed. This means that if X, Y and Θ are random variables satisfying

$$[X|\Theta = \theta] \leq_{\text{st}} [Y|\Theta = \theta]$$

for all $\theta \in \Theta$, then $X \leq_{st} Y$.

4. A further extension of Property 1 above for random convolution is possible. If X_i's and Y_i's are non-negative, M is a non-negative integer valued random variable independent of the X_i's and N is non-negative integer valued random variable and independent of the Y_i's, then

$$X_i \le_{\text{st}} Y_i \Rightarrow \sum_{i=1}^{M} X_i \le_{\text{st}} \sum_{i=1}^{N} Y_i$$

provided $M \le_{\text{st}} N$.
5. The ordering $X \le_{\text{st}} Y$ is closed under shifting and scaling meaning that

$$X \le_{\text{st}} Y \Rightarrow CX \le_{\text{st}} CY$$

and

$$X \le_{\text{st}} Y \Rightarrow X + a \le_{\text{st}} Y + a.$$

More properties of the $\le_{\text{st}}$ ordering will appear in connection with other orderings discussed later. Further properties of $\le_{\text{st}}$ can be found in Muller and Stoyan [432], Scarsini and Shaked [521], Barlow and Proschan [68] and Ma [406].

8.3 Hazard Rate Order

In hazard rate ordering, we compare two distributions by means of the relative magnitude of their hazard rates. The idea behind this comparison is that when the hazard rate becomes larger, the variable becomes stochastically smaller.

Definition 8.2. If X and Y are lifetime random variables with absolutely continuous distribution functions, we say that X is smaller than Y in hazard rate order, denoted by $X \le_{\text{hr}} Y$, if

$$H_X(u) \ge H_Y^*(u),$$

where $H_X(u) = h_X(Q_X(u))$ and $H_Y^*(u) = h_Y(Q_X(u))$ and $h(\cdot)$ denotes the hazard rate function.

Example 8.3. The hazard quantile function of the Pareto II distribution (Table 2.4) is

$$H_X(u) = \frac{c(1-u)^{\frac{1}{c}}}{\alpha}$$

and the hazard rate function of the beta distribution with $R = 1$ is $h_Y(x) = \frac{c}{1-x}$. Hence,

$$H_Y^*(u) = h_Y(Q_X(u)) = h_Y((1-u)^{-\frac{1}{c}} - 1)$$
$$= \frac{c}{2-(1-u)^{-\frac{1}{c}}}.$$

It is easy to check that for $0 < u < 1$, $H_X(u) < H_Y^*(u)$ and so $X \geq_{\text{hr}} Y$.

Some equivalent conditions that ensure hazard rate order are presented in the following theorem.

Theorem 8.6. *X is less than Y in hazard rate order if and only if*

(i) $u^{-1}F_Y(Q_X(1-u))$ *is decreasing in u;*
(ii) $u^{-1}[1-F_X(Q_Y(1-u))]$ *is decreasing in u;*
(iii) $\frac{\overline{F}_Y(x)}{\overline{F}_X(x)}$ *is increasing in x;*
(iv) $\overline{F}_X(x)\overline{F}_Y(y) \geq \overline{F}_X(y)\overline{F}_Y(x)$ *for all* $x \leq y$*;*
(v) $\frac{\overline{F}_X(x+y)}{\overline{F}_X(x)} \leq \frac{\overline{F}_Y(x+y)}{\overline{F}_Y(x)}$ *for all* $x, y \geq 0$*;*
(vi) $(X|X > x) \leq_{st} (Y|Y > x)$.

Proof. (i) From (8.3), we have

$$\frac{\overline{F}_Y(Q_X(1-u))}{u} \text{ is decreasing in } u \Leftrightarrow uf_Y(Q_X(1-u))q_X(1-u) - \overline{F}_Y(Q_X(1-u)) \leq 0$$
$$\Leftrightarrow \frac{f_Y(Q_X(1-u))}{\overline{F}_y(Q_X(1-u))} \leq \frac{1}{uq_X(1-u)}$$
$$\Leftrightarrow h_Y(Q_X(1-u)) \leq H_X(1-u)$$
$$\Leftrightarrow H_Y^*(1-u) \leq H_X(1-u) \text{ for all } 0 < u < 1$$
$$\Leftrightarrow X \leq_{\text{hr}} Y.$$

The proof of (ii) is exactly similar. Result (iii) is obtained from (i) by setting $u = \overline{F}(x)$ and noting that since $u = \overline{F}(x)$ when u is decreasing x is increasing. Notice that (iv) is a consequence of (iii) while (v) is equivalent to (iv) and (vi) to (v).

When different stochastic orders are studied, the implications, if any, between them is also an important aspect. The relationship between $\leq_{\text{st}}$ and $\leq_{\text{hr}}$, e.g., is explained in the following theorem.

Theorem 8.7.

$$X \leq_{hr} Y \Rightarrow X \leq_{st} Y,$$

but not conversely.

Proof.

$$X \le_{\rm hr} Y \Leftrightarrow \frac{\overline{F}_X(x+y)}{\overline{F}_X(x)} \le \frac{\overline{F}_Y(x+y)}{\overline{F}_Y(x)}, \text{ for all } x \le y$$
$$\Rightarrow \overline{F}_X(y) \le \overline{F}_Y(y) \text{ for all } y > 0, \text{ when } x \to 0.$$
$$\Rightarrow X \le_{\rm st} Y.$$

To prove that the converse need not be true, let X be distributed as exponential with $Q_X(u) = -\log(1-u)$ and Y follow distribution with survival function

$$\overline{F}_Y = e^{-x} + e^{-2x} - e^{-3x}, \quad x > 0.$$

Since $\overline{F}_X(x) = e^{-x}$, it is easy to verify that $\overline{F}_X(x) \le \overline{F}_Y(x)$ and so $X \le_{\rm st} Y$. On the other hand,

$$Q(1-u) = \overline{F}^{-1}(u) = -\log u$$

and so

$$u^{-1}F_Y(Q(1-u)) = \frac{F_Y(-\log u)}{u} = 1 + u - u^2.$$

The last expression is increasing for u in $(0, \frac{1}{2}]$ and decreasing for u in $[\frac{1}{2}, 1)$. The hazard rates are therefore not ordered by (i) of Theorem 8.6. Hazard ordering allows definition of certain ageing classes encountered previously in Chap. 4 as the following theorems illustrate.

Theorem 8.8. *The random variable X is IHR (DHR) if and only if any one of the following conditions hold:*

(i) $(X - t|X > t) \ge_{hr} (\le_{hr})(X - s|X > s)$ *for all* $t \le s$*;*
(ii) $X \ge_{hr} (X - t|X > t)$ *for all* $t \ge 0$*;*
(iii) $X + t \le_{hr} X + s,\ t \le s$.

The proof of the theorem rests on the fact that $(X - t|X > t)$ has its hazard rate as $h(x+t)$.

Theorem 8.9. *If* $E(X) < \infty$*, then:*

(a) *X is DMRL* $\Leftrightarrow X \ge_{hr} Z$*;*
(b) *X is IMRL* $\Leftrightarrow X \le_{hr} Z$.

Proof. (a) We see that

$$X \ge_{\rm hr} Z \Leftrightarrow H_X(u) \le H_Z(u) = \frac{1}{M_X(u)}$$
$$\Leftrightarrow H_X(u)M_X(u) \le 1$$
$$\Leftrightarrow 1 - (1-u)H_X(u)M_X'(u) \le 1$$
$$\Leftrightarrow M_X'(u) \le 0 \Leftrightarrow X \text{ is DMRL}.$$

The proof of (b) is obtained by reversing the inequalities in the above argument.

Theorem 8.10. *If $E(X) < \infty$, then:*

(a) $Z \geq_{hr} (Z - t|Z > t) \Leftrightarrow X$ is DMRL;
(b) $Z_{t_1} \geq_{hr} Z_{t_2}$ for $0 < t_1 < t_2 \Leftrightarrow X$ is DMRL.

Proof. By Part (ii) of Theorem 8.8, we see that

$$Z \geq_{\text{hr}} (Z - t|Z > t) \Leftrightarrow Z \text{ is IHR } \Leftrightarrow X \text{ is DMRL}.$$

From proving (b), we use Part (i) of Theorem 8.8 and the same argument as for (a).

Some preservation properties useful in reliability analysis concerning the hazard rate ordering are as follows:

1. For every increasing function $\phi(x)$, $\phi(X) \leq_{\text{hr}} \phi(Y)$, whenever $X \leq_{\text{hr}} Y$;
2. In general, convolution is not preserved under hazard rate ordering. However, if $X_1, X_2, \ldots, X_n$ and $Y_1, Y_2, \ldots, Y_n$ are both independent collections such that $X_i \leq_{\text{hr}} Y_i$, $i = 1, 2, \ldots, n$, and X_i and Y_i are IHR for all i, then
$$\sum_{i=1}^{n} X_i \leq_{\text{hr}} \sum_{i=1}^{n} Y_i.$$
3. If $X_1, X_2, \ldots, X_n$ is a sequence of independent IHR lifetime variables and M and N are discrete positive integer valued random variables such that $M \leq_{\text{hr}} N$ and are independent of the X_i's, then
$$\sum_{i=1}^{M} X_i \leq_{\text{hr}} \sum_{i=1}^{N} X_i.$$
Thus, the ordering '$\leq_{\text{hr}}$' is only conditionally closed under the formation of random convolutions.
4. If X, Y and Θ are random variables such that $X|(\Theta = \theta) \leq_{\text{hr}} Y|(\Theta = \theta')$ for all θ and θ' in the support of Θ, then $X \leq_{\text{hr}} Y$ (Lehmann and Rojo [383]).
5. For $0 < a \leq 1$ and X is IHR, $aX \leq_{\text{hr}} X$ (Kochar [346]).
6. If $X_1, X_2, \ldots, X_n$ are independent, then:
   (a) $X_{k:n} \leq_{\text{hr}} X_{k+1:n}$ (Boland et al. [114, 115]);
   (b) $X_{1:1} \geq_{\text{hr}} X_{1:2} \geq_{\text{hr}} \cdots \geq_{\text{hr}} X_{1:n}$;
   (c) $X_{k:n-1} \geq_{\text{hr}} X_{k:n}$, $k = 1, 2, \ldots, n-1$.

   The results in (b) and (c) are due to Korwar [352] in connection with k-out-of-n system. Proofs of the above properties along with some more general results are given in Sect. 1.B of Shaked and Shantikumar [531].
7. If the hazard rate $h(x)$ of X is such that $xh(x)$ is increasing, then $Y = aX$, $a \geq 1$, satisfies $X \leq_{\text{hr}} Y$.

8.4 Mean Residual Life Order

Let X be a non-negative random variable representing the lifetime of a device with $E(X) = \mu < \infty$. Then, the comparison of the mean residual lives of X and Y by their magnitudes provides a stochastic ordering of the distributions of X and Y. Assume also that $E(Y) < \infty$.

Definition 8.3. X is said to be smaller than Y in mean residual quantile function order if

$$M_X(u) \leq M_Y^*(u),$$

written as $X \leq_{\text{mrl}} Y$, where

$$M_X(u) = m_X(Q_X(u)) \quad \text{and} \quad M_Y^*(u) = m_Y(Q_X(u)).$$

Example 8.4. Let X and Y have distributions with quantile functions

$$Q_X(u) = 1 - (1-u)^{\frac{1}{c}}, \quad c > 0,$$

and

$$Q_Y(u) = 1 - (1-u)^{-\frac{1}{c}} - 1, \quad c > 0,$$

respectively. Then,

$$\overline{F}_Y(x) = (1+x)^{-c}, \quad x > 0.$$

We have

$$M_X(u) = \frac{1}{1-u}\int_u^1 (1-p)q(p)dp = \frac{(1-u)^{\frac{1}{c}}}{c+1},$$

$$M_Y(x) = \frac{1+x}{c-1},$$

and

$$M_Y^*(u) = m_Y(Q_X(u)) = \frac{2-(1-u)^{\frac{1}{c}}}{c-1}, \quad c > 1,$$

$$\begin{aligned} M_X(u) - M_Y^*(u) &= 2c(1-u)^{\frac{1}{c}} - 2(c+1) \\ &= 2c\left\{(1-u)^{\frac{1}{c}} - \frac{c+1}{c}\right\} < 0. \end{aligned}$$

Hence, $X \leq_{\text{mrl}} Y$.

There are several equivalent conditions for the validity of $X \leq_{\text{mrl}} Y$ as presented in the following theorem.

Theorem 8.11. *$X \leq_{mrl} Y$ if and only if any of the following conditions hold:*

(a) $m_X(x) \leq m_Y(x)$ for all $x > 0$;
(b) $\frac{\int_x^\infty \bar{F}_Y(t)dt}{\int_x^\infty \bar{F}_X(t)dt}$ is an increasing function of x, or equivalently

$$\frac{1}{\bar{F}_Y(Q_X(u))}\int_{Q_X(u)}^{\infty} \bar{F}_Y(x)dx \geq \frac{1}{1-u}\int_u^1 (1-p)q_X(p)dp;$$

(c) $\frac{P_X(u)}{P_Y^*(x)}$ is an increasing function of u, when $P_X(u)$ is the partial mean

$$P_X(u) = \int_u^1 (1-p)q(p)dp$$

defined in (6.47) *and*

$$P_Y^*(u) = P_Y(Q_X(u)) = \int_{Q_X(u)}^{\infty} \bar{F}_Y(t)dt.$$

Notice that (a) is the definition of the mean residual life order in the distribution function approach. Differentiating (b) and noting that the derivative is non-negative, we get (a). Setting $x = Q(u)$ in (b), we obtain (c) which is equivalent to (b).

The classes of life distributions induced by $\leq_{\text{mrl}}$ are presented in the following theorem.

Theorem 8.12. *(a) X is DMRL if and only if any one of the following properties hold:*

(i) $X_t \geq_{mrl} X_{t'}$ for $t' \geq t$;
(ii) $X \geq_{mrl} X_t$;
(iii) $X + t \leq_{mrl} X + t'$.

(b) X is DRMRL if and only if any one of the following properties hold:

(i) $X \geq_{mrl} Z$;
(ii) $X_t \geq_{mrl} Z_t$;
(iii) $Z \leq_{mrl} Z_t$.

Part (a) follows readily from the fact that the mean residual life of X_t is $m(x+t)$ and the definition of $\leq_{\text{mrl}}$. To prove (b), recall Definition 4.17. X is said to DRMRL if and only if $e_X(u) \leq M_X(u)$, where (4.24)

$$e(u) = \frac{\int_u^1 [Q(p) - Q(u)](1-p)q(p)dp}{\int_u^1 (1-p)q(p)}.$$

The mean residual functions of X, Z, X_t and Z_t are, respectively, $m(x)$, $e(x)$, $m(x+t)$ and $m^*(x+t)$ (4.23). Hence, (i) implies

$$\begin{aligned} X \geq_{\text{mrl}} Z &\Leftrightarrow m(x) \geq e(x) \\ &\Leftrightarrow M_X(u) \geq e_X(u) \\ &\Leftrightarrow X \text{ is DRMRL.} \end{aligned}$$

Other properties follow similarly.

Regarding the closure properties enjoyed by $\leq_{\text{mrl}}$, some of the important ones are as follows:

1. For every increasing convex function $\phi(x)$, $X \leq_{\text{mrl}} Y$ implies $\phi(X) \leq_{\text{mrl}} \phi(Y)$.
2. The mean residual life order is closed with respect to the formation of mixtures under certain conditions only. If $X|(\Theta = \theta) \leq Y|(\Theta = \theta')$ for all θ, θ' in the support of Θ, then $X \leq_{\text{mrl}} Y$ (Nanda et al. [460]).
3. (X_i, Y_i), $i = 1, 2, \ldots, n$, are independent pairs of IHR random variables such that $X_i \leq_{\text{mrl}} Y_i$ for all i, then (Pellerey [490])
$$\sum_{i=1}^{n} X_i \leq_{\text{mrl}} \sum_{i=1}^{n} Y_i.$$
4. For a sequence $\{X_n\}$, $n = 1, 2, \ldots$, of independent and identically distributed IHR random variables,
$$\sum_{i=1}^{M} X_i \leq_{\text{mrl}} \sum_{i=1}^{N} X_i,$$
where M and N are positive integer valued random variables such that $M \leq_{\text{mrl}} N$ (Pellerey [490]).
5. If X is DMRL and $0 < a \leq 1$, then $aX \leq_{\text{mrl}} X$.
6. Let $X_1, X_2, \ldots X_n$ be independent. If $X_i \leq_{\text{mrl}} X_n$, for $i = 1, 2, \ldots, n-1$, then $X_{n-1:n-1} \leq_{\text{mrl}} X_{n:n}$.
7. Let U be a random variable with mixture distribution function $\alpha F_X(x) + (1 - \alpha) F_Y(x)$, $0 < \alpha < 1$. If $X \leq_{\text{mrl}} Y$, then $X \leq_{\text{mrl}} U \leq_{\text{mrl}} Y$.

The hazard quantile function and the mean residual quantile function are closely related and determine each other. Moreover, the IHR class of life distributions is a subclass of the DMRL class. We now examine how the orderings based on the hazard quantile and mean residual quantile functions imply each other.

Theorem 8.13. *If $X \leq_{hr} Y$, then $X \leq_{mrl} Y$, but the converse need not be true.*

Proof. We have

$$\begin{aligned}
X \leq_{\text{hr}} Y &\Rightarrow H_X(u) \geq H_Y^*(u) \\
&\Rightarrow \int_u^1 \frac{dp}{H_X(p)} \leq \int_u^1 \frac{dp}{H_{Y^*}(p)} \\
&\Rightarrow \int_u^1 (1-p)q_X(p)dp \leq \int_{Q(u)}^{\infty} \frac{\overline{F}_Y(t)dt}{f_Y(t)} \\
&\Rightarrow M_X(u) \leq M_Y^*(u) \\
&\Rightarrow X \leq_{\text{mrl}} Y.
\end{aligned}$$

To prove the second part, let X have standard exponential distribution with

$$Q_X(u) = -\log(1-u)$$

so that $E(X) = 1$, and Y be Weibull with

$$Q_Y(u) = \sigma(-\log(1-u))^{\frac{1}{\lambda}}.$$

The parameters of Y be chosen such that $\lambda > 1$ and $E(Y) < 1$. Since $\lambda > 1$, Y is IHR and hence NBUE. This means that

$$M_Y(u) \leq 1 = E(X) = M_X(u) \text{ for all } 0 < u < 1.$$

Thus, $Y \leq_{\text{mrl}} X$. On the other hand, $H_X(u) = 1$ and

$$h_Y(x) = \frac{\lambda}{\sigma^\lambda} x^{\lambda-1}.$$

This gives

$$H_Y^*(u) = \frac{\lambda}{\sigma^\lambda}(-\log(1-u))^{\lambda-1}$$

or

$$H_X(u) - H_Y^*(u) = 1 - \frac{\lambda}{\sigma^\lambda}(-\log(1-u))^{\lambda-1}.$$

We can see that X and Y are not ordered in hazard rate since

$$H_X(u) \leq H_Y^*(u) \text{ for } u \text{ in } \left(0, 1-\exp\left(\frac{\sigma^\lambda}{\lambda}\right)^{\frac{1}{\lambda-1}}\right)$$

and

$$H_X(u) \geq H_Y^*(u) \text{ in } \left(1-\exp\left(\frac{\sigma^\lambda}{\lambda}\right)^{\frac{1}{\lambda-1}}, 1\right).$$

The above result leads us to seek conditions under which the $\leq_{\text{hr}}$ ordering can be generated from the $\leq_{\text{mrl}}$ ordering.

Theorem 8.14 (Belzunce et al. [86]).

1. $X \leq_{hr} Y \Rightarrow \min(X,Z) \leq_{mrl} \min(Y,Z)$ *for any non-negative random variable* Z *independent of* X *and* Y*;*
2. $X \leq_{hr} Y \Rightarrow 1-e^{-sX} \leq_{mrl} 1-e^{-sY}$, $s>0$.

A result that is helpful in establishing the mrl ordering is stated in the following theorem.

Theorem 8.15. *If* X *and* Y *have finite means,*

$$X \leq_{mrl} Y \Leftrightarrow Z_X \leq_{hr} Z_Y,$$

where Z_X *and* Z_Y *denote the equilibrium random variables corresponding to* X *and* Y*, respectively.*

This result is immediate from the fact that the hazard quantile function of $Z_X\,(Z_Y)$ is the reciprocal of the mean residual quantile function of $X(Y)$. A comparison between the usual stochastic order and the mrl order is even more interesting. Although the mean residual life function determines the distribution uniquely, there is no implication between $\leq_{\text{st}}$ and $\leq_{\text{mrl}}$. This is seen from the following examples furnished by Gupta and Kirmani [241]. Upon choosing

$$F_X(x) = \begin{cases} e^{-x}, & 0 \leq x < 1, \\ e^{-x^{\frac{1}{2}}}, & x \geq 1, \end{cases}$$

and

$$F_Y(x) = e^{-x^{\frac{1}{2}}}, \quad x>0,$$

we see that $F_Y(x) \leq F_X(x)$ or $X \geq_{\text{hr}} Y$. At the same time, $m_X(x)$ and $m_Y(x)$ are not ordered. Secondly, in the counter example in Theorem 8.13, $\overline{F}_X(x)-\overline{F}_Y(x)$ can have both negative and positive signs ruling out either $X \leq_{\text{hr}} Y$ or $X \geq_{\text{hr}} Y$. But, $X \geq_{\text{mrl}} Y$. With additional assumptions on X and Y, implications between the two orders can be established as provided in the following theorem.

Theorem 8.16 (Gupta and Kirmani [241]).

1. *If* $\frac{M_X(u)}{M_Y^*(u)}$ *is increasing in* u*, then*

$$X \leq_{mrl} Y \Rightarrow X \leq_{hr} Y \Rightarrow X \leq_{st} Y;$$

2. *If* $\frac{M_X(u)}{M_Y^*(u)} \geq \frac{E(X)}{E(Y)}$, *then*

$$X \leq_{mrl} Y \Rightarrow X \leq_{st} Y.$$

We have conditions under which the mrl order ensures stochastic equality of X and Y. If $X \geq_{mrl} Y$, $E(Y) > 0$, $E(X) = E(Y)$ and $V(X) = V(Y)$, then X and Y have the same distribution.

For some additional results on mrl ordering, one may refer to Alzaid [35], Ahmed [28], Joag-Dev et al. [298], Fagiouli and Pellerey [190, 192], Hu et al. [288], Zhao and Balakrishnan [602] and Nanda et al. [459].

Another stochastic order that involves the mean residual life is the harmonic mean residual life order defined as follows.

Definition 8.4. X is said to be smaller than Y in harmonically mean residual life order, denoted by $X \leq_{\text{hmrl}} Y$, if and only if

$$\left\{\frac{1}{x}\int_0^x \frac{dt}{m_X(t)}\right\}^{-1} \leq \left\{\frac{1}{x}\int_0^x \frac{dt}{m_Y(t)}\right\}^{-1},$$

or equivalently

$$\int_0^u \frac{q_X(p)dp}{M_X(p)} \geq \int_0^u \frac{q_X(p)dp}{M_Y(Q_X(p))}.$$

Example 8.5. Let X be distributed as Pareto I with $\overline{F}_X(x) = (\frac{x}{\sigma})^{-\alpha_1}$. Then, we have

$$Q_X(u) = \sigma(1-u)^{-\frac{1}{\alpha_1}},$$

$$M_X(u) = \sigma\frac{(1-u)^{-\frac{1}{\alpha_1}}}{\alpha_1 - 1},$$

$$\int_0^u \frac{q_X(p)dp}{M_X(p)} = \frac{\alpha_1 - 1}{\alpha_1}(-\log(1-u)).$$

Assume that Y has Pareto distribution with

$$\overline{F}_Y(x) = \left(\frac{x}{\sigma}\right)^{-\alpha_2},$$

$$M_Y(Q_X(u)) = \sigma\frac{(1-u)^{-\frac{1}{\alpha_1}}}{\alpha_2 - 1},$$

$$\int_0^u \frac{q_X(p)dp}{M_Y^*(p)} = \frac{\alpha_2 - 1}{\alpha_1}(-\log(1-u)).$$

Hence, $X \leq_{\text{hmrl}} Y$ if and only if $\alpha_1 \geq \alpha_2$.

Some equivalent conditions for $X \leq_{\text{hmrl}} Y$ are as follows:

(i) $$\frac{\int_x^\infty \overline{F}_X(t)dt}{E(X)} \leq \frac{\int_x^\infty \overline{G}(t)dt}{E(Y)} \left(\frac{\int_u^1 (1-p)q(p)dp}{E(X)} \leq \frac{\int_{Q_X(u)}^\infty \overline{G}(t)dt}{E(Y)}\right);$$

(ii) $\frac{E\phi(X)}{E(X)} \leq \frac{E\phi(Y)}{E(Y)}$ for all increasing convex functions $\phi(x)$;

(iii) $\frac{P_X(u)}{E(X)} \leq \frac{P_Y^*(u)}{E(Y)}$, where $P_Y^*(u)$ is as in Part (c) of Theorem 8.11.

As a further consequence of the hmrl order, we also have

$$X \leq_{\text{hmrl}} Y \Rightarrow E(X) \leq E(Y)$$

and in addition if Y is NWUE (Kirmani [328, 329]), then

$$V(X) \leq V(Y);$$

(iv) $Z_X \leq_{\text{st}} Z_Y$.

The preservation properties enjoyed by the hmrl order are summarized in the following theorem. Here, all the variables involved X, Y, X_i and Y_i are non-negative. For proofs and other details, we refer the reader to Pellerey [490] and Nanda et al. [460].

Theorem 8.17. *(a) (X_i, Y_i), $i = 1, 2, \ldots, n$, are independent pairs of random variables such that $X_i \leq_{hmrl} Y_i$ for all i. If X_i, Y_i are all NBUE, then*

$$\sum_{i=1}^{n} X_i \leq_{hmrl} \sum_{i=1}^{n} Y_i;$$

(b) (X_n) and (Y_n) are sequences of NBUE independent and identically distributed random variables satisfying $X_n \leq_{hmrl} Y_n$, $n = 1, 2, \ldots$. If M and N are positive integer-valued random variables independent of the sequences $\{X_n\}$ and $\{Y_n\}$ such that $M \leq_{hmrl} N$, then

$$\sum_{i=1}^{M} X_i \leq_{hmrl} \sum_{j=1}^{N} Y_j;$$

(c) Let X, Y and Θ be random variables with $X|(\Theta = \theta) \leq_{hmrl} Y|(\Theta = \theta')$ for all θ and θ' in the support of Θ. Then, $X \leq_{hmrl} Y$;

(d) If X, Y and Θ are random variables such that $X|(\Theta = \theta) \leq_{hmrl} Y|(\Theta = \theta)$ for all θ in the support of Θ along with the additional condition

$$E(Y|\Theta = \theta) = kE(X|\Theta = \theta),$$

where k is independent of θ, then $X \leq_{hmrl} Y$;

(e) If $E(X), E(Y) > 0$ and $E(X) \leq E(Y)$, then $X =_{hmrl} Y$ if and only if $X =_{st} UY$, where U is a Bernoulli variable independent of Y;

(f) If U has mixture distribution

$$F_U(x) = \alpha F_X(x) + (1 - \alpha) F_Y(x), \quad 0 < \alpha < 1,$$

then

$$X \leq_{hmrl} Y \Rightarrow X \leq_{hmrl} U \leq_{hmrl} Y.$$

The DMRL class and NBUE class of life distributions can be characterized by the hmrl order as given in the following theorem.

Theorem 8.18. *(i) X is DMRL* $\Leftrightarrow X_t \geq_{hmrl} X_{t'}$, $t' \geq t \geq 0$;
(ii) X is NBUE $\Leftrightarrow X \leq_{hmrl} Y$, *where Y is independent of X and* $E(Y) > 0$;
(iii) X is NBUE $\Leftrightarrow X + Y_1 \leq_{hmrl} X + Y_2$, *where* Y_1 *and* Y_2, *are independent of X,* $E(Y_i) < \infty$, $i = 1, 2$, *and* $Y_1 \leq_{hmrl} Y_2$.

The results in Parts (ii) and (iii) are due to Lefevre and Utev [381].

Finally, we study the relationships the hmrl order have with some other orders. First of all, by the increasing nature of harmonic averages, we have

$$X \leq_{\text{mrl}} Y \Rightarrow X \leq_{\text{hmrl}} Y.$$

Even otherwise, in terms of quantile functions,

$$\begin{aligned}
X \leq_{\text{mrl}} Y &\Rightarrow M_X(u) \leq M_Y^*(u), \text{ where } M_Y^*(u) = M_Y(Q_X(u)).\\
&\Rightarrow \frac{q_X(u)}{M_X(u)} \geq \frac{q_X(u)}{M_{Y^*}(u)}\\
&\Rightarrow \int_0^u \frac{q_X(p)dp}{M_X(p)} \geq \int_0^u \frac{q_X(p)dp}{M_Y^*(p)dp}\\
&\Leftrightarrow X \leq_{\text{hmrl}} Y.
\end{aligned}$$

The converse need not be true and so the $\leq_{\text{hmrl}}$ order is weaker than the $\leq_{\text{mrl}}$ order. Moreover, neither the usual stochastic order nor the hmrl order imply the other (see Deshpande et al. [173]).

8.5 Renewal and Harmonic Renewal Mean Residual Life Orders

Recall the definition of the renewal mean residual life function (4.23)

$$m^*(x) = \frac{\int_x^\infty (t-x)\overline{F}(t)dt}{\int_x^\infty \overline{F}(t)dt}, \tag{8.6}$$

which is an alternative to the traditional mean residual life function, as it facilitates all the functions and calculations enjoyed by the latter. The quantile-based definition is

$$\begin{aligned} e(u) = m^*(Q(u)) &= \frac{\int_u^1 [Q(p) - Q(u)](1-p)q(p)dp}{\int_u^1 (1-p)q(p)dp} \\ &= \left\{ \int_u^1 (1-p)q(p)dp \right\}^{-1} \int_u^1 \int_p^1 (1-t)q(t)q(p)dtdp. \end{aligned} \tag{8.7}$$

In this section, we discuss the properties of a stochastic order based on the $e(u)$ in (8.7), and these results are taken from Nair and Sankaran [446].

Definition 8.5. The random variable X is said to be less (greater) than Y in renewal mean residual life order, denoted by $X \leq_{\text{rmrl}} Y$, if and only if

$$m_X^*(x) \leq (\geq) m_Y^*(x) \text{ for all } x \geq 0,$$

or equivalently

$$e_X(u) \leq (\geq) e_Y^*(u) \text{ for all } 0 < u < 1,$$

where $e_Y^*(u) = m_Y^*(Q_X(u))$ and $e_X(u) = m_X^*(Q_X(u))$.

Example 8.6. Let X be distributed with quantile function

$$Q_X(u) = 1 - (1-u)^{\frac{1}{3}}$$

and Y have its quantile function as

$$Q_Y(u) = (1-u)^{-\frac{1}{12}} - 1.$$

Then, from (8.7), we have

$$e_X(u) = \frac{(1-u)^{\frac{1}{3}}}{5}.$$

Again, $m_Y^*(x) = \frac{2+x}{10}$ so that

$$e_Y^*(u) = \frac{3-(1-u)^{\frac{1}{3}}}{10}.$$

It is easy to see that $e_X(u) \leq e_Y^*(u)$ for all u, and so $X \leq_{\text{rmrl}} Y$.

Some other conditions that characterize the rmrl order are as follows:

(a) $\frac{\int_x^\infty \int_u^\infty \overline{F}_X(t)dtdu}{\int_x^\infty \int_u^\infty \overline{F}_Y(t)dtdu}$ is increasing in x over $\{x | \int_x^\infty \overline{F}_Y(t)dt > 0\}$;

(b) $(\int_x^\infty \overline{F}_Y(t)dt)(\int_x^\infty \int_u^\infty \overline{F}_X(t)dtdu) \leq (\int_x^\infty \overline{F}_X(t)dt)(\int_x^\infty \int_u^\infty \overline{F}_Y(t)dtdu)$;

(c) $\frac{\int_x^\infty E(X-t)^+dt}{\int_x^\infty E(Y-t)^+dt}$ is decreasing.

By the methods used earlier, the results in (a), (b) and (c) above can also be expressed in terms of quantile functions.

One issue of primary interest is the relationship between the usual mrl order and the rmrl order, which is described in the following theorem.

Theorem 8.19. *If $X \leq_{mrl} Y$, then $X \leq_{rmrl} Y$. But, the converse is not true.*

Proof. For simplicity, we write $Q_X(u) = Q(u)$ throughout the proof. We have

$$\begin{aligned}
X \leq_{\text{mrl}} Y &\Rightarrow \frac{1}{1-u}\int_u^1 (1-p)q(p)dp \leq \frac{1}{\overline{G}(Q(u))}\int_u^1 \overline{G}(Q(p))q(p)dp \\
&\Rightarrow \frac{1-u}{\int_u^1 (1-p)q(p)dp} \geq \frac{\overline{G}Q(u)}{\int_u^1 \overline{G}(Q(p))q(p)dp} \\
&\Rightarrow \frac{d}{du}\log\int_u^1 (1-p)q(p)dp \leq \frac{d}{du}\log\int_u^1 \overline{G}(Q(p))q(p)dp \\
&\Rightarrow \int_p^u \left(\frac{d}{dt}\log\int_t^1 (1-p)q(p)dpdt\right) \leq \int_p^u \left(\frac{d}{dt}\log\int_t^1 \overline{G}(Q(p))q(p)dp\right) \\
&\Rightarrow \frac{\int_u^1 (1-p)q(p)dp}{\int_p^1 (1-t)q(t)dt} \leq \frac{\int_u^1 \overline{G}(Q(p))q(p)dp}{\int_p^1 \overline{G}(Q(t))q(t)dt} \\
&\Rightarrow \frac{\int_p^1\int_u^1 (1-t)q(t)q(u)du}{\int_p^1 (1-t)q(t)dt} \leq \frac{\int_p^1\int_u^1 \overline{G}(Q(t))q(t)q(u)du}{\int_p^1 \overline{G}Q(t)q(t)dt} \\
&\Rightarrow e_X(p) \leq e_Y^*(p) \Leftrightarrow X \leq_{\text{rmrl}} Y.
\end{aligned}$$

To prove the latter part of the theorem, we reconsider Example 8.5 wherein we had established that for the random variables X and Y described therein, $X \leq_{rmrl} Y$. In this case, we also have

$$M_X(u) = \frac{(1-u)^{\frac{1}{3}}}{4}$$

and

$$m_Y(x) = \frac{2+x}{11}$$

giving

$$M_Y^*(u) = \frac{3-(1-u)^{\frac{1}{3}}}{11}.$$

Thus,

$$M_X(u) - M_Y^*(u) = \frac{3}{44}\left\{5(1-u)^{\frac{1}{3}} - 4\right\}$$

which is decreasing in $(0, \frac{61}{125})$ and increasing in $(\frac{61}{125}, 1)$. Hence, X and Y are not ordered in mrl.

Remark. One could see that the $\leq_{\text{rmrl}}$ order is strictly weaker than the $\leq_{\text{mrl}}$ order and consequently generates a larger class of life distributions.

As was done in the mrl order, we consider conditions under which the two orders become equivalent in the following theorem.

Theorem 8.20. *If $\frac{e_X(u)}{e^*_Y(u)}$, is an increasing function of u, then*

$$X \leq_{mrl} Y \Leftrightarrow X \leq_{rmrl} Y.$$

Proof. Since $\frac{e_X(u)}{e^*_Y(u)}$ is an increasing function of u, we have

$$\frac{e'_X(u)}{e_X(u)} \geq \frac{e^{*'}_Y(u)}{e^*_Y(u)}. \tag{8.8}$$

From (4.25), we have

$$M_X(u) = \frac{e_X(u)q_X(u)}{q_X(u) + e'_X(u)}. \tag{8.9}$$

But, by definition, we have

$$\begin{aligned} e^*_Y(u) &= m^*_Y(Q_X(u)) \\ &= \frac{\int_{Q_X(u)}^{\infty}(t-x)\overline{F}_Y(t)dt}{\int_{Q_X(u)}^{\infty}\overline{F}_Y(t)dt} \\ &= \frac{\int_u^1\int_p^1 \overline{F}_Y(Q_X(t))q_X(t)dt}{\int_u^1 \overline{F}_Y(Q_X(p))q_X(p)dp}. \end{aligned}$$

Differentiating and simplifying, we obtain

$$M^*_Y(u) = \frac{e^*_Y(u)q_X(u)}{q_X(u) + e'_X(u)} \tag{8.10}$$

From (8.8), (8.9) and (8.10), whenever $X \leq_{\text{rmrl}} Y$, we must have

$$\begin{aligned} \frac{1}{M_X(u)} &= \frac{1}{e_X(u)} + \frac{e'_X(u)}{e_X(u)} \\ &= \frac{1}{e^*_Y(u)} + \frac{e^{*'}_Y(u)}{e^*_Y(u)} = \frac{1}{M^*_Y(u)} \end{aligned}$$

and so

$$M_X(u) \leq M_Y^*(u) \Leftrightarrow X \leq_{\text{mrl}} Y.$$

The reverse inequality $X \leq_{\text{mrl}} Y \Rightarrow X \leq_{\text{rmrl}} Y$ has already been established in Theorem 8.8 and this completes the proof.

The procedure of taking harmonic averages and then comparing life distributions based on them is also possible with renewal mean residual life functions as described below.

Definition 8.6. X is said to be smaller than Y in harmonic renewal mean residual life, denoted by $X \leq_{\text{hrmrl}} Y$, if and only if

$$\frac{1}{x}\int_0^x \frac{dt}{m_X^*(t)} \leq \frac{1}{x}\int_0^x \frac{dt}{m_Y^*(t)}.$$

An equivalent definition is

$$\int_0^u \frac{q_X(p)dp}{e_X(p)} \geq \int_0^u \frac{q_X(p)dp}{e_Y^*(p)}. \tag{8.11}$$

It can be shown that (8.11) is equivalent to

$$\frac{E[(X-x)^+]^2}{E(X^2)} \leq \frac{E[(Y-x)^+]^2}{E(Y^2)}.$$

The following properties hold for the $\leq_{\text{hrmrl}}$ ordering:

(i) If $\frac{e_X(u)}{e_Y^*(u)}$, is increasing in u, then $X \leq_{\text{hrmrl}} Y \Leftrightarrow X \leq_{\text{hrmrl}} Y$;
(ii) In general,

$$X \leq_{\text{hmrl}} Y \Rightarrow X \leq_{\text{hrmrl}} Y;$$

(iii) $X \leq_{\text{rmrl}} Y \Rightarrow X \leq_{hrmrl} Y$.

The preservation properties and other implications of the rmrl and hrmrl orders have not yet been studied in detail.

8.6 Variance Residual Life Order

Earlier in Sect. 4.3, we have defined the variance residual life of X as

$$\sigma^2(x) = \frac{2}{\overline{F}(x)}\int_x^\infty \int_u^\infty \overline{F}(t)dtdu - m^2(x),$$

or in terms of quantile function as

$$V(u) = \sigma^2 Q(u) = (1-u)^{-1}\int_u^1 M^2(p)dp, \tag{8.12}$$

where $M(u)$ is the mean residual quantile function.

Definition 8.7. We say that X is smaller than Y in variance residual life, denoted by $X \leq_{\text{vrl}} Y$, if and only if any of the following equivalent conditions hold:
(i) $\sigma_X^2(x) \leq \sigma_Y^2(x)$ for all $x > 0$;
(ii) $V_X(u) \leq V_Y^*(u)$ for all $0 < u < 1$, where $V_Y^*(u) = \sigma_Y^2(Q_X(u))$.

For the definition in (i) and properties of the vrl ordering, one may refer to Singh [541].

Connection of the $\leq_{\text{vrl}}$ ordering with the $\leq_{\text{mrl}}$ ordering is presented in the next theorem.

Theorem 8.21. *If $X \leq_{mrl} Y$, then $X \leq_{vrl} Y$.*

Proof. The result easily follows from the fact

$$X \leq_{\text{mrl}} Y \Rightarrow M_X(u) \leq M_Y^*(u)$$

and (8.12).

If $\overline{F}_1$ and $\overline{F}_2$ are survival functions of the equilibrium random variables of X and Y, respectively, Fagiouli and Pellery [192] defined

$$X \leq_{\text{vrl}} Y \text{ if } \frac{\int_x^\infty \overline{F}_1(t)dt}{\int_x^\infty \overline{F}_2(t)dt}$$

is nonincreasing in $x \geq 0$. There has not been much investigation on the preservation properties and other aspects of the vrl order.

8.7 Percentile Residual Life Order

The percentile life ordering was introduced by Joe and Proschan [301] in the context of testing the hypothesis of the equality of two distributions. Earlier, we have defined the αth percentile residual life function for any $0 < \alpha < 1$ as

$$p_\alpha(x) = F^{-1}(1-(1-\alpha)\overline{F}(x)) - x$$

or

$$p_\alpha(u) = p_\alpha(Q(u)) = Q[1-(1-\alpha)(1-u)] - Q(u).$$

Franco-Pereira et al. [202] have discussed some properties of the percentile order.

Definition 8.8. We say that X is smaller than Y in the α-percentile residual life, denoted by $X \leq_{\text{prl}-\alpha} Y$, if and only if

$$p_{\alpha,X}(x) \leq p_{\alpha,Y}(x) \quad (P_{\alpha,X}(u) \leq P^*_{\alpha,Y}(u))$$

for all x (for all u) and $P^*_{\alpha,Y}(u) = p_{\alpha,Y}(Q(u))$.

One specific aspect about the prl order is that, unlike other orderings we have discussed, it is indexed by α which can take any value in $(0,1)$. Moreover, the percentile residual life function $P_\alpha(u)$, for a given α, does not determine the distribution uniquely. If $X \leq_{\text{prl}-\alpha} Y$, then the upper end point of the support of X cannot exceed that of Y, but it is not necessary that a corresponding result hold for the left end point of the supports of the random variables.

Example 8.7. Consider the distribution (Pareto) with quantile function

$$\begin{aligned} Q(u) &= (1-u)^{-\frac{1}{\alpha}}, \quad 0 < u < 1 \\ P_\alpha(u) &= [1 - \{1-(1-\alpha)(1-u)\}]^{-\frac{1}{\alpha}} - (1-u)^{-\frac{1}{\alpha}} \\ &= (1-u)^{-\frac{1}{\alpha}}[(1-u)^{-\frac{1}{\alpha}} - 1]. \end{aligned}$$

Let X and Y be random variables with the above distribution with parameters α_1 and α_2, respectively. Then, we find

$$P_{\alpha,X}(u) - P^*_{\alpha,Y}(u) = (1-u)^{-\frac{1}{\alpha_1}} \left\{ (1-\alpha)^{-\frac{1}{\alpha_1}} - (1-\alpha)^{-\frac{1}{\alpha_2}} \right\},$$

and so

$$X \leq_{\text{prl}-\alpha} Y \text{ for } \alpha_2 \leq \alpha_1.$$

Two useful characterizations of the $\leq_{\text{prl}}$ order, one in terms of quantile functions and the other in terms of distribution functions, are presented in the following theorem both of which are direct consequences of the definition.

Theorem 8.22. *$X \leq_{prl-\alpha} Y$ and only if*

(i) $Q_X(\alpha + (1-\alpha)u) \leq Q_Y(\alpha + (1-\alpha)Q_Y^{-1}(Q_X(u)))$,
(ii) $\frac{\bar{F}_Y(Q_X(u))}{u} \leq \frac{\bar{F}_Y(Q_X(1-\alpha)(u))}{(1-\alpha)u}$ *for all* $0 < u < 1$.

The following relationships exist between the prl order and some other orders we have discussed:

(a) $X \leq_{\text{hr}} Y \Leftrightarrow X \leq_{\text{prl}-\alpha} Y$ for all α in $(0,1)$;
(b) For a specific α, $X \leq_{\text{hr}} Y \Rightarrow X \leq_{\text{prl}-\alpha} Y$. So, the result in (a) is not practically useful;
(c) Percentile life orders do not preserve expectations and as such $\leq_{\text{prl}-\alpha}$ neither implies the usual stochastic order, mean residual life order, and hmrl order, for any α. Further, stochastic order does not imply prl order, or mrl or hmrl orders;
(d) If, for $0 < \beta < 1$, $X \leq_{\text{prl}-\alpha} Y$ for every α in $(0,\beta)$, then $X \leq_{\text{hr}} Y$. Naturally, if $X \leq_{\text{prl}-\alpha} Y$ for all α in $(0,\beta)$, then $X \leq_{\text{prl}-\alpha} Y$ for all α.

Some interesting preservation properties, established by Franco-Pereira et al. [202], are as follows:

1. For an increasing function $\phi(\cdot)$, we have

$$X \leq_{\text{prl}-\alpha} Y \Leftrightarrow \phi(X) \leq_{\text{prl}-\alpha} \phi(Y);$$

2. Let (X_n), (Y_n), $n = 1,2,\ldots$, be two sequences of random variables such that $X_n \to X$ and $Y_n \to Y$ in distribution as $n \to \infty$. If X and Y have continuous distributions with interval support, then for any α, if $X_n \leq_{\text{prl}-\alpha} Y_n$ holds, $n = 1,2,\ldots$, then $X \leq_{\text{prl}-\alpha} Y$;
3. Let X_θ, $\theta \in \Theta$, and Y_θ, $\theta \in \Theta$, be two families of random variables with continuous distributions. If

$$F_W(x) = \int_\Theta F_X(x|\theta)dH(\theta)$$

and

$$F_Z(x) = \int_\Theta F_Y(x|\theta)dH(\theta),$$

where H is some distribution function on Θ and U is a random variable such that

$$X_\theta \leq_{\text{prl}-\alpha} U \leq_{\text{prl}-\alpha} Y_\theta \quad \text{for all } \theta \in \Theta,$$

then

$$W \leq_{\text{prl}-\alpha} Z.$$

In particular, if W has the mixture distribution function

$$F_W = pF_X + (1-p)F_Y$$

for some $0 \leq p \leq 1$, then

$$X \leq_{\text{prl}-\alpha} Y \Rightarrow X \leq_{\text{prl}-\alpha} W \leq_{\text{prl}-\alpha} Y;$$

4. The prl-α order is not closed under the formation of parallel or series systems. However, if X_i, Y_i, $i = 1,2,\ldots,n$, are independent and identically distributed random variables with continuous distributions, satisfying $X_1 \leq_{\text{prl}-\alpha} Y_1$, then
$$\min(X_1,X_2,\ldots,X_n) \leq_{\text{prl}-\beta} (Y_1,Y_2,\ldots,Y_n),$$
where $\beta = 1-(1-\alpha)^n$.

8.8 Stochastic Order by Functions in Reversed Time

Earlier in Sect. 2.4, we have defined and given examples of reliability functions in reversed time like the reversed hazard quantile function and the reversed mean residual quantile function. These functions have also been used in Sect. 4.5 to introduce various ageing classes. It is therefore possible to order life distributions on the basis of their magnitudes, and this is the focus of the present section.

8.8.1 Reversed Hazard Rate Order

Let X and Y be two absolutely continuous random variables with reversed hazard rates
$$\lambda_X(x) = \frac{f_X(x)}{F_X(x)} \text{ and } \lambda_Y(x) = \frac{f_Y(x)}{F_Y(x)},$$
respectively.

Definition 8.9. X is said to be smaller than Y in reversed hazard rate order, denoted by $X \leq_{\text{rh}} Y$, if and only if
$$\lambda_X(x) \leq \lambda_Y(x) \text{ for all } x > 0,$$
or equivalently
$$\Lambda_X(u) \leq \Lambda_Y^*(u) \text{ for all } 0 < u < 1,$$
where $\Lambda_Y^*(u) = \lambda_Y(Q_X(u))$ (see (2.50)).

Some other conditions that characterize the $\leq_{\text{rh}}$ order are presented in the following theorem.

Theorem 8.23. *$X \leq_{rh} Y$ if and only if*

(a) $\frac{Q_Y^{-1}(Q_X(u))}{u} \leq \frac{Q_Y^{-1}Q_X(v)}{v}$ *for all* $0 < u \leq v < 1$;
(b) $\frac{F_Y(x)}{F_X(x)}$ *increases in* x;
(c) $X|(X \leq x) \leq_{st} Y|(Y \leq x)$ *for all* $x > 0$.

Nanda and Shaked [461] have proved a basic relationship between the $\leq_{\text{hr}}$ order and the $\leq_{\text{rh}}$ order as presented in the following theorem, and it simplifies the proofs of many results.

Theorem 8.24. *For two continuous random variables X and Y,*

$$X \leq_{hr} Y \Rightarrow \phi(X) \geq_{rh} \phi(Y)$$

for any continuous function ϕ which is strictly decreasing on (a_1, b_2), where a_1 is the lower end of the support of X and b_2 is the upper end of the support of Y. Furthermore,

$$X \leq_{rh} Y \Rightarrow \phi(X) \leq_{rh} \phi(Y)$$

when ϕ is strictly increasing.

Various properties of the $\leq_{\text{rh}}$ order have been studied by many authors including Kebir [321], Shaked and Wang [533], Kijima [325], Block et al. [111], Hu and He [285], Nanda and Shaked [461], Gupta and Nanda [254], Yu [597], Zang and Li [599] and Brito et al. [120]. There exists a relationship between the $\leq_{st}$ and the $\leq_{\text{rh}}$ orders which is stated in the following theorem.

Theorem 8.25. *If $X \leq_{rh} Y$, then $X \leq_{st} Y$.*

Proof. We observe that

$$\begin{aligned} X \leq_{\text{rh}} Y &\Rightarrow \lambda_X(u) \leq \lambda_Y(Q_X(u)) \Rightarrow \frac{1}{uq(u)} \leq \frac{1}{F_Y(Q_X(u))q_Y(Q_X(u))} \\ &\Rightarrow -\log u \leq -\log F_Y(Q_X(u)) \Rightarrow \frac{1}{u} \leq \frac{1}{F_Y(Q_X(u))} \\ &\Rightarrow Q_X(u) \leq Q_Y(u) \Rightarrow X \leq_{\text{st}} Y, \end{aligned}$$

as required.

The preservation properties enjoyed by the $\leq_{\text{rh}}$ order are as follows:

(i) *Convolution property* Let (X_i, Y_i), $i = 1, 2, \ldots, n$, be n pairs of random variables such that $X_i \leq_{\text{rh}} Y_i$ for all i. If all X_i, Y_i have decreasing reversed hazard rates, then

$$\sum_{i=1}^{n} X_i \leq_{\text{rh}} \sum_{i=1}^{n} Y_i;$$

(ii) *Mixture function* If $X|(\Theta = \theta) \leq_{\text{rh}} Y|(\Theta = \theta')$ for all θ, θ' in the support of Θ, then $X \leq_{\text{rh}} Y$;

(iii) *Order statistics*

(a) If X_i are independent, $i = 1, 2, \ldots, n$, then

$$X_{k:n} \leq_{\text{rh}} X_{k+1:n}, \quad k = 1, 2, \ldots, n-1;$$

(b) If $X_n \leq_{\text{rh}} X_i$ for $i = 1, 2, \ldots, n-1$, then

$$X_{k-1:n-1} \leq_{\text{rh}} X_{k:n}, \quad k = 2, 3, \ldots, n;$$

(c) Let X_i, Y_i be pairs of independent absolutely continuous random variables with $X_i \leq_{\text{rh}} Y_i$, $i = 1, 2, \ldots, n$. If the X_i's and Y_i's are also identically distributed, then

$$X_{k:n} \leq_{\text{rh}} Y_{k:n}, \quad k = 1, 2, \ldots m.$$

Under slightly different conditions, without the assumption of identical distributions for $(X_1, X_2, \ldots, X_n)$ and $(Y_1, Y_2, \ldots, Y_m)$, if $X_i \leq_{\text{rh}} Y_j$ for all i, j, $i = 1, 2, \ldots, n$, $j = 1, 2, \ldots, m$, the result that

$$X_{i:n} \leq_{\text{rh}} Y_{j:m}$$

holds for $i - j \geq \max(0, m-n)$.

8.8.2 Other Orders in Reversed Time

The reversed mean residual life function and the corresponding reversed mean residual quantile function have been defined earlier as

$$r(x) = E[x - X | X \leq x] = \frac{1}{F(x)} \int_0^x F(t) dt$$

and

$$R(u) = r(Q(u)) = \frac{1}{u} \int_0^u pq(p) dp.$$

Nanda et al. [459] introduced an ordering of reversed mean residual life, and their definition and the equivalent version in terms of quantile function are presented in the following theorem.

Definition 8.10. The random variable X is said to be smaller than the random variable Y in reversed mean residual life, denoted by $X \leq_{\text{MIT}} Y$, if and only if

$$r_X(x) \geq r_Y(x) \text{ for all } x,$$

or equivalently

$$R_X(u) \geq R_Y^*(u) \text{ for all } 0 < u < 1,$$

where $R_Y^*(u) = r_Y(Q_X(u))$.

Sometimes, the reversed mean residual life is also called the mean inactivity time and so the corresponding ordering is called the mean inactivity time order or simply the MIT order. The relationship of the MIT order to some other orders has been discussed in the literature; see, e.g., Nanda et al. [462], Kayid and Ahmad [319] and Ahmed et al. [24]. It has been shown that, for $0 < t_1 < t_2$, X is DRHR if and only if

(i) $X_{(t_1)} \leq_{\text{st}} X_{(t_2)}$, $X_{(t)} = t - X|(X \leq t)$ is the inactivity time;
(ii) $X_{(t_1)} \leq_{\text{hr}} X_{(t_2)}$;
(iii) for all positive integers m and n,

$$F^{m+n}(x) \geq F^m\left(\frac{n}{m}x\right) F^n\left(\frac{m}{n}x\right).$$

Further,

$$X \leq_{\text{rh}} Y \Rightarrow X \leq_{\text{MIT}} Y,$$

but the converse need not be true.

Ahmed and Kayid [23] have shown that if $\frac{r_X(x)}{r_Y(x)}$ is an increasing function of x, then the $\leq_{\text{rh}}$ order and the $\leq_{\text{MIT}}$ order are equivalent. Li and Xu [393] have made a comparison of the residual X_t and the inactivity time $X_{(t)}$ of series and parallel systems. Instead of considering the life at a specified time t, Li and Zuo [395] discussed the residual life at a random time Y through the random residual life of the form

$$X_Y = (X - Y)|(X > Y)$$

and the inactivity at the random time of the form

$$X_{(Y)} = (Y - X)|(X \leq Y).$$

Notice that the distribution function of X_Y then becomes

$$\begin{aligned} P(X_Y \leq x) &= P(X - Y \leq x|X > Y) \\ &= \frac{\int_0^\infty [F_X(y+x) - F_X(y)]dF_Y(y)}{\int_0^\infty F_Y(y)dF(y)}. \end{aligned}$$

They then established that X has increasing mean inactivity time if and only if $X \leq_{\text{MIT}} X + Y$ for any Y independent of X. Moreover, if ϕ is a strictly increasing concave function with $\phi(0) = 0$, then

$$X \leq_{\text{MIT}} Y \Rightarrow \phi(X) \leq_{\text{MIT}} \phi(Y).$$

Ortega [474] has some additional results concerning the $\leq_{\text{rh}}$ and $\leq_{\text{MIT}}$ orders presented in the following theorem.

Theorem 8.26. *When X and Y are absolutely continuous random variables,*

$$X \leq_{rh} Y \Leftrightarrow \exp[sX] \leq_{MIT} \exp(sY) \textit{ for all } s > 0.$$

It may be noted that Theorem 8.26 characterizes the $\leq_{\text{rh}}$ order in terms of the $\leq_{\text{MIT}}$ order. Conversely, the reverse characterization is apparent from

$$X \leq_{\text{MIT}} Y \Leftrightarrow \log X^{\frac{1}{s}} \leq_{\text{rh}} \log Y^{\frac{1}{s}} \text{ for all } s > 0.$$

The MIT order is also related to the mrl order as

$$X \leq_{\text{MIT}} Y \Rightarrow \phi(X) \geq_{\text{mrl}} \phi(Y)$$

for any strictly decreasing convex function $\phi : [0,\infty) \to [0,\infty)$.

The following preservation properties of order statistics and convolutions hold in this case.

Theorem 8.27. *(i) Let $(X_1, X_2, \ldots, X_n)$ and $(Y_1, Y_2, \ldots, Y_m)$ be two sets of independent and identically distributed random variable with support $[0,\infty)$. Then,*

$$X_1 \leq_{MIT} Y_1 \Rightarrow X_{k:n} \leq_{rh} Y_{l:m}, \quad k \geq l \text{ and } n-k \leq m-l;$$

(ii) If $X_n \leq_{MIT} X_i$, $i = 1, 2, \ldots, n-1$, then

$$X_{k+1:n} \leq_{rh} X_{k:n-1}, \quad k = 1, 2, \ldots, m-1;$$

also, when $X_1, X_2, \ldots, X_n$ are independent absolutely continuous random variables with $X_i \leq_{MIT} Y_j$ for all i, j, then:

(a) $X_{l:n} \leq_{rh} Y_{l:n}$, $l = 1, 2, \ldots, n$;
(b) $X_{k:n} \leq_{rh} Y_{l:n}$, $k \geq l$, $n \leq m$.

Theorem 8.28. *Let $X = \sum_{i=1}^{N} X_i$ and $Y = \sum_{i=1}^{M} Y_i$, where (X_i, Y_i) are independent pairs of random variables such that X_i has decreasing reversed hazard rate, Y_i also has decreasing reversed hazard rate, and $X_i \geq_{MIT} Y_i$, $i = 1, 2, \ldots$, and $N \geq_{rh} M$, then $X \geq_{MIT} Y$.*

Another function in reversed time for which stochastic orders can be defined is the reversed variance residual life (variance of inactivity time, VIT) given by

$$v(x) = E\left[(x-X)^2|X \le x\right] - r^2(x)$$
$$= \frac{2}{F(x)}\int_0^x\int_0^y F(t)dtdy - r^2(x),$$

or equivalently in quantile form as

$$D(u) = \frac{1}{u}\int_0^u R^2(p)dp$$

(see (2.53)). Mahdy [408] has then defined the following stochastic order.

Definition 8.11. We say that X is smaller than Y in variance inactivity time order, denoted by $X \le_{\mathrm{VIT}} Y$, if and only if

$$\frac{\int_0^x\int_0^t F_X(y)dydt}{F_X(x)} \ge \frac{\int_0^x\int_0^t F_Y(y)dydt}{F_Y(x)}$$

for all $x \ge 0$. In other words,

$$\frac{1}{u}\int_0^u R_X^2(p)dp \ge \frac{1}{F_Y(Q_X(u))}\int_0^u R_X^{*2}(p)dp$$

for all u in $(0,1)$, where $R_Y^*(p) = v_Y(Q_X(p))$

Some properties of the $\le_{\mathrm{VIT}}$ order are as follows:

1. A necessary and sufficient condition for $X \le_{\mathrm{VIT}} Y$ is that
$$\frac{\int_0^x\int_0^t F_X(y)dydt}{\int_0^x\int_0^t F_Y(y)dydt}$$
is an increasing function of x;
2. X has increasing VIT $\Leftrightarrow X \le_{\mathrm{VIT}} X+Y$, where Y is independent of X;
3. If ϕ is strictly increasing and concave with $\phi(0) = 0$, then
$$X \le_{\mathrm{VIT}} \Rightarrow \phi(X) \le_{\mathrm{VIT}} \phi(Y);$$
4. If $X_1,\ldots,X_n$ and $Y_1,\ldots,Y_n$ are independent copies of X and Y, respectively, then
$$\max_{1\le i\le n} X_i \le_{\mathrm{VIT}} \max_{1\le i\le n} Y_i \Rightarrow X \le_{\mathrm{VIT}} Y.$$

8.9 Total Time on Test Transform Order

Recall from (5.6) that the total time on test transform (TTT) of X is defined as

$$T(u) = \int_0^u (1-p)q(p)dp.$$

The role of this function in characterizing life distributions, ageing properties and in various other applications have been described earlier in Chap. 5. Here, $T(u)$ represents the quantile function of a random variable, say X_T, in the support of $[0,\mu]$, where $\mu = E(X)$. In this section, we define and study some properties of an order obtained through the comparison of the TTT's of two random variables; for further details, one may refer to Kochar et al. [349] and Li and Shaked [392].

Definition 8.12. A random variable X is said to be smaller than another random variable Y in total time on test transform order, denoted by $X \leq_{\text{TTT}} Y$, if

$$T_X(u) \leq T_Y(u)$$

for all $u \in (0,1)$.

Example 8.8. Let X be exponential with mean $\frac{1}{4}$, i.e.,

$$Q_X(u) = -4\log(1-u),$$

and Y be uniform with

$$Q_Y(u) = u.$$

Then, we have $T_X(u) = \frac{u}{4}$ and $T_Y(u) = \frac{u(2-u)}{4}$ so that

$$T_X(u) - T_Y(u) = \frac{4}{u}(u-1) < 0 \text{ for all } 0 < u < 1.$$

Hence, $X \leq_{\text{TTT}} Y$.

Some interesting relationships possessed by the $\leq_{\text{TTT}}$ order are presented in the following theorem.

Theorem 8.29. *(i)* $X \leq_{st} Y \Rightarrow X \leq_{TTT} Y$;

(ii) $X \leq_{TTT} Y \Rightarrow aX \leq_{TTT} aY$, $a > 0$;

(iii) $X_T \leq_{st} Y_T \Leftrightarrow X \leq_{TTT} Y$, *where* X_T *denotes the random variable with quantile function* $T(u)$;

(iv) $X \leq_{TTT} Y \Rightarrow X_T \leq_{TTT} Y_T$;

(v) $X \leq_{st} Y \Rightarrow X_T \leq_{st} Y_T$.

Proof. (i) We note that

$$\begin{aligned} T(u) &= \int_0^u (1-p)q(p)dp \\ &= (1-u)Q(u) + \int_0^u Q(p)dp. \end{aligned}$$

Now,

$$\begin{aligned} X \leq_{\text{st}} Y &\Rightarrow Q_X(u) \leq Q_Y(u) \\ &\Rightarrow (1-u)Q_X(u) + \int_0^u Q_X(p)dp \leq (1-u)Q_Y(u) + \int_0^u Q_Y(p)dp \\ &\Rightarrow T_X(u) \leq T_Y(u) \Rightarrow X \leq_{\text{TTT}} Y. \end{aligned}$$

Part (ii) follows from the fact that $Q_{aX}(a) = aQ_X(u)$ and (iii) is obvious from the definitions of the stochastic and TTT orders. To prove Part (iv), we note that the transform of X_T is

$$T_{X_T}(u) = \int_0^u (1-u)t_X(u),$$

where $t_X(u) = T_X'(u)$, the quantile density function of X_T. The last equation, using integration by parts, becomes

$$T_{X_T}(u) = (1-u)T_X(u) + \int_0^u T_X(p)dp.$$

The proof of Part (iv) is then similar to that of (i). Part (v) is a direct consequence of Parts (iii) and (i).

Theorem 8.30. *If X and Y have zero as the common left end point of their supports, then for an increasing concave function ϕ with $\phi(0) = 0$,*

$$X \leq_{TTT} Y \Rightarrow \phi(X) \leq_{TTT} \phi(Y).$$

Theorem 8.31 (Li and Zuo [395]). *Let $\{X_n\}$, $\{Y_n\}$, $n = 1, 2, \ldots$, be two sequences of independent and identically distributed random variables and N be a positive integer valued random variable independent of the X's and Y's. If $X_1 \leq_{TTT} Y_1$, then*

$$\min_{1 \leq i \leq N} X_i \leq_{TTT} \min_{1 \leq i \leq N} Y_i.$$

Extensions of the above results are possible if we consider total time on test transform of order n (TTT $-\, n$) introduced earlier in (5.26). Recall that TTT$-n$ is defined as

$$T_n(u) = \int_0^u (1-p)t_{n-1}(p)dp, \quad n = 1, 2, \ldots,$$

with $T_0(u) = Q(u)$ and $t_n(u) = \frac{dT_n(u)}{du}$, provided $\mu_{n-1} = \int_0^1 T_{n-1}(u)du < \infty$.

Definition 8.13. X is said to be smaller than Y in TTT of order n, written as $X \leq_{\text{TTT}-n} Y$, if and only if $T_{n+1,X} \leq T_{n+1,Y}$ for all u in $(0,1)$. Denote by X_n and Y_n the random variables with quantile functions $T_{n,X}(u)$ and $T_{n,Y}(u)$, respectively.

As in the case of the first order transforms $T(u)$, we have the following relationships:

(i) $X \leq_{\text{TTT}-n} Y \Leftrightarrow X_{n+1} \leq_{\text{st}} Y_{n+1}$;
(ii) $X \leq_{\text{TTT}} Y \Rightarrow X \leq_{\text{TTT}-n} Y$.

If $(X_1, X_2, \ldots X_n)$ and $(Y_1, Y_2, \ldots Y_n)$ are independent copies of X and Y that are identically distributed and $X \leq_{\text{TTT}-n} Y$, then $\min(X_1, X_2, \ldots, X_n) \leq_{\text{TTT}-n} \min(Y_1, Y_2, \ldots, Y_n)$. For further results and other aspects of TTT$-n$ order, we refer the reader to Nair et al. [447].

8.10 Stochastic Orders Based on Ageing Criteria

So far, our attention has focussed on partial orders that compare life distributions on the basis of reliability concepts. In view of the predominant role ageing criteria have in modelling and in the analysis of reliability data, it will be natural to consider similar comparisons that spell out which of the two given distributions is more positively ageing than the other in terms of concepts like IHR, IHRA, NBU, etc. This idea has resulted in some partial orders that are discussed in this section.

We begin with the convex transform order defined by Barlow and Proschan [68].

Definition 8.14. Let X and Y have continuous distributions with $F_X(0) = F_Y(0) = 0$, and $F_Y(x)$ be strictly increasing on an interval support. Then, we say that X is less than Y in convex transform order, denoted by $X \leq_c Y$, if $F_Y^{-1}(F_X(x))$ is a convex function in x on the support of X, assumed to be an interval.

Notice that according to (8.1), $\psi_{F_X,F_Y}(x) = F_Y^{-1}(F(x))$ is the relative inverse function of F_X and F_Y, and it enjoys the properties of ψ mentioned earlier in Sect. 8.1. An immediate consequence of Definition 8.14 is that if Y is exponential, then

$$\psi_{F_X,F_Y}(x) = F_Y^{-1} F_X(x) = -\frac{1}{\lambda} \log(1 - F(x))$$

is convex, which means that

$$\psi'(x) = \frac{1}{\lambda} \frac{f(x)}{\bar{F}(x)} = \frac{1}{\lambda} h(x)$$

is increasing, or X is IHR. It is easy to see that the converse also holds. Thus, we have an equivalent condition for X to be IHR in terms of $\leq_c$ as follows.

Theorem 8.32. *X is IHR if and only if $X \leq_c Y$, where Y is exponential.*

In the above result, Y can have any scale parameter. In general, in terms of distribution function,

$$F_X <_c F_Y \Leftrightarrow F_X(\alpha x) <_c F_Y(\beta x)$$

for all $\alpha, \beta > 0$, and so $<_c$ is unaffected by scaling. Kochar and Wiens [350] have developed an ordering based on IHR from the above facts.

Definition 8.15. We say that X is more IHR than Y if $X \leq_c Y$. Making use of (8.3) and (8.2) and assuming that X and Y have densities, we find

$$\begin{aligned}\frac{d}{dx} F_Y^{-1} F_X(x) &= \frac{f_X(F_X^{-1}(x))}{f_Y(F_Y^{-1}(x))} \\ &= \frac{f_X(Q_X(u))}{f_Y(Q_Y(u))} = \frac{q_Y(u)}{q_X(u)}.\end{aligned}$$

Hence, $X \leq_c Y$ if and only if $\frac{q_Y(u)}{q_X(u)}$ is increasing in u in $[0,1]$.

Theorem 8.33.

$$X \leq_c Y \Leftrightarrow X_T \leq_c Y_T.$$

Proof. From the above discussion, we have seen that $X \leq_c Y$ if and only if the ratio of the quantile density functions $\frac{q_Y(u)}{q_X(u)}$ of X and Y is increasing in u. The quantile density functions of X_T and Y_T are

$$t_{X_T}(u) = (1-u)q_X(u)$$

and

$$t_{Y_T}(u) = (1-u)q_Y(u).$$

Since $\frac{q_Y}{q_X}$ is increasing by hypothesis, $\frac{t_{Y_T}}{t_{X_T}}$ is also increasing by virtue of the fact that $\frac{t_{Y_T}}{t_{X_T}} = \frac{q_Y}{q_X}$. Hence, $X_T \leq_c Y_T$, as required.

There is a preservation property for the order statistics as well as described below.

Theorem 8.34. *Let $\{X_n\}$, $\{Y_n\}$ be two sequences of independent and identically distributed random variables and N be a positive integer valued random variable independent of the X_i's and Y_i's. If $X_1 \leq_c Y_1$, then*

$$\min_{1 \leq i \leq N} X_i \leq_c \min_{1 \leq i \leq N} Y_i \text{ and } \max_{1 \leq i \leq N} X_i \leq_c \max_{1 \leq i \leq N} Y_i.$$

A weaker order than the convex transform order is the star order defined as follows.

Definition 8.16. We say that X is smaller than Y in star order, written as $X \le_* Y$, if and only if $F_Y^{-1}(F_X(x))$ is star-shaped in x.

By definition of star-shaped functions, it means that, for $X \le_* Y$, we should have $\frac{1}{x}F_Y^{-1}(F_X(x))$ increasing in $x \ge 0$. Now,

$$\begin{aligned} x q_Y(F_X(x)) f_X(x) - Q_Y(F_X(x)) &\ge 0 \\ \Rightarrow q_Y(u)\frac{Q_X(u)}{q_X(u)} - Q_Y(u) &\ge 0 \\ \Rightarrow Q_X(u)q_Y(u) - Q_Y(u)q_X(u) &\ge 0 \\ \Rightarrow \frac{Q_Y(u)}{Q_X(u)} \text{ is increasing in } u. \end{aligned}$$

Since $X \le_c Y$ implies $\frac{q_Y}{q_X}$ is increasing, it follows that

$$X \le_c Y \Rightarrow X \le_* Y.$$

The converse need not be true. Bartoszewicz and Skolimowska [78] have shown that

(a) if $X \le_* Y$, $\log Q_Y$ is convex and $\log Q_X$ is concave, then $X \le_c Y$;
(b) if F_X and F_Y are absolutely continuous and $X \le_* Y$, $xf_X(x)$ is increasing and $xg_X(x)$ is decreasing, then $X \le_c Y$.

Assume that Y is exponential with scale parameter λ. Then,

$$X \le_* Y \Rightarrow -\frac{1}{\lambda}\frac{\log(1-u)}{Q(u)}$$

is increasing. Hence, by Definition 4.9, X is IHRA. Thus, the star ordering can be used to define increasing hazard quantile distributions, giving an ordering of IHRA distributions as follows.

Definition 8.17. X is said to be more IHRA than Y if and only if $X \le_* Y$.

The star ordering enjoys properties similar to the convex transform ordering, and they are:

(i) $X \le_* Y \Rightarrow X_T \le_* Y_T$;
(ii) Theorem 8.34 holds when $\le_c$ is replaced by $\le_*$;
(iii) $X \le_* Y \Rightarrow X^p \le_* Y^p$ for any $p \ne 0$.

Ordering life distributions by the NBU property requires the superadditive property which is defined as follows.

Definition 8.18. We say that X is more NBU than Y if $F_Y^{-1}(F_X(x))$ is superadditive in x, i.e., if

$$F_Y^{-1}F_X(x+y) \geq F_Y^{-1}(F_X(x)) + F_Y^{-1}(F_X(y)) \text{ for all } x,y \geq 0. \tag{8.13}$$

This is denoted by $X \leq_{\text{su}} Y$.

To justify the above definition, we note that when Y is exponential, (8.13) becomes

$$-\frac{1}{\lambda}\log(1-F_X(x+y)) \geq -\frac{1}{\lambda}\log(1-F_X(x)) - \frac{1}{\lambda}\log(1-F_Y(x)),$$

or

$$\bar{F}(x+y) \leq \bar{F}(x)\bar{F}(y).$$

Hence, X is NBU by (4.26). Thus, we have the following theorem.

Theorem 8.35. *When Y is exponential, $X \leq_{su} Y \Leftrightarrow X$ is NBU.*

Some other properties of the $\leq_{\text{su}}$ order are:

(a) $X \leq_* Y \Rightarrow X \leq_{\text{su}} Y$;
(b) Theorem 8.34 holds when $\leq_c$ is replaced by $\leq_{\text{su}}$.

A more general result holds for order statistics that involves all three orders discussed in this section in the context of k-out-of-n systems as stated in the following theorem.

Theorem 8.36. *If (X_i, Y_i), $i = 1,2,\ldots,n$, are independent pairs of random variables with the property $X_i \leq_c (\leq_*, \leq_{su})Y_i$ for all i, and X_i's and Y_i's are identically distributed, then*

$$X_{k:n} \leq_c (\leq_*, \leq_{su})Y_{k:n}, \quad k = 1,2,\ldots,n.$$

The orderings with respect to other ageing criteria discussed below are due to Kochar and Weins [350] and Kochar [347].

Definition 8.19. We say that X is more decreasing mean residual life than Y, denoted by $X <_{\text{DMRL}} Y$, if

$$\frac{M_X(u)}{M_Y(u)} \text{ is nonincreasing in } u.$$

Since the reciprocal of the hazard quantile function of Z is the mean residual quantile function of X, an equivalent condition for $X \leq_{\text{DMRL}} Y$ is that

$$\frac{H_{Z,X}(u)}{H_{Z,Y}(u)} \text{ is non-decreasing in } u,$$

where $H_{Z,X}$ is the hazard quantile function of the equilibrium distribution of X. Observe that the definition

$$M_X(u) = m_X(Q_X(u)) = \frac{1}{1-u}\int_u^1 (1-p)q_X(p)dp$$

is the mean residual quantile of X, and similarly

$$M_Y(u) = m_Y(Q_Y(u)) = \frac{1}{1-u}\int_u^1 (1-p)q_Y(p)dp.$$

Theorem 8.37. *If Y is exponential, then*

$$X \leq_{DMRL} Y \Leftrightarrow X \text{ is DMRL.}$$

The proof is immediate upon substituting $M_Y(u) = \frac{1}{\lambda}$ in Definition 8.19.

Theorem 8.38.

$$X \leq_{DMRL} Y \Leftrightarrow \frac{\mu_Y - T_Y(u)}{\mu_X - T_X(u)} \text{ is increasing in } u.$$

Proof. We have

$$\begin{aligned} X \leq_{\text{DMRL}} Y &\Leftrightarrow \frac{M_X(u)}{M_Y(u)} \text{ is increasing} \\ &\Leftrightarrow \frac{\int_u^1 (1-p)q_X(p)dp}{\int_u^1 (1-p)q_Y(p)dp} \text{ is increasing} \end{aligned}$$

The proof is completed simply by noting that $\int_u^1 (1-p)q_X(p)dp = \mu - T(u)$.

Theorem 8.39.

$$X \leq_c Y \Rightarrow X \leq_{DMRL} Y.$$

In other words, the IHR order implies the DMRL order.

Definition 8.20. X is said to be smaller than Y in NBUE order (X is more NBUE than Y) if and only if

$$\frac{M_X(u)}{M_Y(u)} \leq \frac{\mu_X}{\mu_Y} \text{ for all } u \text{ in } [0,1],$$

and we denote it by $X \leq_{\text{NBUE}} Y$.

Two equivalent conditions for the $\leq_{NBUE}$ order are:

(a) $\frac{H_{Z,X}(u)}{H_{Z,Y}(u)} \geq \frac{\mu_Y}{\mu_X}$;
(b) $\frac{T_X(u)}{T_Y(u)} \geq \frac{\mu_X}{\mu_Y}$.

Theorem 8.40. *Let Y be an exponential random variable. Then,*

$$X \leq_{NBUE} Y \Leftrightarrow X \text{ is NBUE.}$$

Proof. Since $M_Y(u) = \mu_Y = \frac{1}{\lambda}$, the definition of $\leq_{\text{NBUE}}$ gives the desired result.

Theorem 8.41. *If X and Y have supports of the form $[0,a)$, then:*

(i) $X \leq_{DMRL} Y \Rightarrow X \leq_{NBUE} Y$;
(ii) $X \leq_{*} Y \Rightarrow X \leq_{NBUE} Y$.

The proof of Part (i) is straightforward from the definitions of the two orderings. To prove Part (ii), we note that

$$\begin{aligned} X \leq_* Y &\Rightarrow X_T \leq_* Y_T \\ &\Rightarrow \frac{T_Y(u)}{T_X(u)} \text{ is increasing in } u \\ &\Rightarrow \frac{T_Y(u)}{T_X(u)} \leq \frac{T_Y(1)}{T_X(1)} = \frac{\mu_Y}{\mu_X} \\ &\Rightarrow X \leq_{\text{NBUE}} Y. \end{aligned}$$

The characterization of the class of distributions for which $X \leq_{\text{su}} Y$ implies $X \leq_{\text{NBUE}} Y$ remains open.

Definition 8.21. We say that F is more NBUHR (new better than used in hazard rate) if $\frac{d}{dx}\psi_{F_X,F_Y}(x) \geq \psi'(0)$, and is denoted by $X \leq_{\text{NBUHR}} Y$.

From this definition, we see that

$$\frac{d}{dx}\psi(x) = \frac{d}{dx}F_Y^{-1}F(x) = \frac{H_X(u)}{H_Y(u)}$$

from the discussion following Definition 8.15. Hence,

$$X \leq_{\text{NBUHR}} Y \Leftrightarrow \frac{H_X(u)}{H_X(0)} > \frac{H_Y(u)}{H_Y(0)},$$

using which we obtain the interpretation in the following theorem.

Theorem 8.42. *If Y is exponential, then $X \leq_{NBUHR} Y \Leftrightarrow X$ is NBUHR.*

Proof. We observe that

$$\begin{aligned} X \leq_{\text{NBUHR}} Y &\Leftrightarrow \frac{d}{dx}\psi(x) \geq \psi'(0) \\ &\Leftrightarrow \frac{H_X(u)}{\lambda} \geq \frac{H_X(0)}{\lambda} \\ &\Leftrightarrow X \text{ is NBUHR} \end{aligned}$$

by Definition 4.6.

A similar definition for the NBUHRA order can be provided as follows.

Definition 8.22. X is more NBUHRA (new better than used in hazard rate average than Y), denoted by $X \leq_{\text{NBUHRA}} Y$, if and only if

$$\psi(x) \geq x\psi'(0).$$

We then have

$$X \leq_{\text{NBUHRA}} Y \Rightarrow X \text{ is NBUHRA}$$

and

$$X \leq_{\text{NBU}} Y \Rightarrow X \leq_{\text{NBUHRA}} Y \Rightarrow X \leq_{\text{NBUHRA}} Y.$$

8.11 MTTF Order

Earlier in Sect. 4.2, we have defined the mean time to failure (MTTF) in an age replacement model as (see (4.19)).

$$M(T) = \frac{1}{F(T)} \int_0^T \overline{F}(t)dt.$$

Another formulation of MTTF is

$$\mu(u) = M(Q(u)) = \frac{1}{u} \int_0^u (1-p)q(p)dp.$$

Now, a comparison of life distributions by the magnitude of MTTF is possible by considering an appropriate stochastic order.

Definition 8.23. A lifetime random variable X is smaller than another lifetime random variable Y in MTTF order, denoted by $X \leq_{\text{MTTF}} Y$, if and only if $\mu_X(u) \leq \mu_Y^*(u)$ for all u in $(0,1)$ (or equivalently, $M_X(T) \leq M_Y(T)$ for all $T > 0$), where $\mu_Y^*(u) = M_Y(Q_X(u))$.

First, we discuss the relationship of the MTTF order with other stochastic orders discussed earlier.

Theorem 8.43. *If $X \leq_{st} Y$, then $X \leq_{MTTF} Y$, but the converse is not always true.*

The proof of this result and a counter example are given in Asha and Nair [39]. Resulting from Theorem 8.43, we have the following chain of implications:

$$X \leq_{\text{hr}} Y \Rightarrow X \leq_{\text{st}} Y \Rightarrow X \leq_{\text{MTTF}} Y$$

$$\Uparrow$$

$$X \leq_{\text{rh}} Y.$$

Two other basic reliability orders are $\leq_{\text{mrl}}$ and $\leq_{\text{MIT}}$, comparing the mean residual life and the mean inactivity time. As already seen, the hr order implies the mrl order and the hr order also implies the MTTF order. Hence, the point of interest is to know whether there exist any implications between the $\leq_{\text{mrl}}$ and the $\leq_{\text{MTTF}}$ orders. By taking

$$f_Y(x) = \frac{1}{2}\exp\left(-\frac{x}{2}\right)$$

and

$$f_X(x) = xe^{-x}, \quad x > 0,$$

we see that $X \geq_{\text{MTTF}} Y$, but $X \leq_{\text{mrl}} Y$.

Conditions under which the $\leq_{\text{st}}$ and the $\leq_{\text{mrl}}$ orders have implications with the $\leq_{\text{MTTF}}$ order are of interest. These are presented in the next theorem. The conditions can be stated in terms of quantiles by setting $x = Q(u)$ as usual.

Theorem 8.44. *(a) If $\frac{\int_0^x F_X(t)dt}{\int_0^x F_Y(t)dt}$ is decreasing, then $X \geq_{MTTF} Y \Rightarrow X \geq_{st} Y$;*
(b) If $\frac{m_X(x)}{m_Y(x)}$ is decreasing, then $X \geq_{mrl} Y \Rightarrow X \geq_{MTTF} Y$.

A similar result holds for the MIT order as well. It has been mentioned earlier that if $\frac{r_X(x)}{r_Y(x)}$ is an increasing function of x, then the $\leq_{\text{rh}}$ and the $\leq_{\text{MIT}}$ orders are equivalent. Accordingly, when $\frac{r_X(x)}{r_Y(x)}$ is decreasing,

$$X \geq_{\text{MIT}} Y \Rightarrow X \geq_{MTTF} Y.$$

Further, if $X \geq_{\text{st}} Y$, then $X \geq_{\text{MTTF}} Y \Rightarrow X \geq_{\text{hmrl}} Y$. Returning to decreasing mean time to failure as an ageing concept (see Sect. 4.3), we have a stochastic order comparison based on DMTTF as follows.

Definition 8.24. X has more DMTTF than Y if $\frac{\mu_X(u)}{\mu_Y(u)}$ is decreasing in u for all $0 \leq u \leq 1$, and we denote it by $X \leq_{\text{DMTTF}} Y$.

Suppose Y is exponential. Then, $\mu_Y(u) = \frac{1}{\lambda}$ and so in this particular case, we have

$$X \geq_{\text{DMTTF}} Y \Leftrightarrow X \text{ is DMTTF}.$$

Two other properties of this ordering are as follows:

1. $X \geq_{\text{DMRL}} Y \Rightarrow X \leq_{\text{DMTTF}} Y$;
2. $X \leq_{\text{NBUE}} Y \Leftrightarrow \frac{\mu_X(u)}{\mu_x} \geq \frac{\mu_Y(u)}{\mu_Y}$.

8.12 Some Applications

When X represents a continuous lifetime with distribution function $F(x)$, the proportional reversed hazard model is represented by a non-negative absolutely continuous random variable U whose distribution function is

$$F_U(x) = [F_X(x)]^{\theta},$$

where θ is a positive real number (see Example 1.3). When $F(x)$ is strictly increasing, $F_X(x) = u$ gives the quantile function of U as

$$Q_U(\theta) = Q_X(u^{\frac{1}{\theta}}).$$

For this model, the reversed hazard rates of U and X are proportional, i.e., $\lambda_U(x) = \theta\lambda_X(x)$ or $\Lambda_U^*(u) = \theta\Lambda_X(u)$, where

$$\Lambda_U^*(u) = \lambda_U(Q_X(u)).$$

Gupta et al. [239] and Di Crecenzo [177] have studied the order relationship between X and U and also between two random variable X and Y and their proportional reversed hazard models U and V. Let

$$\mathscr{H}(x) = -\log F_X(x) = \int_x^{\infty} \lambda(t)dt$$

be the cumulative reversed hazard rate of X.

Theorem 8.45. *Let $[\mathscr{H}(x)]^{-1}$ be star-shaped (antistarshaped). Then:*

(i) If $\theta < 1$, $\theta X \leq_{st} U(\theta X) \geq_{st} U$;
(ii) If $\theta > 1$, $\theta X \geq_{st} U(\theta X) \leq_{st} U$.

Theorem 8.46. *(i) $X \leq_{st} Y \Leftrightarrow U \leq_{st} V$;*
(ii) $X \leq_{rh} Y \Leftrightarrow U \leq_{rh} V$;
(iii) $X \leq_{hr} Y$ and $\theta > 1 \Leftrightarrow U \leq_{hr} V$.

Gupta and Nanda [254] have considered X_i, $i=1,2$, with distribution functions $F_i(x)$ and U_i as proportional reversed hazards models of X_i with distribution functions $[F_i(x)]^{\theta_i}$, $i=1,2$.

Theorem 8.47. *$\theta_1 \geq \theta_2$ and $X_1 \geq_{rh} X_2 \Rightarrow Y_1 \geq_{rh} Y_2$.*

In particular, if

$$S_i(x) = 1 - e^{-(\frac{x}{\sigma_i})^{\lambda}},$$

then $X_1 \geq_{\mathrm{rh}} X_2$ if and only if $\sigma_1 \geq \sigma_2$ (>0), irrespective of the value of λ. Similarly, for the exponentiated Weibull distribution with

$$F_i(x) = [1 - e^{-(\frac{x}{\sigma_i})^{\alpha}}]^{\theta},$$

$X_1 \geq_{\mathrm{rh}} X_2$ if and only if $\sigma_1 \geq \sigma_2$. If $X_1, X_2, \ldots$ are independent and identically distributed random variables and N is geometric with $P(N=n) = p(1-p)^{n-1}$, $n=1,2,\ldots$, independent of the X_i's, then the sum

$$S_N = X_1 + \cdots + X_N$$

is said to be a geometric compound. It is easy to see that S_N belongs to the random convolution discussed earlier. Hu and Lin [284] have given several characterizations of the exponential distribution using stochastic orders, some of which are presented in the following theorem.

Theorem 8.48. *1. If F, the common distribution function of the X_i's, is NWU and $pS_N \leq_{st} T\min(X_1,\ldots,X_T)$, then F is exponential, where T is an integer valued random variable. If F is NBU and $T\min(X_1\ldots X_T) \leq_{st} pS_N$, then F is exponential;*
2. If $pS_N \leq_{st} X_1$, then F is exponential;
3. In the renewal process $(S_n)_{n=1}^{\infty}$, $S_n = \sum_{k=1}^{n} X_k$ and $r(t) = S_{N(t)+1} - t$ is the residual life at time t, if F is NBU and $pS_N \leq_{st} r(t)$, then F is exponential.

Nanda et al. [458] have discussed stochastic orderings in terms of the proportional mean residual life model. Let X be a non-negative random variable with absolutely continuous distribution function and finite mean and V be another non-negative random variable with the same properties. Then, we say that V is the proportional mean residual life model (PMRLM) of X if

$$m_V(x) = cm_X(x),$$

where $m_X(x)$ is as usual the mean residual life function. An equivalent condition is

$$M_V^*(u) = cM_X(u),$$

where $M_V^*(u) = m_v(Q_X(u))$. For this model, we have the following properties:

(i) $X \leq_{\text{hr}} (\geq) V$ if $c > (<1)$;
(ii) Let $X \leq_{\text{st}} Y$. If either (a) $c < 1$ and

$$\frac{m_Y(x)}{\mu_Y} \geq \frac{m_X(x)}{\mu_X},$$

or (b) $c > 1$ and

$$\frac{m_Y(x)}{\mu_Y} \leq \frac{m_X(x)}{\mu_X},$$

then $V_X \leq_{\text{st}} V_Y$, where $V_X(V_Y)$ is the PMRLM corresponding to $X(Y)$;
(iii) $X \leq_{\text{hr}} (\geq_{\text{hr}}) Y$ and $c < 1 \Rightarrow V_X \leq_{\text{hr}} (\geq_{\text{hr}}) V_Y$;
(iv) $X \leq_{\text{mrl}} (\geq_{\text{mrl}}) Y \Leftrightarrow V_X \leq_{\text{mrl}} (\geq_{\text{mrl}}) V_Y$;
(v) $X \leq_{\text{hmrl}} (\geq_{\text{hmrl}}) Y \Leftrightarrow V_X \leq_{\text{hmrl}} (\geq_{\text{hmrl}}) V_Y$.

The preservation of stochastic orders among weighted distributions has been discussed in Misra et al. [417]. Let X_1 and Y_1 be weighted versions of X and Y defined as

$$F_{X_1}(x) = \frac{\int_0^x w_1(t) f_X(t) dt}{EW_1(X)}$$

and

$$F_{Y_1}(x) = \frac{\int_0^x w_2(t) f_Y(t) dt}{EW_2(Y)}.$$

We then have the following results.

Theorem 8.49. *(i) If $X \leq_{st} Y$, $w_1(\cdot)$ is decreasing and $w_2(\cdot)$ is increasing, then $X_1 \leq_{st} Y_1$;*
(ii) If X and Y have a common support, $X \leq_{hr} Y$ and $w(x) = w_1(x) = w_2(x)$ is increasing, then $X_1 \leq_{hr} Y_1$;
(iii) If in (ii) $w(\cdot)$ is decreasing and $X \leq_{rh} Y$, then $X_1 \leq_{rh} Y_1$;
(iv) Let $X \leq_{hr} Y$ ($X \leq_{rh} Y$), $w_2(x)$ is increasing ($w_1(x)$ is decreasing) and $\frac{w_2(x_1)}{w_1(x_1)}$ is increasing on the intersection of the supports, then $X_1 \leq_{hr} Y_1$ ($X_1 \leq_{rh} Y_1$) provided that $l_1 \leq l_2$, $u_1 \leq u_2$, where (l_1, u_1) and (l_2, u_2) are the supports of X_1 and Y_1, respectively.

Yu [597] has discussed stochastic comparisons between exponential family of distributions and their mixtures with respect to various stochastic orders. Members of this family have been frequently used in reliability analysis and for this reason we present some results relevant in this regard. The exponential family is expressed by the probability density function

$$f(x,\theta) = a(x)e^{b(\theta)x}h(\theta),$$

where the support is $(0,\infty)$. Let

$$g(x) = \int f(x;t)d\mu(t)$$

be the mixture of $f(x,\theta)$. Then we have the order relations, between X and Y, the random variables corresponding to $f(x;\theta)$ and $g(x)$, as follows:

(a) $X \leq_{\text{st}} Y$ ($X \leq_{\text{hr}} Y$) if and only if $\int h(t)d\mu(t) \leq h(\theta)$;
(b) $X \leq_{\text{rh}} Y$ if and only if

$$b(\theta) \leq \frac{\int b(t)h(t)d\mu(t)}{\int h(t)d\mu(t)}.$$

Let $X = \sum_{i=1}^{\infty} \beta_i X_i$, where X_i is gamma $(\alpha_i, 1)$ independently and $\beta_i > 0$. The order relations between X and Y which is gamma $(\sum_{i=1}^{n} \alpha_i, \beta)$ have been discussed by many authors. When X_i's are independent exponential with different scale parameters (i.e., when $\alpha_i = 1$), Boland et al. [114] have established that

$$\beta \leq \frac{n}{\sum_{i=1}^{n} \beta_i^{-1}} \Rightarrow X \leq_{\text{rh}} Y$$

and Bon and Paltanea [117] have extended this result to

$$Y \leq_{\text{st}} X \Leftrightarrow Y \leq_{\text{hr}} X \Leftrightarrow \beta \leq \left(\prod_{i=1}^{n} \beta_i\right)^{\frac{1}{n}}.$$

Yu [597] has further established that

$$Y \leq_{\text{st}} X (Y \leq_{\text{hr}} X) \text{ if and only if } \beta \leq \left(\prod_{i=1}^{n} \beta_i^{\alpha_i}\right)^{\sum_i^n \frac{1}{\alpha_i}},$$

$$Y \leq_{\text{rh}} X \text{ if and only if } \beta \leq \frac{\sum_1^n \alpha_i}{\sum_i^n \frac{\alpha_i}{\beta_i}}.$$

These results are useful in developing bounds for the hazard rate of X through simpler hazard rate of Y.

If X and Y are lifetime variables with cumulative hazard functions $\mathscr{H}_X(x)$ and $\mathscr{H}_Y(x)$, Sengupta and Deshpande [526] have defined X to be ageing faster than Y if and only if $\mathscr{H}_1\mathscr{H}_2^{-1}$ is superadditive, i.e.,

$$\mathscr{H}_1\mathscr{H}_2^{-1}(x+y) \geq \mathscr{H}_1\mathscr{H}_2^{-1}(x) + \mathscr{H}_1\mathscr{H}_2^{-1}(y).$$

Abraham and Nair [13] have proposed a relative ageing factor

$$B(x,y) = \frac{\mathscr{H}^{-1}(\mathscr{H}(x) + \mathscr{H}(y)) - x}{y}$$

between a new component and an old component that survived up to time x. They then defined an order $X \leq_{B:\mathrm{NBU}} Y$ by the relation $B_X(x,y) \leq B_Y(x,y)$ for all $x,y > 0$. They provided the result that

$$B_X(x,y) \leq B_Y(x,y) \Leftrightarrow X \text{ is NBU},$$

where the NBU part arises from the fact that Y is exponential. The relative ageing defined by the superadditive order now becomes

$$X \leq_{\mathrm{su}} Y \Leftrightarrow X \leq_{B:\mathrm{NBU}} Y.$$

Thus, an ageing criterion is prescribed in terms of $B(x,y)$ to assess the concept of 'X ageing faster than Y'.

If X is a random variable with survival function $\overline{F}(x)$ and Z has survival function $\overline{F}_2(x) = [\overline{F}(x)]^{\theta}$, $\theta > 0$, then $F_Z(x)$ is called the proportional hazards model corresponding to X. The terminology is evident from the fact that $h_Z(x) = \theta h_X(x)$. There are other interpretations also for Z. If $\theta < 1$, Z represents the lifetime of a component in which the original lifetime of the component X is subjected imperfect repair procedure, where θ is the probability of a minimal repair. If $\theta = n$, obviously we have $(\overline{F}(x))^n$ as the survival function of a series system consisting of n independent and identical components whose lifetimes are distributed as X. Franco-Pereira et al. [202] have shown that if X and Y are continuous random variables on interval supports, the α-percentile life order satisfies

$$X \leq_{\mathrm{prl}-\alpha} Y \Rightarrow Z_X \leq_{\mathrm{prl}-\beta} Z_Y,$$

where $\beta = 1 - (1-\alpha)^{\theta}$ and Z_X (Z_Y) is the proportional hazards model corresponding to $X(Y)$.

Extensions of some of the stochastic orders discussed above as well as a variety of applications of all these stochastic orders can be found in Kayid et al. [320], Aboukalam and Kayid [11], Li and Shaked [388], Boland et al. [115], Navarro and Lai [467], Zhang and Li [599], Hu and Wei [286], and Da et al. [164] and the references contained therein.

Chapter 9
Estimation and Modelling

Abstract Earlier in Chaps. 3 and 7, several types of models for lifetime data were discussed through their quantile functions. These will be candidate distributions in specific situations. The selection of one of them or a new one is dictated by how well it can justify the data generating mechanisms and satisfy well other criteria like goodness of fit. Once the question of an initial choice of the model is resolved, the problem is then to test its adequacy against the observed data. This is accomplished by first estimating the parameters of the model and then carrying out a goodness-of-fit test. This chapter addresses the problem of estimation as well as some other modelling aspects.

In choosing the estimates, our basic objective is to get estimated values that are as close as possible to the true values of the model parameters. One method is to seek estimate that match the basic characteristics of the model with those in the sample. This includes the method of percentiles and the method of moments that involve the conventional moments, *L*-moments and probability weighted moments. These methods of estimation are explained along with a discussion of the properties of these estimates. In the quantile form of analysis, the method of maximum likelihood can also be employed. The approach of this method, when there is no tractable distribution function, is described. Many functions required in reliability analysis are estimated by nonparametric methods. These include the quantile function itself and other functions such as quantile density function, hazard quantile function and percentile residual quantile function. We review some important results in these cases that furnish the asymptotic distribution of the estimates and the proximity of the proposed estimates to the true values.

9.1 Introduction

In Chaps. 3 and 7, we have seen several types of models, specified by their quantile functions, that can provide adequate representations of lifetime data. These will be candidate distributions in specific real situations. The selection of one of

N.U. Nair et al., *Quantile-Based Reliability Analysis*, Statistics for Industry and Technology, DOI 10.1007/978-0-8176-8361-0_9,

them or a new one is dictated by finding out how well it can justify the data generating mechanism and satisfy well other criteria like goodness of fit. Perhaps, the most important requirement in all modelling problems is that the chosen lifetime distribution captures the failure patterns that are inherent in the empirical data. Often, the features of the failure mechanism are assessed from the data with the aid of the ageing concepts discussed earlier in Chap. 5. For instance, it could be the shape of the hazard or mean residual quantile function, assessed from a plot of the observed failure times. Based on this preliminary knowledge, the choice of the distribution can be limited to one from the corresponding ageing class discussed in Chap. 4. Once the question of a suitable model is resolved as an initial choice, the problem then is to test its adequacy against the observed failure times. This is accomplished by first estimating the parameters of the distribution and then carrying out a goodness-of-fit test. Alternatively, nonparametric methods can also be employed to infer various reliability characteristics. In this chapter, we address both general parametric methods and nonparametric procedures from a quantile-based perspective.

Our basic objective in estimation is to find estimates that are as close as possible to the true values of the model parameters. There are different criteria which ensure proximity between the estimate and the true parameter value and accordingly different approaches can be prescribed that meet the desired criteria. One method is to seek estimates by matching the basic characteristics of the chosen model with those in the sample. This includes the method of percentiles, method of moments involving conventional moments, *L*-moments and probability weighted moments, and then identifying basic characteristics such as location, dispersion, skewness, kurtosis and tail behaviour. A second category of estimation procedures are governed by optimality conditions that renders the difference between the fitted model and the observed data as small as possible or that provides estimates which are most probable. In the following sections, we describe various methods of estimation as well as the properties of these estimates.

9.2 Method of Percentiles

Recall from Chap. 1 that the pth percentile in a set of observations is the value that has $100p\,\%$ of values below it and $100(1-p)\,\%$ values above it. Let $X_1, X_2, \ldots, X_n$ be a random sample from a population with distribution function $F(x;\Theta)$, or equivalently quantile function $Q(u;\Theta)$, where Θ is a vector of parameters consisting of one or more elements. The sample observations are arranged in order of magnitude with $X_{r:n}$ being the rth order statistic. Then, the sample (empirical) distribution function is defined as

$$F_n(x) = \begin{cases} 0, & x \le X_{1:n} \\ \frac{i}{n}, & X_{i-1:n} < x \le X_{i:n} \text{ for } i = 1,2,\ldots n-1 \\ 1, & x \ge X_{n:n}. \end{cases}$$

Obviously, $F_n(x)$ is the fraction of the sample observations that does not exceed x. The empirical (sample) quantile function then becomes

$$Q_n(p) = F_n^{-1}(p) = \inf[x|F_n(x) \geq p] \tag{9.1}$$

which is a step function with jump $\frac{1}{n}$. Notice that (9.1) can be interpreted as a function ξ_p such that the number of observations $\leq \xi_p$ is $\geq [np]$ and the number of observations $\geq \xi_p$ is $\geq [n(1-p)]$. Thus, e.g.,

$$\begin{aligned} \xi_p &= X_{[np]+1:n} \quad \text{if } np \text{ is not an integer} \\ &= X_{[np]:n} \quad \text{if } np \text{ is an integer.} \end{aligned}$$

In practice, some of the other methods of calculating ξ_p are as follows:

(i) Set $p(n+1) = k+a$, n being the sample size, k an integer, and $0 \leq a < 1$. Then,

$$\xi_p = \begin{cases} X_{k:n} + a(X_{k+1:n} - X_{k:n}), & 0 < k < n \\ X_{1:n}, & k = 0 \\ X_{n:n}, & k = n \end{cases}$$

(see Sect. 3.2.1);

(ii) In some software packages, the setting is $1 + p(n-1) = k + a$;

(iii) Calculate np. If it is not an integer, round it up to the next higher integer k and $X_{k:n}$ is the value. If np is an integer k, take

$$\xi_p = \frac{1}{2}[X_{k:n} + X_{k+1:n}].$$

The value $X_{[np]+1:n}$ is popular as it assures the monotonic nature of ξ_p in the sense that if x is the p-quantile and y is the q-quantile with $p < q$, then $y < x$.

Some properties of ξ_p as an estimate of $Q(p)$ are described in the following theorems.

Theorem 9.1. *If there is a unique value of $Q(p)$ satisfying*

$$P(X \leq Q(p)) \geq p \quad \text{and} \quad P(X \geq Q(p)) \geq 1-p,$$

then $\xi_p \to Q(p)$ as $n \to \infty$ with probability 1.

Theorem 9.2. *Let $F(x)$ have a density $f(x)$ which is continuous. If $Q(p)$ is unique and $f(Q(p)) > 0$, then*

(i) $\sqrt{n}(\xi_p - Q(p)) = n^{\frac{1}{2}}\{p - F_n(Q(p))\}[f(Q(p))]^{-1} + O(n^{-\frac{1}{4}}(\log n)^{\frac{3}{4}})$;

(ii) $\sqrt{n}(\xi_p - Q(p))$ *is asymptotically distributed as*

$$N\left(0, \frac{p(1-p)}{n[f(Q(p))]^2}\right).$$

In particular, the asymptotic distribution of the sample median is normal with mean as the population median and variance $\frac{[f(M)]^{-2}}{4n}$, where $M = Q(\frac{1}{2})$ is the population median. For proofs of the above theorems and further results on the asymptotic behaviour of ξ_p, one may refer to Bahadur [44], Kiefer [324], Serfling [527] and Csorgo and Csorgo [160]. Sometimes, the following bound may be useful in evaluating the bias involved in estimating $Q(p)$ by ξ_p. For $0 < p < 1$ and unique $Q(p)$, for all n and every $\varepsilon > 0$,

$$P[|\xi_p - Q(p)| > \varepsilon] \leq 2\exp[-2n\delta^2],$$

where $\delta = \min(F(Q(p)+\varepsilon) - p, p - F(Q(p)-\varepsilon))$.

A multivariate generalization of Theorem 9.2 is as follows; see Serfling [527].

Theorem 9.3. *Let $0 < p_1 < \cdots < p_k < 1$. Suppose F has a density f in the neighbourhood of $Q(p_1),\ldots,Q(p_k)$ and f is positive and continuous at $Q(p_1)\ldots Q(p_k)$. Then, $(\xi_{p_1}, \xi_{p_2}, \ldots, \xi_{p_k})$ is asymptotically normal with mean vector $(Q(p_1),\ldots,Q(p_k))$ and covariance $\frac{1}{n}\sigma_{ij}$, where*

$$\sigma_{ij} = \frac{p_i(1-p_j)}{f(Q(p_i))f(Q(p_j))}, \quad i \leq j,$$

and $\sigma_{ij} = \sigma_{ji}$ for $i > j$.

Since the order statistic $X_{k:n}$ is equivalent to the sample distribution function $F_n(x)$, the sample quantile may be expressed as

$$\xi_p = \begin{cases} X_{[np]:n}, & np \text{ is an integer} \\ X_{[np]+1:n}, & np \text{ is not an integer.} \end{cases}$$

By inverting this relation, we get

$$X_{k:n} = \xi_{\frac{k}{n}}, \quad 1 \leq k \leq n,$$

and so any discussion of order statistics could be carried out in terms of sample quantiles and vice versa.

Bahadur [44] has given representations for sample quantiles and order statistics. His results with subsequent modifications are of the following form:

1. If F is twice differentiable at $Q(p)$, $0 < p < 1$, with $q(p) > 0$, then

$$\xi_p = Q(p) + [p - F_n(Q(p))]q(p) + R_n,$$

where, with probability 1,

$$R_n = O(n^{-\frac{3}{4}}(\log n)^{\frac{3}{4}}), \quad n \to \infty.$$

Alternatively,

$$n^{\frac{1}{2}}(\xi_p - Q(p)) = n^{\frac{1}{2}}\{p - F_n(Q(p))\}q(p) + O(n^{-\frac{1}{4}}(\log n)^{\frac{3}{4}}),$$

$n \to \infty$;

2. $n^{\frac{1}{2}}(\xi_p - Q(p))$ and $n^{\frac{1}{2}}[p - F_n(Q(p))]q(p)$ each converge in distribution to $N(0, p(1-p)q^2(p))$;
3. Writing $Y_i = Q(p) + [p - I(X_i \leq Q(p))]q(p)$, $i = 1, 2, \ldots$,

$$\xi_p = \frac{1}{n}\sum_{i=1}^{n} Y_i + O(n^{-\frac{3}{4}}(\log n)^{\frac{3}{4}}), \quad n \to \infty,$$

or with probability 1. ξ_p is asymptotically the mean of the first n values of Y_i. Thus, we have a representation of the sample quantile as a sample mean. Consider a sequence of order statistics $X_{k_n:n}$ for which $\frac{k_n}{n}$ has a limit. Provided

$$\frac{k_n}{n} = p + \frac{k}{n^{\frac{1}{2}}} o\left(\frac{1}{n^{\frac{1}{2}}}\right), \quad n \to \infty,$$

$n^{\frac{1}{2}}(X_{k_n:n} - \xi_p)$ converge to $kq(p)$ with probability 1, and $n^{\frac{1}{2}}(X_{k_n:n} - Q(p))$ converge in distribution to $N(kq(p), p(1-p)q^2(p))$.

In other words, $X_{k_n:n}$ and ξ_p are roughly equivalent as estimates of $Q(p)$. Rojo [510] considered the problem of estimation of a quantile function when it is more dispersed than distribution function, based on complete and censored samples. Rojo [511] subsequently developed an estimator of quantile function under the assumption that the survival function is increasing hazard rate on the average (IHRA). The estimator of the quantile function in the censored sample case is also given. He has shown that estimators of $Q(u)$ are uniformly strongly consistent.

The percentiles of the population are, in general, functions of the parameter Θ in the model. In the percentile method of estimation, we choose as many percentile points as there are model parameters. Equating these percentile points of the population with the corresponding sample percentiles and solving the resulting equations, we obtain the estimate of Θ. This method ensures that the model fits exactly at the specific points chosen. Since the method does not specify which percentiles are to be chosen in a practical situation, some judgement is necessary in the choice of the percentile points. Issues such as the interpretation of the model parameters and the purpose for which the model is constructed could be some of the guidelines. Shapiro and Gross [535] pointed out that it is a good practice to choose percentiles where inferences are drawn and not to estimate a percentile where interpolation is required between two highest or two lowest values. As a general practice, they recommended using $p = 0.05$ and $p = 0.95$ for moderate samples and $p = 0.01$ and $p = 0.99$ for somewhat larger samples. In the case of two parameters, one of the above two sets, when there are three parameters augment these by the median $p = 0.50$, and when there are four parameters, the two extreme points along with the quartiles $p = 0.25$ and $p = 0.75$ are their recommendations. The percentile estimators are generally biased, less sensitive to the outliers, and may not guarantee that the mean and variance of the approximating distribution correspond to the sample values.

Example 9.1. Suppose we have a sample of 100 observations from the loglogistic distribution with

$$Q(u) = \frac{1}{\alpha}\left(\frac{u}{1-u}\right)^{\frac{1}{\beta}}, \quad \alpha, \beta > 0.$$

Choosing the values $p = 0.05$ and $p = 0.95$ for matching the population and sample quantiles, we look at the order statistics $X_{5:100}$ and $X_{95:100}$. Then, we have the equations

$$\frac{1}{\alpha}\left(\frac{0.05}{0.95}\right)^{\frac{1}{\beta}} = X_{5:100}, \tag{9.2}$$

$$\frac{1}{\alpha}\left(\frac{0.95}{0.05}\right)^{\frac{1}{\beta}} = X_{95:100}, \tag{9.3}$$

yielding

$$\frac{1}{\alpha^2} = X_{5:100} \times X_{95:100}$$

$$\text{or} \quad \hat{\alpha} = [X_{5:100} \times X_{95:100}]^{-\frac{1}{2}}.$$

Upon substituting $\hat{\alpha}$ in either (9.2) or (9.3) and solving for β, we obtain the estimate of β as

$$\hat{\beta} = \left(\frac{\log 19}{\log \hat{\alpha} X_{5:100}}\right).$$

Instead of using percentiles as such, various quantile-based descriptors of the distribution such as median (M), interquantile range (IQR), measures of skewness (S) and Kurtosis (T), mentioned earlier in Sect. 1.4, may also be matched with the corresponding measures in the sample. The idea is that the fitted distribution has approximately the same distributional characteristics as the observed one. The number of equations should be the same as the number of parameters and the characteristics are so chosen that all the parameters are represented. If there is a parameter representing location (scale), it is a good idea to equate it to the median M (IQR). Some results concerning the asymptotic distributions involving the statistics are relevant in this context. See Sects. 1.4 and 1.5 for the definitions of various measures and Chap. 3 in which percentile method is applied for various quantile functions.

Theorem 9.4. *The sample interquartile range*

$$iqr = \frac{1}{2}(\xi_{\frac{3}{4}} - \xi_{\frac{1}{4}})$$

is asymptotically normal

$$N\left(\frac{1}{2}\left(Q\left(\frac{3}{4}\right)-Q\left(\frac{1}{4}\right)\right), \frac{1}{64n}\left(\frac{3}{f(Q(\frac{3}{4}))} - \frac{2}{f(Q(\frac{1}{4}))f(Q(\frac{3}{4}))} + \frac{3}{f^2(Q(\frac{1}{4}))}\right)\right).$$

Note that $\mathrm{IQR} = \frac{1}{2}\left(Q(\frac{3}{4}) - Q(\frac{1}{4})\right)$, and that iqr is strongly consistent for IQR.

Theorem 9.5. *The sample skewness s and the sample Moor's kurtosis t possess the following properties:*

$$(s,t) \text{ is consistent for } (S,T)$$

and

$$\sqrt{n}(s-S, t-T)^* \text{ has asymptotic bivariate normal distribution}$$

with mean $(0,0)^*$ *and dispersion matrix* $\phi'(c)A(\phi'(c))^*$, *where*

$$A = (\sigma_{ij}), \quad \sigma_{ij} = \frac{i(8-j)}{64 f(E_i) f(E_j)}, \quad i \le j,$$

$$\phi(c) = (S,T)^*, \quad E_i = \frac{i}{8}, \quad i,j = 1,3,5,7.$$

and $*$ *denotes the transpose.*

Some other nonparametric estimators of $Q(u)$ have been suggested in literature. Kaigh and Lachenbruch [308] suggested consideration of a subsample of size k without replacement from a complete sample of size n. Then, by defining the total sample estimator of $Q(u)$ as the average of all possible subsamples of size k, they arrived at

$$\hat{Q}_1(p) = \sum_{r=1}^{k} w_r X_{r:n},$$

where

$$w_r = \frac{\binom{r-1}{j-1}\binom{n-r}{k-j}}{\binom{n}{k}} \text{ and } j = (k+1)p.$$

The choice of k is such that the extreme order statistics should have negligible weight. A somewhat different estimator is proposed in Harrell and Davis [262] of the form

$$\hat{Q}_2(p) = \sum_{r=1}^{n} w_r X_{r:n},$$

where

$$w_r = I_{\frac{r}{n}}(p(n+1),(1-p)(n+1)) - I_{\frac{r-1}{n}}(p(n+1),(1-p)(n+1))$$

with $I_x(a,b)$ being the incomplete beta function. Both $\hat{Q}_1(p)$ and $\hat{Q}_2(p)$ have asymptotic normal distribution. Specific cases of estimation of the exponential and Weibull quantile functions have been discussed by Lawless [378] and Mann and Fertig [410].

9.3 Method of Moments

The method of moments is also a procedure that matches the sample and population characteristics. We consider three such characteristics here, viz., the conventional moments, L moments and probability weighted moments.

9.3.1 Conventional Moments

In this case, either the raw moments $\mu'_r = E(X^r)$ or the central moments $\mu_r = E(X - \mu)^r$ are equated to the same type of sample moments. When μ'_r is used, we construct the equations

$$\mu'_r = \frac{1}{n}\sum_{i=1}^{n} X_i^r = m'_r, \quad r = 1,2,3,\ldots$$

where $X_1, X_2, \ldots, X_n$ are independent and identically distributed and the number of such equations is the same as the number of parameters in the distribution. The sample moments

$$m'_r = \int_0^\infty x^r dF_n(x)$$

as the estimate of μ'_r have the following properties:

(i) m'_r is strongly consistent for μ'_r;
(ii) $E(m'_r) = \mu'_r$, i.e., the estimates are unbiased;
(iii) $V(m'_r) = \frac{\mu'_{2r} - \mu'^2_r}{n}$;
(iv) If $\mu_{2r}' < \infty$, the random vector $n^{-1}(m'_1 - \mu'_1, m'_2 - \mu'_2, \ldots, m'_n - \mu'_n)$ converges in distribution to a n-variate normal distribution with mean vector $(0,0,\ldots,0)$ and covariance matrix $[\sigma_{ij}]$, $i,j = 1,2,\ldots,n$, where $\sigma_{ij} = \mu'_{i+j} - \mu'_i\mu'_j$.

On the other hand, if we use the central moments, the equations to be considered become

$$\mu_r = \frac{1}{n}\sum_{i=1}^{n}(X_i - \bar{X})^r = m_r, \quad \bar{X} = m'_1.$$

The statistic m_r estimates μ_r with the following properties:

(a) m_r is strongly consistent for μ_r;
(b) $E(m_r) = \mu_r + \frac{\frac{1}{2}r(r-1)\mu_{r-2}\mu_2 - r\mu_r}{n} + O(n^{-2})$;
(c) $V(m_r) = \frac{1}{n}(\mu_{2r} - \mu_r^2 - 2r\mu_{r-1}\mu_{r+1} + r^2\mu_2\mu_{r-1}^2) + O(\frac{1}{n^2})$;
(d) If $\mu_{2r} < \infty$, the random vector $n^{\frac{1}{2}}(m_2 - \mu_2, \ldots, m_r - \mu_r)$ converges in distribution to $(r-1)$ dimensional normal distribution with mean vector $(0,0,\ldots,0)$ and covariance matrix $[\sigma_{ij}]$, $i,j = 2,3,\ldots,r$, where

$$\sigma_{ij} = \mu_{i+j+2} - \mu_{i+1}\mu_{j+1} - (i+1)\mu_i\mu_{j+2} - (j+1)\mu_{i+2}\mu_j + (i+1)(j+1)\mu_i\mu_j\mu_2.$$

In general, m_r gives biased estimator of μ_r. The correction factor required to make them unbiased and the corresponding statistics are

$$M_2 = \frac{n}{n-1}m_2,$$

$$M_3 = \frac{n}{(n-1)(n-2)}m_3,$$

$$M_4 = \frac{n(n^2-2n+3)}{(n-1)(n-2)(n-3)}m_4 - \frac{3n(2n-3)}{(n-1)(n-2)(n-3)}m_2^2.$$

Occasionally, the parameters are also estimated by matching μ'_1, μ_2, β_1 and β_2 with the corresponding sample values. For example, the estimation of parameters of the lambda distributions is often done in this manner.

Example 9.2. Consider the generalized Pareto model (Table 1.1) with

$$Q(u) = \frac{b}{a}[(1-u)^{-\frac{a}{a+1}} - 1].$$

In this case, the first two moments are

$$\mu_1' = \int_0^1 Q(p)dp = b$$

and

$$\mu_2' = \int_0^1 Q^2(p)dp = \frac{2b^2}{1-a}.$$

Hence, we form the equations

$$b = \frac{1}{n}\sum_{i=1}^{n} X_i = \bar{X},$$

$$\frac{2b^2}{1-a} = \frac{1}{n}\sum X_i^2,$$

and solve them to obtain the moment estimates of a and b as

$$\hat{a} = 1 - \frac{2n\bar{X}^2}{\sum X_i^2} \quad \text{and} \quad \hat{b} = \bar{X}.$$

9.3.2 L-Moments

In the method of L-moments, the logic is the same as in the case of usual moments except that we equate the population L-moments with those of the sample and then solve for the parameters. Here again, the number of equations to be considered is the same as the number of parameters to be estimated. Thus, we consider r equations

$$L_r = l_r, \quad r = 1,2,\ldots,$$

where

$$L_r = \int_0^1 \sum_{k=0}^{r-1} (-1)^{r-k} \binom{r}{k} \binom{r+k}{k} u^k Q(u)du$$

and

$$l_r = \frac{1}{n}\sum_{j=0}^{r-1} p_{rj} \left(\sum_{r=1}^{n} \frac{(r-1)_{(j)}}{(n-1)_{(j)}} \right) \tag{9.4}$$

with

$$p_{ij} = \frac{(-1)^{i-1-j}(i+j-1)!}{(j!)^2(i-j-1)!},$$

when the model to be fitted contains r parameters. The expressions for the first four L-moments L_1 through L_4 are given in (1.34)–(1.37) (or equivalently (1.38)–(1.41)). Next, we have the sample counterparts as

$$\begin{aligned}
l_1 &= \frac{1}{n}\sum_{i=1}^{n} X_{(i)} = \bar{X}, \\
l_2 &= \frac{1}{n(n-1)}\sum_{i=1}^{n}(2i-1-n)X_{i:n}, \\
l_3 &= \frac{1}{n(n-1)(n_2)}\sum_{i=1}^{n}\{6(i-1)(i-2)-6(i-1)(n-2) \\
&\quad +(n-1)(n-2)\}X_{i:n}, \\
l_4 &= \frac{1}{n(n-1)(n_2)(n-3)}\sum_{i=1}^{n}\{20(i-1)(i-2)(i-3)-30(i-1)(i-2)(n-3) \\
&\quad +12(i-1)(n-2)(n-3)-(n-1)(n-2)(n-3)\}X_{i:n}.
\end{aligned}$$

Regarding properties of l_r as estimates of L_r, we note that l_r is unbiased, consistent and asymptotically normal (Hosking [276]). Elamir and Seheult [187] have obtained expressions for the exact variances of the sample L-moments. They have used an equivalent representation of (9.4) in the form

$$l_r = \sum_{k=0}^{r-1} p^*_{r-1,k} b_k, \tag{9.5}$$

where

$$p^*_{r-1,k} = (-1)^{r-k}\binom{r}{k}\binom{r+k}{k}$$

and

$$b_k = \frac{1}{n_{(k+1)}}\sum_{i=1}^{n}(i-1)_{(k)}X_{i:n}.$$

For a sample of size n, (9.5) is also expressible in the vector form

$$l = bC^T$$

with $l = (l_1, l_2, \ldots, l_n)$ and $b = (b_0, b_1, \ldots, b_{k-1})$ and C is a triangular matrix with entries $p^*_{r-1,k}$. So,

$$V(l) = C\Theta C^T, \quad \Theta = V(b). \tag{9.6}$$

As special cases, we have

$$V(l_1) = \frac{\sigma^2}{n}, \tag{9.7}$$

where σ^2 is the population variance and

$$\begin{aligned} V(l_2) =& \frac{1}{n(n-1)}\Big[\frac{4(n-2)}{3}\{E(Y_{3:3}^2) + E(Y_{1:3}Y_{2:3}) + E(Y_{2:3}Y_{3:3})\} \\ & - 2(n-3)E(Y_{1:2}Y_{2:2}) - 2(n-2)E(Y_{2:2}^2) + (n-1)E(Y_{1:1}^2) \\ & - 2(2n-3)E(Y_{2:2}^2) + E(Y_{1:1})\{4(2n-3)E(Y_{2:2} - 5(n-1)E(Y_{1:1})\}\Big]. \end{aligned} \tag{9.8}$$

In the case of the first four sample moments,

$$C = \begin{pmatrix} 1 & 0 & 0 & 0 \\ -1 & 2 & 0 & 0 \\ -1 & -6 & 6 & 0 \\ -1 & 12 & -30 & 20 \end{pmatrix}$$

along with

$$\begin{aligned} V(l_1) &= \theta_{00}, \\ V(l_2) &= 4\theta_{11} - 4\theta_{01} + \theta_{00}, \end{aligned}$$

were used to derive the expressions in (9.7) and (9.8). Furthermore,

$$\operatorname{Cov}(l_1, l_2) = \frac{1}{3n}[E(Y_{3:3} - Y_{2:3})^2 - E(Y_{2:3} - Y_{1:3})^2]$$

and

$$\begin{aligned} \operatorname{Cov}(l_1, l_r) = \frac{1}{n}\Big[&\int_0^1 u^2 P_{r-1}^*(u)du - \int_0^1 \int_0^u [uvF(P_{r-1}^*(u))]' dudv \\ & - (-1)^r \int_0^1 \int_0^v uv[(1-G)P_{r-1}^*(u)(1-G)\Big]' dudv, \end{aligned}$$

where $G(y) = v$ and $F(x) = u$. Also, if we define

$$\theta_{kl} = \operatorname{Cov}(b_k, b_l), \tag{9.9}$$

then

$$\hat{\theta}_{kl} = b_k b_l - \frac{1}{n_{(n+l+2)}} \sum_{1 \le i < j \le n} [(i-1)_{(k)}(j-k-2)_{(l)}$$
$$+(i-1)_{(l)}(j-l-2)_{(k)}]X_{i:n}X_{j:n}$$

is a distribution-free unbiased estimator of (9.9). In the above discussion, $Y_{r:n}$ denotes the conceptual order statistics of the population. The expression for $V(l_2)$ is equivalent to the estimate of Nair [455]. In finding the variance of the ratio of two sample L-moments, the approximation

$$V\left(\frac{X}{Y}\right) \doteq \left[\frac{V(X)}{E(X)^2} + \frac{V(Y)}{E(Y)^2} - \frac{2\mathrm{Cov}(X,Y)}{E(X)E(Y)}\right]\left(\frac{E(X)}{E(Y)}\right)^2$$

is useful. Thus, approximate variances of the sample L-skewness and kurtosis can be obtained. The sample L-moment ratios are consistent but not unbiased.

A more general sampling scheme involving censoring of observations has been discussed recently. Let $T_1, T_2, \ldots, T_n$ be independent and identically distributed lifetimes following distribution function $F(x)$. Assume that lifetimes are censored on the right by independent and identically distributed random variables $Y_1, Y_2, \ldots, Y_n$ having common distribution function $H(x)$. Further, let Y_i's be independent of the T_i's. Thus, we observe only the right censored data of the form $X_i = \min(T_i, Y_i)$. Define indicator variables

$$\Delta_i = \begin{cases} 1 & \text{if } T_i \le Y_i \\ 0 & \text{if } T_i > Y_i \end{cases}$$

so that $\Delta_i = 1$ indicates T_i is uncensored and $\Delta_i = 0$ indicates T_i is censored. The distribution of each X_i is then

$$G(x) = 1 - (1 - F(x))(1 - H(x)).$$

To estimate the distribution function of the censored samples (X_i, Δ_i), the Kaplan–Meir [311] product limit estimator is popular.

The estimator of the survival function is

$$S_n(t) = \prod_{j:x_{(j)} \le t} \left(\frac{n-j}{n-j+1}\right)^{\Delta(j)}, \quad t \le X_{(n)},$$

where $X_{(1)} \le \cdots \le X_{(n)}$ are ordered X_i's and $\Delta(j)$ is the censoring status corresponding to $X_{(j)}$.

The estimator of the rth L-moment for right censored data (Wang et al. [576]) is

$$\hat{L}_r = \sum_{j=1}^{n} X_{(j)} U_{j(r)},$$

where

$$U_{j(r)} = \sum_{k=0}^{j-1} r^{-1} \binom{r-1}{k} \left[B_{p,q}(1 - S_n(X_{(j)})) - B_{p,q}(1 - S_n(X_{(j-1)})) \right]$$

with $X_{(0)} = 0$, $p = r - k$ and $q = k + 1$.

Now, let $T = \min(X, C)$, where X denotes the failure time and C denotes the noninformative censoring time. For a constant $0 \leq u_0 < 1$ and $Q(u_0) < T^*$, where T^* is the minimum of the upper most support points of the failure time X and the censoring time C, suppose that $F''(x)$ is bounded on $[0, Q(u_0) + \Delta]$, $\Delta > 0$ and $\inf_{0 \leq u \leq u_0} f(Q(u)) > 0$. Then (Cheng [145]),

(i) with probability one

$$\sup_{0 \leq u \leq u_0} |\hat{Q}(u) - Q(u)| = O(n^{-\frac{1}{2}} (\log\log n)^{\frac{1}{2}}),$$

(ii)

$$\sup_{0 \leq u \leq u_0} |n^{\frac{1}{2}} f(Q(u))(\hat{Q}(u) - Q(u) - G_n(u)| = O(n^{-\frac{1}{3}} (\log n)^{\frac{3}{2}}),$$

where $G_n(u)$ is a sequence of identically distributed Gaussian process with zero mean and covariance function

$$\text{Cov}\,(G_n(u_1), G_n(u_2)) = (1 - u_1)(1 - u_2) \int_0^n \frac{dt}{(1-t)^2 [1 - H(F^{-1}(t))]},$$

with $u_1 \leq u_2$, $H(x) = 1 - (1 - F(x)(1 - G(x)))$ and $G(x)$ is the distribution function of the censoring time C.

Under the above regularity conditions, Wang et al. [576] have shown that, as $n \to \infty$,

(i) $\hat{L}_r = L_r + o(n^{-\frac{1}{2}} (\log\log n)^{\frac{1}{2}})$;

(ii) $n^{\frac{1}{2}} (\hat{L}_r - L_r)$, $r = 1, 2, \ldots, n$ converges in distribution to multivariate normal $(0, \Sigma)$, where Σ has its elements as

$$\Sigma_{rs} = \iint_{x \leq y} \frac{P^*_{r-1}(x) P^*_{s-1}(y) + P^*_{s-1}(x) P^*_{r-1}(y)}{f(Q(x)) f(Q(y))} \text{Cov}(G_n(x), G_n(y)) dx dy,$$

with

$$P^*_{r-1}(u) = \sum_{k=0}^{r-1} (-1)^{r-1-k} \binom{r-1}{k} \binom{r+k-1}{k} u^k$$

is the $(r-1)$th shifted Legendre polynomial;

(iii) the vector

$$n^{\frac{1}{2}}(\hat{L}_1 - L_1, \hat{L}_2 - L_2, \hat{\tau}_3 - \tau_3, (\hat{\tau}_3 - \tau_3)\dots(\hat{\tau}_m - \tau_m))$$

converges in distribution to multivariate normal $(0,\Lambda)$, where Λ has its elements as

$$\Lambda_{rs} = \begin{cases} \Sigma_{rs}, & r \leq 2,\ s \leq 2 \\ \frac{(\Sigma_{rs} - \tau_r \Sigma_{2s})}{L_2}, & r \geq 2,\ s \leq 2 \\ \frac{\Sigma_{rs} - \tau_r \Sigma_{2s} - \tau_s \Sigma_{2r} + \tau_r \tau_s \Sigma_{22}}{L_2^2}, & r \geq 3, s \geq 3. \end{cases}$$

There are several papers that deal with L-moments of specific distributions and comparison of the method of L-moments with other methods of estimation. Reference may be made, e.g., to Hosking [277], Pearson [488], Guttman [256], Gingras and Adamowski [217], Hosking [278], Sankarasubramonian and Sreenivasan [517], Chadjiconstantinidis and Antzoulakos [131], Hosking [280], Karvanen [312], Ciumara [150], Abdul-Moniem [3], Asquith [40] and Delicade and Goria [169]. Illustration of the method of L-moments with some real data can be seen in Sect. 3.6 for different models.

9.3.3 Probability Weighted Moments

Earlier in Sect. 1.4, we defined the probability weighted moment (PVM) of order (p,r,s) as

$$M_{p,r,s} = E(X^p F^r(X) \bar{F}^s(X))$$

which is the same as

$$\begin{aligned} M_{p,r,s} &= \int_0^\infty x^p F^r(x) \bar{F}^s(x) f(x) dx \\ &= \int_0^1 [Q(u)]^p u^r (1-u)^s du, \end{aligned}$$

provided that $E(|X|^P) < \infty$. Commonly used quantities are

$$\beta_r = \int_0^1 [Q(u)] u^r du$$

and

$$\alpha_r = \int_0^1 [Q(u)](1-u)^r du,$$

which are the special cases $M(1,r,0)$ and $M(1,0,r)$. Since

$$\alpha_r = \sum_{s=0}^{r} \binom{r}{s} (-1)^s \beta_s \text{ and } \beta_r = \sum_{s=0}^{r} \binom{r}{s} (-1)^s \alpha_s,$$

characterization of a distribution with finite mean by α_r or β_r is interchangeable. A natural estimate of α_r (also called the nonparametric maximum likelihood estimate) based on ordered observations $X_{1:n}, X_{2:n}, \ldots, X_{n:n}$ is

$$\hat{\alpha}_r = \int_0^{\alpha} x(1-F_n(x))^r dF_n(x) = \frac{1}{n} \sum_{i=1}^{n} X_{i:n} \left(1 - \frac{i}{n}\right)^r.$$

The asymptotic covariance of the estimator is

$$\begin{aligned}\sigma_{rs} &= \text{Cov}\,(\hat{\alpha}_r, \hat{\alpha}_s) \\ &= \frac{1}{n} \iint_{x<y} [1-F(x)]^r [1-F(y)]^s F(x)(1-F(x)) dxdy.\end{aligned}$$

Similarly, the estimate of β_r is

$$\hat{\beta}_r = \int_0^{\infty} x[F_n(x)]^r dF_n(x) = \frac{1}{n} \sum_{i=1}^{n} X_{i:n} \left(\frac{i}{n}\right)^r,$$

with asymptotic covariance

$$\text{Cov}\,(\hat{\beta}_r, \hat{\beta}_s) = \frac{1}{n} \iint_{x<y} [F(x)]^r [F(y)]^s F(x)[1-F(y)] dy.$$

Landwehr and Matalas [373] have shown that

$$b_r = \frac{1}{n} \sum_{i=1}^{n} \frac{(i-1)(i-2)\ldots(i-r)}{(n-1)(n-2)\ldots(n-r)} X_{i:n} \tag{9.10}$$

is an unbiased estimator of β_r. Similarly, for α_r, we have the estimator

$$a_r = \frac{1}{n} \sum_{i=1}^{n} \frac{(n-i)(n-i-1)(n-i+r-1)}{(n-1)(n-2)\ldots(n-r)}.$$

Hosking [278] has developed estimates based on censored samples. In Type I censoring from a sample of size n, m are observed and $(n-m)$ are censored above a known threshold T so that m is a random variable with binomial distribution. The estimate based on the uncensored sample of m values is (9.10) with m replacing n and when the $n-m$ censored values are replaced by T, the estimate is given by

$$\bar{b}_r = \frac{1}{n}\left[\sum_{j=1}^{m} \frac{(i-1)(i-2)\dots(i-r)}{(n-1)(n-2)\dots(n-r)} X_{i:n} + \sum_{i=m+1}^{m} \frac{(i-1)\dots(i-r)}{(n-1)\dots(n-r)} T\right].$$

Assume that $F(t) = v$. Conditioned on the achieved value of m, the uncensored values are a random sample of size m from the distribution with quantile function $Q(uv)$, $0 < u < 1$. The population PWM's for this distribution are

$$\beta_{r,m} = \frac{1}{v^{r+1}} \int_0^v u^r Q(u)du.$$

On the other hand, the completed sample is of size n from the distribution with quantile function

$$Q_1(u) = \begin{cases} Q(u), & 0 < u < v \\ Q(v), & v \le u < 1 \end{cases}$$

and hence its PWM's are

$$\begin{aligned} \beta_{r,n} &= \int_0^1 u^r Q_1(u)du \\ &= \int_0^v u^r Q(u)du + \frac{1-v^{r+1}}{r+1} Q(v). \end{aligned}$$

The asymptotic distributions in this case are derived as in the case of the usual PWM's. Furrer and Naveau [205] have examined the small-sample properties of probability weighted moments.

As in the case of the other two moments, we equate the sample and population probability weighted moments for the estimation of the parameters. In such cases, it is useful to adopt the formulas

$$\bar{b}_r = \frac{1}{n} \sum_{i=1}^{n} x_{(i)} p_{(i)}^r$$

and

$$\bar{a}_r = \frac{1}{n} \sum_{i=1}^{n} x_{(i)} (1 - p_{(i)})$$

with $p(i)$ as some ordered suitably chosen probabilities. Gilchrist [215] has prescribed the choice for $p_{(i)}$ as $\frac{i}{n+1}$, or inverse of the beta function $(0.5, i, n-i+1)$, or $p_i = \frac{i-0.5}{n}$.

Example 9.3. Consider the Govindarajulu distribution with

$$Q(u) = \sigma((\beta+1)u^{\beta} - \beta u^{\beta+1})$$

$$\beta_r = \sigma \int_0^1 [(\beta+1)u^{\beta} - \beta u^{\beta+1}]u^r du,$$
$$= \frac{\sigma(2\beta+r+2)}{(\beta+r+1)(\beta+r+2)}.$$

Since we have only two parameters in this case to estimate, the equations to be considered are

$$\beta_r = \bar{b}_r, \quad r = 0, 1,$$

or

$$\frac{2\sigma}{\beta+2} = \bar{X} = \bar{b}_0$$

and

$$\frac{\sigma(2\beta+3)}{(\beta+2)(\beta+3)} = \frac{1}{n(n+1)} \sum iX_{(i)} = \bar{b}_1.$$

Upon solving these equations, we obtain the estimates

$$\hat{\beta} = \frac{3\bar{b}_0 - 6\bar{b}_1}{2(\bar{b}_1 - \bar{b}_0)}, \quad \hat{\sigma} = \frac{\bar{b}_0(\bar{b}_0 + 2\bar{b}_1)}{4(\bar{b}_0 - \bar{b}_1)}.$$

9.4 Method of Maximum Likelihood

We proceed by writing the likelihood function based on a random sample $x_1, \ldots, x_n$ as

$$L(\theta) = f(x_1;\theta)f(x_2;\theta)\ldots f(x_n;\theta).$$

Taking $x_i = Q(u_i;\theta)$, we then have

$$L(\theta) = f(Q(u_1,\theta))f(Q(u_2,\theta))\ldots f(Q(u_n,\theta))$$
$$= [q(u_1;\theta)q(u_2;\theta)\ldots q(u_n;\theta)]^{-1}.$$

The estimate of θ is the solution that maximizes $L(\theta)$, or equivalently, we have to solve the equation

$$\frac{d\log L}{d\theta} = -\frac{\sum_i d\log q(u_i;\theta)}{d\theta} = 0$$

for θ. Notice that, in practice, the calculation of $L(\theta)$ requires the derivation of u_i from the equation $x_i = Q(u_i;\theta)$. If $u_i = F(x_i;\theta)$ is explicitly available, a direct solution of the u_i's are possible from the observed x_i in the fit of $F(x)$ after substituting the estimated values for the parameters. Otherwise, one has to use some numerical method to extract u_i. The observations are ordered. When U is a uniform random variable, X and $Q(U)$ have identical distributions. Let $\hat{Q}(u)$ be a fitted quantile function and $x_{(r)} = Q(u_{(r)})$. If ψ_0 is an initial estimate of u for a given x value, using the first two terms of the Taylor expansion, we have

$$Q(\psi) = Q(u_0) + (u - u_0)Q'(u_0)$$

as an approximation. Solving for u, we get

$$u \doteq u_0 + \frac{x - Q(u_0)}{q(u_0)}. \tag{9.11}$$

In practical problems, the initial value could be $u_{(r)} = \frac{r}{n+1}$. With this value, (9.11) is used iteratively until x differs from $Q(u)$ by ε, a small pre-set tolerance value, in a trial. Gilchrist [215] has provided a detailed discussion on the subject and an example of the estimation of the parameters of the generalized lambda distribution and layout for the calculations. The properties of the maximum likelihood estimates, though widely known, is included here for the sake of completeness.

Theorem 9.6. *Let $X_1, X_2, \ldots X_n$ be independent and identically distributed with distribution function $F(x:\theta)$ where θ belongs to an open interval Θ in R, satisfying the following conditions:*

(a)

$$\frac{\partial \log f(x;\theta)}{\partial\theta}, \frac{\partial^2 \log f(x;\theta)}{\partial\theta}, \frac{\partial^3 \log f(x;\theta)}{\partial\theta}$$

exist for all x;

(b) for each $\theta_0 \in \Theta$, there exist functions $g_i(x)$ in the neighbourhood of θ_0 such that

$$\left|\frac{\partial f}{\partial\theta}\right| \le g_1(x), \quad \left|\frac{\partial^2 f}{\partial\theta^2}\right| \le g_2(x), \quad \left|\frac{\partial^3 \log f}{\partial\theta^3}\right| \le g_3(x)$$

for all x, and

$$\int g_1(x)dx < \infty, \quad \int g_2(x)dx < \infty, \quad Eg_3(X)dx < \infty$$

in the neighbourhood of θ_0;

(c) $0 < E\left(\frac{\partial \log f(X;\theta)}{\partial \theta}\right)^2 < M < \infty$ *for each* θ*, then with probability 1, the likelihood equations*

$$\frac{\partial L}{\partial \theta} = 0$$

admit a sequence of solutions $\{\hat{\theta}_n\}$ *with the following properties:*

(i) $\hat{\theta}_n$ *is strongly consistent for* θ*;*
(ii) $\hat{\theta}_n$ *is asymptotically distributed as*

$$N\left(\theta, \frac{1}{nE\left(\frac{\partial \log f(X;\theta)}{\partial \theta}\right)^2}\right).$$

When θ *contains more than one element, then also the sequence of vector values* $(\hat{\theta}_n)$ *satisfies consistency and asymptotic normality* $(\theta, \frac{1}{nI(\theta)})$ *where* $I(\theta)$*, called the information matrix, has its elements as*

$$E\left(\frac{\partial \log f(X;\theta)}{\partial \theta_i}, \frac{\partial \log f(X;\theta)}{\partial \theta_i}\right)$$

where $\theta = (\theta_1, \ldots, \theta_K)$ *and* $I(\theta)$ *has order* $K \times K$*.*

9.5 Estimation of the Quantile Density Function

The quantile density function $q(u)$ is a vital component in the definitions of reliability concepts like hazard quantile function, mean residual quantile function and total time on test transforms. Moreover, it appears in the asymptotic variances of different quantile-based statistics. Babu [43] has pointed out that the estimate of the bootstrap variance of the sample median needs consistent estimates of $q(u)$.

Assume that $q(p) \geq 0$. Then,

$$\frac{Q(v) - Q(p)}{v - p} = q(p) + o(1)$$

as $v \to p$. Thus, to get an approximation for $q(p)$, it is enough to consider $Q(v) - Q(p)$ for $v \geq u$ near u. As $Q(p)$ is not known, it is replaced by $Q_n(p)$. Since all quantiles $\frac{Q_n(v)-Q(p)}{v-p}$ are close to $q(p)$, a linear combination of these two also will be near $q(p)$. With this as the motivating point, Babu [43] has provided the following results.

Let h be a function on the positive real line such that $h(y)e^y$ is a polynomial of degree not exceeding k ($k \geq 2$ is an integer) and

$$\int_0^\infty h(y)y^j dy = \begin{cases} 1, & j = 0,1 \\ 0, & j = 2,3,\dots k. \end{cases}$$

Then,

$$\sigma^2(x) = \int_0^x \int_0^x h(p)h(v)\min(p,v)dpdv$$

and

$$\sigma^2 = \int_0^\infty \int_0^\infty h(p)h(v)\min(p,v)dudv.$$

Defining $J = (p-\varepsilon, p+\varepsilon) \subset (0,1)$, for independent variables $X_1, \dots, X_n$ with common distribution function $F(x)$, we have the following two results.

Theorem 9.7. *If $F(x)$ is k times continuously differentiable at $Q(p)$ for $p \in J$ such that $f(x)$ at $Q(p)$ is positive and $E(X^2) < \infty$, then uniformly in $x > 0$,*

$$n^{2\beta} E\left(\frac{D(x,n)}{q(p)} - 1\right)^2 = \sigma^2(x) + o(1) + n^{2\beta}(1 - L(x,n))^2,$$

where

$$D(x,n) = n^\delta \int_0^x [Q_n(p + vn^{-\delta}) - Q_n(p)]h(v)dv,$$

$$L(x,n) = n^\delta b_1^{-1} \int_0^x b(vn^{-\delta})h(v)dv,$$

$$\delta = (2k-1)^{-1},\ \beta = \frac{1-\delta}{2},\ b_j = \frac{1}{j!}\frac{d^jQ(p)}{dp^j} \text{ at } p \in J, \text{ and } b(p) = \sum_{j=1}^k b_j p^j.$$

Theorem 9.8. *Let $f_i(x) = \int_x^\infty y^i h(y)dy$ and f_j and f_1 do not have common positive roots for any $2 \leq j \leq k-1$. If the jth derivative of Q at p is non-zero, for $2 \leq j \leq k-1$, then*

$$n^{2\beta} E\left(\frac{D(x,n)}{q(p)} - 1\right)^2 \geq \sigma^2 + o(1)$$

and that the equality occurs at $x = \log n$. As a result, $D(\log n, n)$ is an efficient estimator of $q(u)$ in the mean square sense among the class of estimators $\{D(x,n)|x > 0\}$. Also, $n^\beta(D(\log n,n)(\frac{1}{q(p)} - 1))$ is asymptotically distributed as $N(0,\sigma^2)$.

A histogram type estimator of the form

$$A_n(p) = \frac{Q_n(p+\alpha_n) - Q_n(p-\alpha_n)}{2\alpha_n}, \quad \alpha_n > 0,$$

has been discussed by Bloch and Gastwirth [108] and Bofinger [113]. Its asymptotic distribution is presented in the following theorem.

Theorem 9.9 (Falk [193]). *Let $0 < p_1 < \cdots < p_r < 1$ and $Q(p)$ be twice differentiable near p_j with bounded second derivative, $j = 1,2,\ldots,r$. Then, if $\alpha_{nj} \to 0$ and $n\alpha_{jn} \to \infty$, $j = 1,2,\ldots r$,*

$$(2n\alpha_{jn})^{\frac{1}{2}} \left\{ A_{jn}(p_j) - \frac{Q(p_j+\alpha_{jn}) - Q(p_j-\alpha_{jn})}{2\alpha_{jn}} \right\}_{j=1}^{r},$$

where

$$A_{jn}(p_j) = \frac{Q_n(p_j+\alpha_{jn}) - Q_n(p_j-\alpha_{jn})}{2\alpha_{jn}},$$

converges in distribution to $\prod_{j=1}^{r} N(0,q^2(p_j))$, with Π denoting the product measure.

In the above result, if we further assume that $n\alpha_{jn}^3 \to 0$, then $\frac{Q(p_j+\alpha_{jn})-Q(p_j-\alpha_{jn})}{2\alpha_{jn}}$ can be replaced $q(p_j)$. Moreover, if Q is three times differentiable near p with bounded third derivative which is continuous at p, an optimal bandwidth α_n^* in the sense of mean square is

$$\alpha_n^* = \left\{ \frac{\frac{6}{2^{\frac{1}{2}}} q(p)}{q^3(p)} \right\}^{\frac{2}{5}} n^{-\frac{1}{5}}.$$

Falk [193] considered a kernel estimator of $q(p)$ defined by

$$\hat{k}_n(F_n) = \int_0^1 Q_n(x)\alpha_n^{-2} h\left(\frac{p-x}{\alpha_n}\right) dx,$$

where h is a real valued kernel function with bounded support and $\int h(x)dx = 0$. Notice that $k_n(p)$ is a linear combination of order statistics of the form $\sum_{i=1}^{n} C_{in} X_{i:n}$, where

$$C_{in} = \frac{1}{\alpha_n^2} \int_{\frac{i-1}{n}}^{\frac{i}{n}} h\left(\frac{p-x}{\alpha_n}\right) dx, \quad i = 1,2,\ldots,n.$$

Some key properties of the above kernel estimator are presented in the following theorem.

Theorem 9.10. *Let $0 < p_1 < \cdots < p_r < 1$ and Q be twice differentiable near p_j with bounded second derivative, $j = 1, 2, \ldots, r$. Then, if h_j has the properties of h given above and $\alpha_{jn} \to 0$, $n\alpha_{jn}^2 \to \infty$,*

$$\left[(n\alpha_{jn})^{\frac{1}{2}}(\hat{k}_{jn}(p_j) - k_{jn}(p_j))\right]_{j=1}^{r}$$

converges to $\prod_{j=1}^{r} N(0, q^2(p_j)) \int H_j^2(y)dy$, where $\prod$ is the product measure,

$$H_j(y) = \int_{-\infty}^{y} h_j(x)dx,$$

$$\hat{k}_{nj}(p_j) = \int_0^1 Q_n(x)\alpha_{jn}^{-2} h_j\left(\frac{p_j - x}{\alpha_{jn}}\right)dx,$$

and

$$\hat{k}_{nj}(p_j) = \int_0^1 Q(x)\alpha_{jn}^{-2} h_j\left(\frac{p - x}{\alpha_{jn}}\right)dx.$$

Using additional conditions $\int xh(x)dx = -1$ and $n\alpha_n^3 \to 0$, $k_n(p)$ in Theorem 9.4 can be replaced by $q(p)$. Further, if Q is differentiable $(m+1)$ times with bounded derivatives which are continuous at p, with $n\alpha_n^{3m+1} \to 0$, the approximate bias of $\hat{k}_n(p)$ becomes

$$k_n(p) - q(p) = o(\alpha_n^m)$$

and the optimal bandwidth that minimizes the mean squared error $E(\hat{k}_n(p) - q(p))^2$ is

$$\alpha_n^* = \frac{(m+1)!(\int H^2(y)dy)^{\frac{1}{2}} q(p)}{Q^{m+1}(p) \int x^{m+1} h(x)dx}.$$

Mean squared error of the kernel quantile density estimator is compared with that of the estimate $\tilde{g}(u)$ in Jones [305], where $\tilde{g}(u)$ is the reciprocal of the kernel density estimator given by

$$\tilde{g}(u) = \frac{1}{\hat{f}(Q_n(u))}.$$

It is proved that the former estimator is better than the latter one in terms of the mean squared error.

Estimation of $q(p)$ in a more general framework and for different sampling strategies has been discussed by Xiang [590], Zhou and Yip [603], Cheng [143] and Buhamra et al. [123].

9.6 Estimation of the Hazard Quantile Function

The hazard quantile function in reliability analysis, as described in the preceding chapters, plays a key role in describing the patterns of failure and also in the selection of the model. Sankaran and Nair [515] have provided the methodology for the nonparametric estimation of the hazard function, by suggesting two estimators, with one based on the empirical quantile density function and the other based on a kernel density approach. The properties of the kernel-based estimator and comparative study of the two estimators have also provided by them. Recall that the hazard quantile function is defined as

$$H(p) = [(1-p)q(p)]^{-1}, \quad 0 < p < 1.$$

Suppose that the lifetime X is censored by a non-negative random variable Z. We observe (T,Δ), where $T = \min(X,Z)$ and $\Delta = I(X \leq Z)$, with

$$T(X \leq Z) = \begin{cases} 1, & X \leq Z \\ 0, & X > Z. \end{cases}$$

If $G(x)$ and $L(x)$ are the distribution functions of Z and T, respectively, under the assumption that Z and X are independent, we have

$$1 - L(x) = (1-F(x))(1-G(x)).$$

Let (T_i,Δ_i), $i = 1,2,\ldots,n$, be independent and identically distributed and each (T_i,Δ_i) has the same distribution as (T,Δ). This framework includes time censored observations if all the Z_i's are fixed constants, a Type I censoring when all Z_i's are the same constant, and Type II censoring if $Z_i = X_{r:n}$ for all i. The first estimator proposed by Sankaran and Nair [515] is

$$\hat{H}(p) = \frac{1}{[1-F_n(Q_n(p))]q_n(p)},$$

where

$$q_n(p) = n(T_{j:n} - T_{j-1:n}), \quad \frac{j-1}{n} \leq p \leq \frac{j}{n},$$

and $T_{0:n} \equiv 0$. From Parzen [486], it follows that $q_n(p)$ is asymptotically exponential with mean $q(p)$. Thus, $q_n(p)$ is not a consistent estimator of $q(p)$ nor $\hat{H}(p)$ is for $H(p)$.

A second estimator has been proposed by considering a real valued function $K(\cdot)$ such that

(i) $K(x) \geq 0$ for all x and $\int K(x)dx = 1$,
(ii) $K(x)$ has finite support, i.e., $K(x) = 0$ for $|x| > c$ for some constant $c > 0$,
(iii) $K(x)$ is symmetric about zero,
(iv) $K(x)$ satisfies the Lipschitz condition

$$|K(x) - H(y)| \leq M|x - y|$$

for some constant M. Further, let $\{h_n\}$ be a sequence of positive numbers such that $h_n \to 0$ as $n \to \infty$. Define a new estimator as

$$H_n(p) = \frac{1}{h_n}\int_0^1 \frac{1}{[1-F_n(Q_n(t))]q_n(t)} K\Big(\frac{t-p}{h_n}\Big)dt \tag{9.12}$$

$$= \frac{1}{h_n}\sum_{i=1}^{u} \frac{1}{[1-F_n(T_{i:n})]^n[T_{i:n}-T_{i-1:n}]}\int_{S_{i-1:n}}^{S_{i:n}} K\Big(\frac{t-p}{h_n}\Big)dt, \tag{9.13}$$

where

$$S_{i:n} = \begin{cases} 0, & i = 0 \\ F_n(T_{i:n}), & i = 1,2,\ldots,n-1 \\ 1, & i = n. \end{cases}$$

When $S_{i:n} - S_{i-1:n}$ is small, by the first mean value theorem, (9.13) is approximately equal to

$$H_n^*(p) = \frac{1}{h_n}\sum_{i=1}^{n} \frac{S_{i:n}-S_{i-1:n}}{[1-F_n(T_{i:n})]n(T_{i:n}-T_{i-1:n})} K\Big(\frac{S_{i:n}-p}{h_n}\Big).$$

When no censoring is present, $S_{i:n} - S_{i-1:n} = \frac{i}{n}$ for all i. When heavy censoring is present, $S_{i:n} - S_{i-1:n}$ is large for $i = n$ so that $H_i^*(p)$ need not be a good approximation for $H_n(p)$.

When F is continuous and $K(\cdot)$ satisfies Conditions (i)–(iv) given above, the estimator $H_n(p)$ is uniformly strongly consistent and for $0 < p < 1$, $(\sqrt{n}H_n(p) - H(p))$ is asymptotically normal with mean zero and variance

$$\sigma^2(p) = \frac{n}{(h(n))^2}E\Big[\int_0^1 Q_n(t)dM'(t,p) + \int_0^1 F_n(Q_n(t))\frac{M(t,p)}{(1-t)}q(t)dt\Big]^2.$$

A simulation was carried out in order to make a small sample comparison of $H_n(p)$ and $\hat{H}(p)$ in terms of mean squared error. The random censorship model with $F(t) = 1 - e^{\lambda t}$ was used by varying λ. Observations were censored with the uniform distribution $U(0,1)$ with probability 0.3, so that 30 % of the observations were censored. As a choice of the kernel function, the triangular density

$$K(x) = (1 - |x|)I(|x| \leq 1)$$

was used. The ratios of the mean squared error of $\hat{H}(p)$ to that of $H_n(p)$ were compared in the study, which revealed the following points:

(a) $H_n(p)$ gave reasonable performance for $h_n \leq 0.50$;
(b) When $0.05 < h_n < 0.50$, for each value of p, there is a range of widths h_n for which $H_n(p)$ has smaller mean square error. For large h_n values $h_n = 0.15$ gives the smallest discrepancy between the two estimators;
(c) The two estimators $\hat{H}(p)$ and $H_n(p)$ do not perform well when p becomes large. The method of estimation has also been illustrated with a real data by Sankaran and Nair [515].

9.7 Estimation of Percentile Residual Life

As we have seen in the preceding chapters, the residual life distribution plays a fundamental role in inferring the lifetime remaining to a device given that it has survived a fixed time in operation. The percentiles of the residual life quantile function are the percentile residual life defined in (2.19) and its quantile form in (2.19). Classes of lifetime distributions based on monotone percentile residual life functions have been discussed in Sect. 4.3. The $(1-p)$th percentile life function, according to the definition in (2.19), is

$$P(x) = Q[1 - p(1 - F(x))] - x, \quad x > 0. \tag{9.14}$$

As before, assume that $X_{1:n} \leq \cdots \leq X_{n:n}$ are the ordered observations in a random sample of size n from the distribution with quantile function $Q(p)$. The sample analogue of (9.14) is then

$$p_n(x) = Q_n[1 - p(1 - F_n(x)_)] - x. \tag{9.15}$$

Csorgo and Csorgo [160] have discussed the asymptotic distribution of $p_n(x)$ for different cases consisting of $p_n(x)$, (a) as a stochastic process in x for fixed $0 < p < 1$, (b) as a stochastic process in p for a fixed $x > 0$, and (c) a two-parameter process in (p, x). Using the density function of X, define

$$g(x) = [f(P(x)) + x]^{-1},$$

$$r_n(x) = \frac{\sqrt{n}}{g(x)}[p_n(x) - P(x)],$$

$$H(x) = B[1 - p(1 - F(x)) - pB(F(x))],$$

where B is a Brownian bridge over $(0,1)$. For a fixed $x > 0$, $H(x)$ is distributed as $N(0, p(1-p)(1-F(x)))$. For a fixed p, provided $q(p)$ is positive and continuous at $1 - p(1 - F(x))$, $r_n(x)$ is asymptotically $N(0, p(1-p)(1-F(x)))$. In addition, if $f(x) > 0$ on $(Q(1-p), \infty)$ and some $r > 0$,

$$\sup_{x>Q(1-p)} \left[F(x)(1-F(x)) \frac{|f'(x)|}{f^2(x)} \right] \leq \gamma$$

and $f(x)$ is ultimately non-increasing as $x \to \infty$, then almost surely

$$\sup_{0<x<\infty} |r_n(x) - H(x)| = o(\delta_n)$$

with

$$\delta_n = \left(\frac{\log \log n}{n} \right)^{\frac{1}{4}} (\log n)^{\frac{1}{2}} \text{ as } n \to \infty.$$

A smooth version of the empirical estimator has been studied by Feng and Kulasekera [196]. Following this, as $n \to \infty$, Alam and Kulasekera [33] have established that under the above assumptions,

$$r_n(x) = \frac{1}{g(x)} \int_{-\infty}^{\infty} r_n(u) g(u) K\left(\frac{u-x}{\lambda} \right) \frac{du}{\lambda}$$

is asymptotically normal as

$$\bar{H}(x) = \frac{1}{g(x)} \int_{-\infty}^{\infty} H(u) g(u) K\left(\frac{u-x}{\lambda} \right) \frac{du}{\lambda}$$

and also

$$|\bar{r}_n(x) - \bar{H}(x)| = o(\delta_n)$$

almost surely. Since

$$E\bar{H}(x) = 0$$

and

$$\begin{aligned} E[\bar{H}(x)]^2 &= \frac{1}{g^2(x)} \int_{-\infty}^{\infty} \int_{-\infty}^{\infty} \mu(u,w) g(u) g(w) K\left(\frac{u-x}{\lambda} \right) K\left(\frac{w-x}{\lambda} \right) \frac{du\,dw}{\lambda^2} \\ &= \nu_\lambda(x), \end{aligned} \tag{9.16}$$

the asymptotic normal distribution has mean zero and variance $\nu_\lambda(x)$. It is easy to see that the function $K(\cdot)$ denotes the kernel, which the authors assume to be a probability density function centred at the origin. The efficiency of the Csorgo's estimator $r_n(x)$ relative to that of $\bar{r}_n(x)$ is now

$$e = \frac{\nu_\lambda(x)}{p(1-p)(1-F(x))}.$$

The term $\mu(u,w)$ occurring in (9.16) is

$$\begin{aligned}\mu(u,w) &= EG(u)G(w) \\ &= p(1-p)+(p+p^2)(F(u)\wedge F(w)) \\ &\quad -p\{F(u)\wedge((1-p)(1-F(w)))+F(w)\wedge((1-p)(1-F(u)))\}.\end{aligned}$$

Assuming that

$$\int_{-\infty}^{\infty} g(u)K\Big(\frac{p-x}{\lambda}\Big)\frac{du}{\lambda}<\infty,$$

we have

$$\sup_{0<x<\infty} E(r_n^2(x)-v^*(x))=o(n^{-\frac{1}{4}})$$

and

$$\sup_{0<x<\infty} E(\bar{r}_n^2(x)-v_\lambda^*(x))=o(n^{-\frac{1}{4}})$$

as $n\to\infty$. In the above results,

$$v^*(x)=p[1-p(1-F(x))][1-F(x)]$$

and

$$\begin{aligned}v_\lambda^*(x)) = \Big(\frac{p}{\lambda g(x)}\Big)^2\int_{-\infty}^{\infty}\int_{-\infty}^{\infty}&[F(u)F(w)][1-F(u)F(w)]g(u)g(w) \\ &\times K\Big(\frac{u-x}{\lambda}\Big)K\Big(\frac{w-x}{\lambda}\Big)dudw+v_\lambda(x).\end{aligned}$$

The asymptotic value of the normalized difference between $\bar{r}_n(x)$ and the empirical estimator is given by

$$\frac{n[\mathrm{MSE}(p_n(x))-\mathrm{MSE}(\bar{p}_n(x))]}{\lambda g^2(x)}=v_0^*(x),$$

where

$$\bar{p}_n(x)=\int_{-\infty}^{\infty} p_n(u)K\Big(\frac{(u-t)}{\lambda}\Big)\frac{du}{\lambda}.$$

Alam and Kulasekera [33] have also presented a Monte Carlo study when the underlying distribution is exponential and Weibull using uniform distribution over $[-1,1]$ as the kernel. It has been observed through this study that the kernel estimator provides better results for moderate sample sizes and chosen values of λ.

More properties of $P_n(x)$ have been given by Csorgo and Mason [162], Aly [34] and Csorgo and Viharos [163].

Pereira et al. [491] have studied properties of the class of distributions with decreasing percentile residual life (DPRL). They introduced a nonparametric estimator of $P(x)$ based on the fact that

$$P(x) \text{ is DPRL} \Leftrightarrow P(x) = \inf_{y \le x} P(y).$$

Thus, the estimator of $P(x)$ is given by

$$\tilde{P}(x) = I_{(x,\infty)}(X_{n:n}) \inf_{y \le x} P_n(y), \tag{9.17}$$

where $I_{(x,\infty)}$ denotes the indicator function of the indicated interval. Note that $\tilde{P}(x)$ is the largest decreasing function that lies below the empirical $P_n(x)$. In practice, the estimator $\tilde{P}(x)$ can be computed easily in the following way. When $X_{1:n} \le \cdots \le X_{n:n}$ are the ordered observations in a random sample of size n from the distribution $F(x)$, find the number of distinct values in the sample, say k. Let $Y_1 < \cdots < Y_k$ be the resulting ordered values with no ties. Then, the estimate $\tilde{P}(x)$ is given by

$$\tilde{P}(x) = \begin{cases} P_n(Y_1-) + Y_1 - x & \text{if } x < Y_1 \\ \min\{P_n(Y_1-), P_n(Y_2-), P_n(Y_j-), P_n(Y_{j+1}-) + Y_{j+1} - x\} & \text{if } Y_j \le x < Y_{j+1} \\ 0 & \text{if } x > Y_k. \end{cases}$$

The strong uniform consistency of the estimator $\tilde{P}(x)$ is presented in the following theorem.

Theorem 9.11. *Let X be a random variable having DPRL property. If the distribution function $F(x)$ of X has a continuous positive density function $f(x)$ such that $\inf_{0 \le p \le 1} f_x(Q(p)) > 0$, then $\tilde{P}(x)$ is a strongly uniformly consistent estimator of $P(x)$.*

Note that in order to estimate $p(x)$ under the condition that it increases, an estimator that is a modification of the estimator given in (9.15) can be obtained. It is also strongly uniformly consistent.

9.8 Modelling Failure Time Data

In this and in the subsequent sections, we consider various aspects of the process of modelling lifetime data using distributions. As a problem solving activity, the statistical concepts expressed in terms of quantile functions offer new perspectives that are not generally available in the distribution function approach or at least provides an alternative approach with possibly different interpretations with almost equivalent results. Several factors have to be considered while constructing a model. Generally, the model builder will have some information about the phenomenon under consideration or will be able to extract some features from a preliminary assessment of the observations. The background information about the variables and possible distributions along with the necessary level of details required for the analysis are crucial points. Choice of the appropriate model also depends on the data available to ensure its adequacy and the method of estimation of the parameters. Finally, model parsimony is an attractive feature that prefers a simpler model to a more complex one. For example, models with lesser number of parameters or functional forms that have simpler structure (like constancy or linearity as against nonlinear) will be easier to build and analyse. Qualities such as tractability of the model, ease of analysis and interpretation are often prime considerations. This should be consistent with the ability of the model to represent the essential features of the life distribution that are inherent in the observations. In practice, there are three essential steps in building a model. They are identification of the appropriate model, fitting the model and finally checking its adequacy.

9.9 Model Identification

The procedure involved in model choice is to try out possible candidates and choose the best among them. We have seen in previous chapters (see Table 1.1 and the review of bathtub models in Chap. 7) that a plethora of lifetime distributions have been proposed to represent lifetimes in the distribution function approach. This adds to the complexity of determining the potential initial choice. A generalized version may fit in many practical situations, but more parsimonious solution that render easy analysis and interpretations may exist. The problem is somewhat of a lesser degree when quantile functions are used. We have the generalized lambda distribution (Sect. 3.2.1) or the generalized Tukey lambda family (Sect. 3.2.2) that can take care of a wide variety of practical problems in view of their ability to provide reasonable approximations to many continuous distributions. See the discussion on the structural properties of the two quantile functions. The reliability properties of the models, methods of estimating the parameters and examples of fitting them (Sect. 3.6) have been described in detail in Chap. 3. When there are multiple solutions that give models which fit the data, the one which captures the observed features of the reliability characteristics more closely may be preferred. The reliability characteristic may be the hazard quantile function, mean residual quantile function or any other for which the fitted model is put to good use. We can also make use of other models including those suggested in the distribution function approach with tractable quantile functions. A look at the admissible range of skewness and kurtosis values for the proposed distribution will indicate if it covers the distributional shape that fits the observations. The skewness and kurtosis coefficients of the sample values have to be within the ranges prescribed for the chosen model.

Another useful method to arrive at a realistic model is to compare any of the basic reliability functions that uniquely determines the life distribution, with its sample counterpart. The hazard quantile function, mean residual quantile function, etc. can be used for this purpose. To use the hazard quantile function, recall its definition

$$H(u) = \frac{1}{(1-u)q(u)}.$$

Let $X_{1:n} \le X_{2:n} \le \cdots \le X_{n:n}$ be an ordered set of observations on failure times. Then, the quantile function of the distribution of $X_{r:n}$ is (1.26)

$$Q_r(u_r) \equiv Q(I^{-1}(r, n-r+1)).$$

If U has a uniform distribution, then X, where $x = Q(p)$, has quantile function $Q(u)$. Hence, the ordered U_r, say $u_{(r)}$, leads to $x_{r:n} = Q(u_{(r)})$. So, as an approximation, either the mean

$$EQ(U_{(n)}) = Q\left(\frac{r-0.5}{n}\right),$$

or the median

$$M_r = Q(I^{-1}(0.5, r, n-r+1)),$$

or equivalently

$$u^*_{(r)} = I^{-1}(0.5, r, n-r+1), \quad u^*_{(r)} = Q^{-1}(M_r) = F(M_r),$$

can be used. Gilchrist [215] refers to the function I^{-1} as BETAINV and points out that it is a crucial standard function in most spreadsheets and statistical software. The empirical quantile density function $\hat{q}(u)$ can be obtained from the data, u_r, from the median probability. Thus, from the above formula, $q(u^*_{(r)})$ can be plotted against u^*_r. Once a graph of

$$\hat{H}(u) = \frac{1}{(1-u^*_r)q(u^*_r)}$$

is obtained, its functional form can be obtained by comparing the plot with one of the hazard quantile function forms. Several such forms are available from Table 2.4, Chaps. 3 and 7.

A third alternative in model identification is to start with a simple model and then modify it to accommodate the features of the data. Various properties of quantile functions described in Sect. 1.2 can assist in this regard. For example, the power distribution has an increasing $H(u)$, while the Pareto II has a decreasing hazard quantile function. The product of these two is the power-Pareto distribution discussed in Chap. 3, with a highly flexible form for $H(u)$. We can also use various kinds of transformations to arrive at new models from the initially assumed one. An excellent discussion of these methods along with various illustrations is available in Gilchrist [215].

9.10 Model Fitting and Validation

Once the data is collected and a specific model form is assumed, the next goal is to estimate the parameters. One of the methods of estimation discussed in the preceding sections of this chapter can be employed for this purpose. The only remaining step in the model building process is to ascertain whether the model with the estimated parameters describes the data adequately. This is called model validation. Since the parameters have been estimated from the data using some optimality criteria, the reproducibility of the model will be enhanced if its validation is made by another set of data if it exists. When the data is large, part of it can be used for identification and fitting while the remaining for validation. Sometimes, cross-validation is made use of wherein part of the data used for fitting and the remaining part of the data used for validation are interchanged and the two acts are repeated.

There are graphical methods to ascertain the goodness of fit. One is the *Q-Q* plot and the other is the box plot mentioned in Chap. 1. The *Q-Q* plots were illustrated in the modelling of real data using lambda distribution, the power-Pareto and the Govindarajulu distributions in Chap. 3; see, for example, Figs. 3.7–3.9. An advantage of the *Q-Q* plot is that it can be used to specifically compare the tail areas. We can consider the plots comparing $x_{(r)}$ with its values at 0.90 or 0.95 or 0.05, using the median rankits

$$\hat{Q}(I^{-1}(0.05, r, n-r+1)),\ \hat{Q}(I^{-1}(0.025, r, n-r+1)),\ \hat{Q}(I^{-1}(0.975, r, n-r+1)).$$

The recent work of Balakrishnan et al. [52] constructing optimal plotting points based on Pitman closeness and its performance as a good of fit and comparison with other plotting points is of special interest here. A second method is to apply some goodness-of-fit tests like chi-square. Suppose there are n observations and are divided into m groups each containing the same number of observations. Take $u_j = \frac{j}{m}$, $u_0 = 0$, $u_m = 1$, $r = 0, \ldots, m-1$. If $p_j = \hat{Q}(u_r)$ and f_j is the frequency of observations in (p_{r-1}, p_r), the expected value of f_r is $\frac{n}{m}$ for all value of r. Then, the statistic

$$\sum \left[\frac{f_r - (\frac{n}{m})^2}{(\frac{n}{m})} \right]$$

has approximately a chi-square distribution with $n-1$ degrees of freedom. This scheme is more easier to apply than the conventional chi-square procedure. For elaborate details on chi-squared tests and their power properties, one may refer to the recent book by Voinov et al. [575]. The general references on different forms of goodness-of-fit tests of D'Agostino and Stephens [165] and Huber-Carol et al. [289] will also provide valuable information in this regard. See also Gilchrist [214,215] for methods of estimators when quantile functions are used in modelling statistical data.

References

1. Aarset, M.V.: The null distribution of a test of constant versus bathtub failure rate. Scand. J. Stat. **12**, 55–62 (1985)
2. Abdel-Aziz, A.A.: On testing exponentiality against RNBRUE alternative. Contr. Cybern. **34**, 1175–1180 (2007)
3. Abdul-Moniem, I.B.: L-moments and TL-moments estimation of the exponential distribution. Far East J. Theor. Stat. **23**, 51–61 (2007)
4. Abouammoh, A., El-Neweihi, E.: Closure of NBUE and DMRL under the formation of parallel systems. Stat. Probab. Lett. **4**, 223–225 (1986)
5. Abouammoh, A.M., Ahmed, A.N.: The new better than used failure rate class of life distributions. Adv. Appl. Probab. **20**, 237–240 (1988)
6. Abouammoh, A.M., Ahmed, A.N., Barry, A.M.: Shock models and testing for the mean inactivity time. Microelectron. Reliab. **33**, 729–740 (1993)
7. Abouammoh, A.M., Ahmed, R., Khalique, A.: On renewal better than used classes of ageing. Stat. Probab. Lett. **48**, 189–194 (2000)
8. Abouammoh, A.M., Kanjo, A., Khalique, A.: On aspects of variance residual life distributions. Microelectron. Reliab. **30**, 751–760 (1990)
9. Abouammoh, A.M., Khalique, A.: Some tests for mean residual life criteria based on the total time on test transform. Reliab. Eng. **19**, 85–101 (1997)
10. Abouammoh, A.M., Qamber, I.S.: New better than renewal-used classes of life distribution. IEEE Trans. Reliab. **52**, 150–153 (2003)
11. Aboukalam, F., Kayid, M.: Some new results about shifted hazard and shifted likelihood ratio orders. Int. Math. Forum **31**, 1525–1536 (2007)
12. Abraham, B., Nair, N.U.: On characterizing mixtures of life distributions. Stat. Paper. **42**, 387–393 (2001)
13. Abraham, B., Nair, N.U.: A criterion to distinguish ageing patterns. Statistics **47**, 85–92 (2013)
14. Abraham, B., Nair, N.U., Sankaran, P.G.: Characterizations of some continuous distributions by properties of partial moments. J. Kor. Stat. Soc. **36**, 357–365 (2007)
15. Abromowitz, M., Stegun, I.A.: Handbook of Mathematical Functions: Formulas, Graphs and Mathematical Tables. Applied Mathematics Series, vol. 55. National Bureau of Standards, Washington, DC (1964)
16. Abu-Youssef, S.E.: A moment inequality for decreasing (increasing) mean residual life distribution with hypothesis testing applications. Stat. Probab. Lett. **57**, 171–177 (2002)
17. Adamidis, K., Dimitrakopoulou, T., Loukas, S.: On an extension of the exponential geometric distribution. Stat. Probab. Lett. **73**, 259–269 (2005)

N.U. Nair et al., *Quantile-Based Reliability Analysis*, Statistics for Industry and Technology, DOI 10.1007/978-0-8176-8361-0,

18. Adamidis, K., Loukas, S.: A lifetime distribution with decreasing failure rate. Stat. Probab. Lett. **39**, 35–42 (1998)
19. Adatia, A., Law, A.G., Wang, Q.: Characterization of mixture of gamma distributions via conditional moments. Comm. Stat. Theor. Meth. **20**, 1937–1949 (1991)
20. Agarwal, A., Kalla, S.L.: A generalized gamma distribution and its applications to reliability. Comm. Stat. Theor. Meth. **25**, 201–210 (1996)
21. Ahmad, I.A.: Moments inequalities of ageing families of distributions with hypothesis testing applications. J. Stat. Plann. Infer. **92**, 121–132 (2001)
22. Ahmad, I.A.: Some properties of classes of life distributions with unknown age. Stat. Probab. Lett. **69**, 333–342 (2004)
23. Ahmad, I.A., Kayid, M.: Characterization of the RHR and MIT orderings and the DRHR and IMIT classes of life distributions. Probab. Eng. Inform. Sci. **19**, 447–461 (2005)
24. Ahmad, I.A., Kayid, M., Pellerey, F.: Further results involving the MIT order and the IMIT class. Probab. Eng. Inform. Sci. **19**, 377–395 (2005)
25. Ahmad, I.A., Li, X., Kayid, M.: The NBUT class of life distributions. IEEE Trans. Reliab. **54**, 396–401 (2005)
26. Ahmad, I.A., Mugadi, A.R.: Further moment inequalities of life distributions with hypothesis testing applications, the IFRA, NBUC and DMRL classes. J. Stat. Plann. Infer. **120**, 1–12 (2004)
27. Ahmad, K.E., Jaheen, Z.F., Mohammed, H.S.: Finite mixture of Burr type XII distribution and its reciprocal: properties and applications. Stat. Paper. **52**, 835–845 (2009)
28. Ahmed, A.N.: Preservation properties of the mean residual life ordering. Stat. Paper. **29**, 143–150 (1988)
29. AL-Hussaini, E.K., Sultan, K.S.: Reliability and hazard based on finite mixture models. In: Balakrishnan, N., Rao, C.R. (eds.) Advances in Reliability, vol. 20, pp. 139–183. North-Holland, Amsterdam (2001)
30. Al-Ruzaiza, A.C., Hendi, M.I., Abu-Youssef, S.E.: A note on moment inequality for HNBUE property with hypothesis testing applications. J. Nonparametr. Stat. **15**, 267–272 (2003)
31. Al-Wasel, I.A., El-Bassiouny, A.H., Kayid, M.: Some results on NBUL class of life distributions. Appl. Math. Sci. **1**, 869–881 (2007)
32. Al-Zahrani, B., Stoyanov, J.: Moment inequalities for DVRL distributions, characterization and testing for exponentiality. Stat. Probab. Lett. **78**, 1792–1799 (2008)
33. Alam, K., Kulasekera, K.B.: Estimation of the quantile function of residual lifetime distribution. J. Stat. Plann. Infer. **37**, 327–337 (1993)
34. Aly, E.E.A.A.: On some confidence bands for percentile residual life function. J. Nonparametr. Stat. **2**, 59–70 (1992)
35. Alzaid, A.A.: Mean residual life ordering. Stat. Paper. **29**, 35–43 (1988)
36. Alzaid, A.A.: Ageing concerning of items of unknown age. Stoch. Model. **10**, 649–659 (1994)
37. Arnold, B.C., Balakrishnan, N., Nagaraja, H.N.: A First Course in Order Statistics. Wiley, New York (1992)
38. Arnold, B.C., Brockett, P.L.: When does the β-th percentile residual life function determine the distribution. Oper. Res. **31**, 391–396 (1983)
39. Asha, G., Nair, N.U.: Reliability properties of mean time to failure in age replacement models. Int. J. Reliab. Qual. Saf. Eng. **17**, 15–26 (2010)
40. Asquith, W.H.: L-moments and TL-moments of the generalized lambda distribution. Comput. Stat. Data Anal. **51**, 4484–4496 (2007)
41. Avinadav, T., Raz, T.: A new inverted hazard rate function. IEEE Trans. Reliab. **57**, 32–40 (2008)
42. Azevedo, C., Leiva, V., Athayde, E., Balakrishnan, N.: Shape and change point analyses of the Birnbaum-Saunders-t hazard rate and associated estimation. Comput. Stat. Data Anal. **56**, 3887–3897 (2012)
43. Babu, G.J.: Efficient estimation of the reciprocal of the density quantile function at a point. Stat. Probab. Lett. **4**, 133–139 (1986)
44. Bahadur, R.R.: A note on quantiles in large samples. Ann. Math. Stat. **37**, 577–580 (1966)

45. Bain, L.J.: Analysis of linear failure rate life testing distributions. Technometrics **16**, 551–560 (1974)
46. Bain, L.J.: Statistical Analysis of Reliability and Life-Testing Models. Marcel Dekker, New York (1978)
47. Balakrishnan, N.: Order statistics from the half logistic distribution. J. Stat. Comput. Simulat. **20**, 287–309 (1985)
48. Balakrishnan, N. (ed.): Handbook of the Logistic Distribution. Marcel Dekker, New York (1992)
49. Balakrishnan, N., Aggarwala, R.: Relationships for moments of order statistics from the right-truncated generalized half logistic distribution. Ann. Inst. Stat. Math. **48**, 519–534 (1996)
50. Balakrishnan, N., Balasubramanian, K.: Equivalence of Hartley-David-Gumbel and Papathanasiou bounds and some further remarks. Stat. Probab. Lett. **16**, 39–41 (1993)
51. Balakrishnan, N., Cohen, A.C.: Order Statistics and Inference: Estimation Methods. Academic, Boston (1991)
52. Balakrishnan, N., Davies, K., Keating, J.P., Mason, R.L.: Computation of optimal plotting points based on Pitman closeness with an application to goodness-of-fit for location-scale families. Comput. Stat. Data Anal. **56**, 2637–2649 (2012)
53. Balakrishnan, N., Kundu, D.: Hybrid censoring: Models, inferential results and applications (with discussions). Comput. Stat. Data Anal. **57**, 166–209 (2013)
54. Balakrishnan, N., Leiva, V., Sanhueza, A., Vilca, F.: Scale-mixture Birnbaum-Saunders distributions: characterization and EM algorithm. SORT **33**, 171–192 (2009)
55. Balakrishnan, N., Malik, H.J., Puthenpura, S.: Best linear unbiased estimation of location and scale parameters of the log-logistic distribution. Comm. Stat. Theor. Meth. **16**, 3477–3495 (1987)
56. Balakrishnan, N., Rao, C.R.: Order Statistics: Theory and Methods. Handbook of Statistics, vol. 16. North-Holland, Amsterdam (1998)
57. Balakrishnan, N., Rao, C.R.: Order Statistics - Applications. Handbook of Statistics, vol. 17. North-Holland, Amsterdam (1998)
58. Balakrishnan, N., Saleh, H.M.: Relations for moments of progressively Type-II censored order statistics from log-logistic distribution with applications to inference. Comm. Stat. Theor. Meth. **41**, 880–906 (2012)
59. Balakrishnan, N., Sandhu, R.: Recurrence relations for single and product moments of order statistics from a generalized half logistic distribution, with applications to inference. J. Stat. Comput. Simulat. **52**, 385–398 (1995)
60. Balakrishnan, N., Sarabia, J.M., Kolev, N.: A simple relation between the Leimkuhler curve and the mean residual life function. J. Informetrics **4**, 602–607 (2010)
61. Balakrishnan, N., Wong, K.H.T.: Approximate MLEs for the location and scale parameters of the half-logistic distribution with Type-II right-censoring. IEEE Trans. Reliab. **40**, 140–145 (1991)
62. Balakrishnan, N., Zhu, X.: On the existence and uniqueness of the maximum likelihood estimates of parameters of Birnbaum-Saunders distribution based on Type-I, Type-II and hybrid censored samples. Statistics, DOI: 10.1080/02331888.2013.800069 (2013)
63. Balanda, K.P., MacGillivray, H.L.: Kurtosis: a critical review. Am. Stat. **42**, 111–119 (1988)
64. Barlow, R.E.: Geometry of the total time on test transforms. Nav. Res. Logist. Q. **26**, 393–402 (1979)
65. Barlow, R.E., Bartholomew, D.J., Bremner, J.M., Brunk, H.D.: Statistical Inference Under Order Restrictions. Wiley, New York (1972)
66. Barlow, R.E., Campo, R.: Total time on test process and applications to fault tree analysis. In: Reliability and Fault Tree Analysis, pp. 451–481. SIAM, Philadelphia (1975)
67. Barlow, R.E., Doksum, K.A.: Isotonic tests for convex orderings. In: The Sixth Berkeley Symposium in Mathematical Statistics and Probability I, Statistical Laboratory of the University of California, Berkeley, pp. 293–323, 1972
68. Barlow, R.E., Proschan, F.: Statistical Theory of Reliability and Life Testing. Holt, Rinehart and Winston, New York (1975)

69. Barlow, R.E., Proschan, F.: Statistical Theory of Reliability and Life Testing. To Begin with, Silver Spring (1981)
70. Barlow, R.E., Proschan, F.: Mathematical Theory of Reliability. SIAM, Philadelphia (1996)
71. Barreto-Souza, W., Cribari-Neto, F.: A generalization of the exponential Poisson distribution. Stat. Probab. Lett. **79**, 2493–2500 (2009)
72. Barriga, G.D., Cribari-Neto, F., Cancho, V.G.: The complementary exponential power lifetime model. Comput. Stat. Data Anal. **55**, 1250–1258 (2011)
73. Bartoszewicz, J.: Stochastic order relations and the total time on test transforms. Stat. Probab. Lett. **22**, 103–110 (1995)
74. Bartoszewicz, J.: Tail orderings and the total time on test transforms. Applicationes Mathematicae **24**, 77–86 (1996)
75. Bartoszewicz, J.: Application of a general composition theorem to the star order of distributions. Stat. Probab. Lett. **38**, 1–9 (1998)
76. Bartoszewicz, J., Benduch, M.: Some properties of the generalized TTT transform. J. Stat. Plann. Infer. **139**, 2208–2217 (2009)
77. Bartoszewicz, J., Skolimowska, M.: Preservation of classes of life distributions and stochastic orders under weighting. Stat. Probab. Lett. **76**, 587–596 (2006)
78. Bartoszewicz, J., Skolimowska, M.: Preservation of stochastic orders under mixtures of exponential distributions. Probab. Eng. Inform. Sci. **20**, 655–666 (2006)
79. Basu, A.P., Bhattacharjee, M.C.: On weak convergence within HNBUE family of life distributions. J. Appl. Probab. **21**, 654–660 (1984)
80. Basu, A.P., Ebrahimi, N.: On the k-th order harmonic new better than used in expectation distributions. Ann. Inst. Stat. Math. **36**, 87–100 (1984)
81. Bawa, V.S.: Optimal rules for ordering uncertain prospects. J. Financ. Econ. **2**, 95–121 (1975)
82. Bebbington, M., Lai, C.-D., Zitikis, R.: Bathtub curves in reliability and beyond. Aust. New Zeal. J. Stat. **49**, 251–265 (2007)
83. Bebbington, M., Lai, C.-D., Zitikis, R.: A flexible Weibull extension. Reliab. Eng. Syst. Saf. **92**, 719–726 (2007)
84. Bebbington, M., Lai, C.-D., Zitikis, R.: A proof of the shape of the Birnbaum-Saunders hazard rate function. Math. Sci. **33**, 49–56 (2008)
85. Bebbington, M., Lai, C.-D., Zitikis, R.: Modelling human mortality using mixtures of bathtub shaped failure distributions. J. Theor. Biol. **245**, 528–538 (2007)
86. Belzunce, F., Navarro, J., Ruiz, J.M., Aguila, Y.D.: Some results on residual entropy function. Metrika **59**, 147–161 (2004)
87. Belzunce, F., Orlege, E., Ruiz, J.M.: A note on replacement policy comparisons from NBUC lifetime of the unit. Stat. Paper. **46**, 509–522 (2005)
88. Bennet, S.: Log-logistic models for survival data. Appl. Stat. **32**, 165–171 (1983)
89. Bergman, B.: Crossings in the total time on test plot. Scand. J. Stat. **4**, 171–177 (1977)
90. Bergman, B.: On age replacement and total time on test concept. Scand. J. Stat. **6**, 161–168 (1979)
91. Bergman, B.: On the decision to replace a unit early or late: A graphical solution. Microelectron. Reliab. **20**, 895–896 (1980)
92. Bergman, B., Klefsjö, B.: The TTT-transforms and age replacements with discounted costs. Nav. Res. Logist. Q. **30**, 631–639 (1983)
93. Bergman, B., Klefsjö, B.: The total time on test and its uses in reliability theory. Oper. Res. **32**, 596–606 (1984)
94. Bergman, B., Klefsjö, B.: Burn-in models and TTT-transforms. Qual. Reliab. Eng. Int. **1**, 125–130 (1985)
95. Bergman, B., Klefsjö, B.: The TTT-concept and replacements to extend system life. Eur. J. Oper. Res. **28**, 302–307 (1987)
96. Bergman, B., Klefsjö, B.: A family of test statistics for detecting monotone mean residual life. J. Stat. Plann. Infer. **21**, 161–178 (1989)
97. Bhattacharjee, A., Sengupta, D.: On the coefficient of variations of the $\mathscr{L}$ and $\bar{\mathscr{L}}$ classes. Stat. Probab. Lett. **27**, 177–180 (1996)

98. Bhattacharjee, M.C., Kandar, R.: Simple bounds on availability in a model with unknown life and repair distributions. J. Stat. Plann. Infer. **8**, 129–142 (1983)
99. Bhattacharya, G.K., Fries, A.: Fatigue failure models—Birnbaum-Saunders vs. inverse Gaussian. IEEE Trans. Reliab. **31**, 439–440 (1982)
100. Bigerelle, M., Najjar, D., Fournier, B., Rupin, N., Iost, A.: Application of lambda distributions and bootstrap analysis to the prediction of fatigue lifetime and confidence intervals. Int. J. Fatig. **28**, 233–236 (2005)
101. Birnbaum, Z.W.: On random variables with comparable peakedness. Ann. Math. Stat. **19**, 76–81 (1948)
102. Birnbaum, Z.W., Esary, J.D., Marshall, A.W.: A stochastic characterization of wear out for components and systems. Ann. Math. Stat. **37**, 816–825 (1966)
103. Birnbaum, Z.W., Esary, J.D., Saunders, S.C.: Multicomponent systems and structures and their reliability. Technometrics **12**, 55–57 (1961)
104. Birnbaum, Z.W., Saunders, S.C.: A statistical model for life length of materials. J. Am. Stat. Assoc. **53**, 151–160 (1958)
105. Birnbaum, Z.W., Saunders, S.C.: Estimation of a family of life distributions with application to fatigue. J. Appl. Probab. **6**, 328–347 (1969)
106. Birnbaum, Z.W., Saunders, S.C.: A new family of life distributions. J. Appl. Probab. **6**, 319–327 (1969)
107. Blazej, P.: Preservation of classes of life distributions under weighting with a general weighting function. Stat. Probab. Lett. **78**, 3056–3061 (2008)
108. Bloch, D.A., Gastwirth, J.L.: On a simple estimate of the reciprocal of the density function. Ann. Math. Stat. **39**, 1083–1085 (1968)
109. Block, H.W., Li, Y., Savits, T.H.: Initial and final behaviour of failure rate functions for mixtures and systems. J. Appl. Probab. **40**, 721–740 (2003)
110. Block, H.W., Li, Y., Savits, T.H., Wang, J.: Continuous mixtures with bathtub shaped failure rates. J. Appl. Probab. **45**, 260–270 (2008)
111. Block, H.W., Savits, T.H., Singh, H.: The reversed hazard rate function. Probab. Eng. Inform. Sci. **12**, 69–90 (1998)
112. Block, H.W., Savits, T.H., Wondmagegnehu, E.T.: Mixtures of distributions with increasing failure rates. J. Appl. Probab. **40**, 485–504 (2003)
113. Bofinger, E.: Estimation of a density function using order statistics. Aust. J. Stat. **17**, 1–7 (1975)
114. Boland, P.J., Shaked, M., Shanthikumar, J.G.: Stochastic ordering of order statistics. In: Balakrishnan, N., Rao, C.R. (eds.) Handbook of Statistics 16, Order Statistics: Theory and Methods, pp. 89–103. North-Holland, Amsterdam (1998)
115. Boland, P.J., Singh, H., Cukic, B.: Stochastic orders in partition and random testing of software. J. Appl. Probab. **39**, 555–565 (1999)
116. Bon, J., Illayk, A.: A note on some new renewal ageing notions. Stat. Probab. Lett. **57**, 151–155 (2002)
117. Bon, J., Paltanea, E.: Ordering properties of convolution of exponential random variables. Lifetime Data Anal. **5**, 185–192 (1999)
118. Bosch, G.: Model for failure rate curves. Microelectron. Reliab. **19**, 371–375 (1979)
119. Boyan, C.: Renewal and non homogeneous Poisson process generated by distribution with periodic failure rates. Stat. Probab. Lett. **17**, 19–25 (1993)
120. Brito, G., Zequeira, R.I., Valdes, J.G.: On hazard and reversed hazard rate orderings in two component series systems with active redundancies. Stat. Probab. Lett. **81**, 201–206 (2011)
121. Brogan, A.J., Stidham, S., Jr.: A note on separation in mean-lower partial moments portfolio optimization with fixed and moving targets. IEEE Trans. Reliab. **37**, 901–906 (2005)
122. Bryson, M.C., Siddiqui, M.M.: Some criteria for aging. J. Am. Stat. Assoc. **64**, 1472–1483 (1969)
123. Buhamra, S.S., Al-Kandari, N.M., Ahmed, S.E.: Nonparametric inference strategies for the quantile functions under left truncation and right censoring. J. Nonparametr. Stat. **21**, 1–10 (2007)

124. Cai, J., Wu, Y.: A note on preservation of NBUC class under the formation of parallel system with dissimilar components. Microelectron. Reliab. **37**, 359–360 (1997)
125. Campo, R.A.: Probabilistic optimality in long-term energy sales. IEEE Trans. Power Syst. **17**, 237–242 (2002)
126. Canfield, R.V., Borgman, L.E.: Some distributions of time to failure for reliability applications. Technometrics **17**, 263–268 (1975)
127. Cao, J., Wang, Y.: The NBUC and NWUC classes of life distributions. J. Appl. Probab. **28**, 473–479 (1991)
128. Cao, R., Lugosi, G.: Goodness of fit tests based on the kernel density estimator. Scand. J. Stat. **32**, 599–616 (2005)
129. Carrasco, J.M.F., Ortega, E.M.M., Cordeiro, G.M.: A generalized modified Weibull distribution for lifetime modelling. Comput. Stat. Data Anal. **53**, 450–462 (2008)
130. Castillo, X., Sieworek, D.P.: Workload, performance, and reliability of digital computing systems. Proc. FTCS **11**, 84–89 (1981)
131. Chadjiconstantinidis, S., Antzoulakos, D.L.: Moments of compound mixed Poisson distribution. Scand. Actuarial J. **3**, 138–161 (2002)
132. Chan, P.K.W., Downs, T.: Two criteria for preventive maintenance. IEEE Trans. Reliab. **27**, 272–273 (1968)
133. Chandra, M., Singpurwalla, N.D.: The Gini index, Lorenz curve and the total time on test transform. Technical Report, George Washington University, Washington, DC, 1978
134. Chandra, M., Singpurwalla, N.D.: Relationships between some notions which are common to reliability theory and economics. Math. Oper. Res. **6**, 113–121 (1981)
135. Chandra, N.K., Roy, D.: Some results on reversed hazard rates. Probab. Eng. Inform. Sci. **15**, 95–102 (2001)
136. Chang, D.S., Tang, L.C.: Reliability bounds and critical time of bathtub shaped distributions. IEEE Trans. Reliab. **42**, 464–469 (1993)
137. Chang, D.S., Tang, L.C.: Percentile bounds and tolerance limits for the Birnbaum-Saunders distribution. Comm. Stat. Theor. Meth. **23**, 2853–2863 (1994)
138. Chaudhury, G.: Coefficient of variation of $\mathscr{L}$-class of life distributions. Comm. Stat. Theor. Meth. **22**, 331–344 (1993)
139. Chaudhury, G.: A note on the $\mathscr{L}$-class of life distributions. Sankhyā Ser. A **57**, 158–160 (1995)
140. Chen, G., Balakrishnan, N.: The infeasibility of probability weighted moments estimation of some generalized distributions. In: Balakrishnan, N. (ed.) Recent Advances in Life-Testing and Reliability, pp. 565–573. CRC Press, Boca Raton (1995)
141. Chen, Z.: Statistical inference about shape parameters of the exponential power distribution. Stat. Paper. **40**, 459–465 (1999)
142. Chen, Z.: A new two-parameter lifetime distribution with bathtub shape or increasing failure rate function. Stat. Probab. Lett. **49**, 155–161 (2000)
143. Cheng, C.: Almost sure uniform error bounds of general smooth estimator of quantile density functions. Stat. Probab. Lett. **59**, 183–194 (2002)
144. Cheng, K., Lam, Y.: Reliability bounds on HNUBUE life distributions with known first two moments. Eur. J. Oper. Res. **132**, 163–175 (2001)
145. Cheng, K.F.: On almost sure representation for quantiles of the product limit estimator with applications. Sankhyā **46**, 426–443 (1984)
146. Chhikara, R.S., Folks, J.L.: The inverse Gaussian distribution as a lifetime model. Technometrics **19**, 461–464 (1977)
147. Chong, K.M.: On characterization of exponential and geometric distributions by expectations. J. Am. Stat. Assoc. **72**, 160–161 (1977)
148. Chukova, S., Dimitrov, B.: On distributions having the almost lack of memory property. J. Appl. Probab. **29**, 691–698 (1992)
149. Chukova, S., Dimitrov, B., Khalil, Z.: A class of probability distributions similar to the exponential. Can. J. Stat. **21**, 260–276 (1993)

150. Ciumara, R.: L-moment evaluation of identically and nonidentically Weibull distributed random variables. In: Proceedings of the Romanian Academy of Sciences, vol. A-8 (2007)
151. Cleroux, P., Dubuc, S., Tilquini, C.: The age replacement problem with minimal repair costs. Oper. Res. **27**, 1158–1167 (1979)
152. Cobb, L.: The multimodal exponential formulas of statistical catastrophe theory. In: Tallie, C., Patil, G.P., Baldessari, B. (eds.) Statistical Distributions in Scientific Work, vol. 4, pp. 87–94. D. Reidel, Dordrecht (1981)
153. Cobb, L., Koppstein, P., Chen, N.H.: Estimation and moment recursion relations for multimodal distributions of the exponential families. J. Am. Stat. Assoc. **78**, 124–130 (1983)
154. Cohen, A.C.: Truncated and Censored Samples: Theory and Applications. Marcel Dekker, New York (1991)
155. Consul, P.C.: Some characterizations of the exponential class of distributions. IEEE Trans. Reliab. **40**, 290–295 (1995)
156. Cooray, K.: Generalization of the Weibull distribution: The odd Weibull family. Stat. Model. **6**, 265–277 (2006)
157. Cox, D.R.: Renewal Theory. Methuen & Co., London (1962)
158. Cox, D.R., Oakes, D.: Analysis of Survival Data. Chapman and Hall, London (1984)
159. Crow, E.L., Shimizu, K. (eds.): Lognormal Distributions: Theory and Applications. Marcel Dekker, New York (1988)
160. Csorgo, M., Csorgo, S.: Estimation of percentile residual life. Oper. Res. **35**, 598–606 (1987)
161. Csorgo, M., Yu, H.: Estimation of total time on test transforms for stationary observations. Stoch. Proc. Appl. **68**, 229–253 (1997)
162. Csorgo, S., Mason, D.M.: Bootstrapping empirical functions. Ann. Stat. **17**, 1447–1471 (1989)
163. Csorgo, S., Viharos, L.: Confidence bands for percentile residual lifetimes. J. Stat. Plann. Infer. **30**, 327–337 (1992)
164. Da, G., Ding, W., Li, X.: On hazard rate ordering of parallel systems with two independent components. J. Stat. Plann. Infer. **140**, 2148–2154 (2010)
165. D'Agostino, R.G., Stephens, M.A.: Goodness-of-Fit Techniques. Marcel Dekker, New York (1986)
166. Dallas, A.C.: A characterization using conditional variances. Metrika **28**, 151–153 (1981)
167. Davis, H.T., Feldstein, M.: The generalised Pareto law as a model for progressively censored data. Biometrika **66**, 299–306 (1979)
168. de Gusmao, F.R.S., Ortega, E.M.M., Cordeiro, G.M.: The generalized inverse Weibull distribution. Stat. Paper. **52**, 591–619 (2009)
169. Delicade, P., Goria, M.N.: A small sample comparison of maximum likelihood moments and L moment methods for the asymmetric exponential power distribution. Comput. Stat. Data Anal. **52**, 1661–1673 (2008)
170. Denuit, M.: s-convex extrema, Taylor type expansions and stochastic approximations. Scand. Actuarial J. **1**, 45–67 (2002)
171. Derman, C., Lieberman, G.J., Ross, S.M.: On the use of replacements to extend system life. Oper. Res. **32**, 616–627 (1984)
172. Deshpande, J.V., Kochar, S.C., Singh, H.: Aspects of positive aging. J. Appl. Probab. **23**, 748–758 (1986)
173. Deshpande, J.V., Singh, H., Bagai, I., Jain, K.: Some partial orders describing positive ageing. Comm. Stat. Stoch. Model. **6**, 471–481 (1990)
174. Desmond, A.F.: On relationship between two fatigue life models. IEEE Trans. Reliab. **35**, 167–169 (1986)
175. Dhillon, B.: Life distributions. IEEE Trans. Reliab. **30**, 457–460 (1981)
176. Diaz-Garcia, J.A., Leiva, V.: A new family of life distributions based on elliptically contoured distributions. J. Stat. Plann. Infer. **128**, 445–457; Erratum, **137**, 1512–1513 (2005)
177. DiCresenzo, A.: Some results on the proportional reversed hazards model. Stat. Probab. Lett. **50**, 313–321 (2000)

178. Dimitrakopoulou, T., Adamidis, K., Loukas, S.: A life distribution with an upside down bathtub-shaped hazard function. IEEE Trans. Reliab. **56**, 308–311 (2007)
179. Dimitrov, B., Khalil, Z., El-Saidi, M.A.: On probability distribution with accumulation of failure rates in periodic random environment. Environmetrics **7**, 17–26 (1998)
180. Dohi, T., Kiao, N., Osaki, S.: Solving problems of a repairable limit using TTT concept. IMA J. Math. Appl. Bus. Ind. **6**, 101–115 (1995)
181. Dudewicz, E.J., Karian, A.: The extended generalized lambda distribution (EGLD) system for fitting distribution to data with moments, II: Tables. Am. J. Math. Manag. Sci. **19**, 1–73 (1996)
182. Dupuis, D.J., Mills, J.E.: Robust estimation of the Birnbaum-Saunders distribution. IEEE Trans. Reliab. **47**, 88–95 (1998)
183. Ebrahimi, N., Spizzichino, F.: Some results on normalised total time on test and spacings. Stat. Probab. Lett. **36**, 231–243 (1997)
184. El-Arishi, S.: A conditional probability characterization of some discrete probability distributions. Stat. Paper. **46**, 31–45 (2005)
185. El-Bassiouny, A.H., Sarhan, A.H., Al-Garian, M.: Testing exponentiality against NBUFR (NWUFR). Appl. Math. Comput. **149**, 351–358 (2004)
186. Elabatal, I.: Some ageing classes of life distributions at specific age. Int. Math. Forum **2**, 1445–1456 (2007)
187. Elamir, E.A.H., Seheult, A.H.: Trimmed L-moments. Comput. Stat. Data Anal. **43**, 299–314 (2003)
188. Erto, P.: Genesis, properties and identification of the inverse Weibull lifetime model. Statistica Applicato **1**, 117–128 (1989)
189. Esary, J.D., Marshall, A.W., Proschan, F.: Shock models and wear process. Ann. Probab. **1**, 627–649 (1973)
190. Fagiouli, E., Pellerey, F.: Mean residual life and increasing convex comparison of shock models. Stat. Probab. Lett. **20**, 337–345 (1993)
191. Fagiouli, E., Pellerey, F.: New partial orderings and applications. Nav. Res. Logist. **40**, 829–842 (1993)
192. Fagiouli, E., Pellerey, F.: Moment inequalities for sums of DMRL random variables. J. Appl. Probab. **34**, 525–535 (1997)
193. Falk, M.: On the estimation of the quantile density function. Stat. Probab. Lett. **4**, 69–73 (1986)
194. Falk, M.: On MAD and comedians. Ann. Inst. Stat. Math. **45**, 615–644 (1997)
195. Feaganes, J.R., Suchindran, C.M.: Weibull regression with unobservable heterogeneity, an application. In: ASA Proceedings of Social Statistics Section, pp. 160–165. American Statistical Association, Alexandria (1991)
196. Feng, Z., Kulasekera, K.B.: Nonparametric estimation of the percentile residual life function. Comm. Stat. Theor. Meth. **20**, 87–105 (1991)
197. Filliben, J.J.: Simple and robust linear estimation of the location parameter of a symmetric distribution. Ph.D. thesis, Princeton University, Princeton (1969)
198. Finkelstein, M.S.: On the reversed hazard rate. Reliab. Eng. Syst. Saf. **78**, 71–75 (2002)
199. Fishburn, P.C.: Mean risk analysis with risk associated with below-target returns. Am. Econ. Rev. **67**, 116–126 (1977)
200. Fournier, B., Rupin, N., Bigerelle, M., Najjar, D., Iost, A.: Application of the generalized lambda distributions in a statistical process control methodology. J. Process Contr. **16**, 1087–1098 (2006)
201. Fournier, B., Rupin, N., Bigerelle, M., Najjar, D., Iost, A., Wilcox, R.: Estimating the parameters of a generalized lambda distribution. Comput. Stat. Data Anal. **51**, 2813–2835 (2007)
202. Franco-Pereira, A.M., Lillo, R.E., Romo, J., Shaked, M.: Percentile residual life orders. Appl. Stoch. Model. Bus. Ind. **27**, 235–252 (2011)
203. Freimer, M., Mudholkar, G.S., Kollia, G., Lin, C.T.: A study of the generalised Tukey lambda family. Comm. Stat. Theor. Meth. **17**, 3547–3567 (1988)

204. Fry, T.R.L.: Univariate and multivariate Burr distributions. Pakistan J. Stat. **9**, 1–24 (1993)
205. Furrer, R., Naveau, P.: Probability weighted moments properties for small samples. Stat. Probab. Lett. **77**, 190–195 (2007)
206. Galton, F.: Statistics by inter-comparison with remarks on the law of frequency error. Phil. Mag. **49**, 33–46 (1875)
207. Galton, F.: Enquiries into Human Faculty and Its Development. MacMillan, London (1883)
208. Galton, F.: Natural Inheritance. MacMillan, London (1889)
209. Ganter, W.A.: Comment on reliability of modified designs, a Bayes analysis of an accelerated test of electronic assembles. IEEE Trans. Reliab. **39**, 520–522 (1990)
210. Gaver, D.P., Acar, M.: Analytical hazard representations for use in reliability, mortality and simulation studies. Comm. Stat. Simulat. Comput. **8**, 91–111 (1979)
211. Ghai, G.L., Mi, J.: Mean residual life and its association with failure rate. IEEE Trans. Reliab. **48**, 262–266 (1999)
212. Ghitany, M.E.: The monotonicity of the reliability measures of the beta distribution. Appl. Math. Lett. 1277–1283 (2004)
213. Ghitany, M.E., El-Saidi, M.A., Khalil, Z.: Characterization of a general class of life testing models. J. Appl. Probab. **32**, 548–553 (1995)
214. Gilchrist, W.G.: Modelling with quantile functions. J. Appl. Stat. **24**, 113–122 (1997)
215. Gilchrist, W.G.: Statistical Modelling with Quantile Functions. Chapman and Hall/CRC Press, Boca Raton (2000)
216. Gilchrist, W.G.: Regression revisited. Int. Stat. Rev. **76**, 401–418 (2011)
217. Gingras, D., Adamowski, K.: Performance of flood frequency analysis. Can. J. Civ. Eng. **21**, 856–862 (1994)
218. Giorgi, G.M.: Concentration index, Bonferroni. In: Encyclopedia of Statistical Sciences, vol. 2, pp. 141–146. Wiley, New York (1998)
219. Giorgi, G.M., Crescenzi, M.: A look at the Bonferroni inequality measure in a reliability framework. Statistica **61**(4), 571–583 (2001)
220. Glaser, R.E.: Bathtub related failure rate characterizations. J. Am. Stat. Assoc. **75**, 667–672 (1980)
221. Glaser, R.E.: The gamma distribution as a mixture of exponential distributions. Am. Stat. **43**, 115–117 (1989)
222. Gohout, W., Kunhert, I.: NBUFR closure under formation of coherent systems. Stat. Paper. **38**, 243–248 (1997)
223. Gore, A.P., Paranjpe, S.A., Rajarshi, M.B., Gadgul, M.: Some methods of summarising survivorship in nonstandard situations. Biometrical J. **28**, 577–586 (1986)
224. Govindarajulu, Z.: A class of distributions useful in life testing and reliability with applications to nonparametric testing. In: Tsokos, C.P., Shimi, I.N. (eds.) Theory and Applications of Reliability, vol. 1, pp. 109–130. Academic, New York (1977)
225. Greenwich, M.: A unimodal hazard rate function and its failure distribution. Statistische Hefte **33**, 187–202 (1992)
226. Greenwood, J.A., Landwehr, J.M., Matalas, N.C., Wallis, J.R.: Probability weighted moments. Water Resour. Res. **15**, 1049–1054 (1979)
227. Groeneveld, R.A., Meeden, G.: Measuring skewness and kurtosis. The Statistician **33**, 391–393 (1984)
228. Guess, F., Proschan, F.: Mean residual life: Theory and Applications. In: Krishnaiah, P.R., Rao, C.R. (eds.) Handbook of Statistics, vol. 7, pp. 215–224. North-Holland, Amsterdam (1988)
229. Gumbel, E.J.: The maxima of the mean largest value and of the range. Ann Math. Stat. **25**, 76–84 (1954)
230. Gupta, A.K., Nadarajah, S. (eds.): Handbook of Beta Distribution and Its Applications. Marcel Dekker, New York (2004)
231. Gupta, P.L., Gupta, R.C.: On the moments of residual life in reliability and some characterization results. Comm. Stat. Theor. Meth. **12**, 449–461 (1983)

232. Gupta, P.L., Gupta, R.C.: The monotonicity of the reliability measures of the beta distribution. Appl. Math. Lett. **13**, 5–9 (2000)
233. Gupta, R.C.: On the characterization of survival distributions in reliability by properties of their renewal densities. Comm. Stat. Theor. Meth. **8**, 685–697 (1979)
234. Gupta, R.C.: On the monotonic properties of residual variance and their applications in reliability. J. Stat. Plann. Infer. **16**, 329–335 (1987)
235. Gupta, R.C.: Variance residual life function in reliability studies. Metron **54**, 343–345 (2006)
236. Gupta, R.C.: Role of equilibrium in reliability studies. Probab. Eng. Inform. Sci. **21**, 315–334 (2007)
237. Gupta, R.C., Akman, H.O., Lvin, S.: A study of log-logistic model in survival analysis. Biometrical J. **41**, 431–433 (1999)
238. Gupta, R.C., Bradley, D.M.: Representing mean residual life in terms of failure rate. Math. Comput. Model. **1**, 1–10 (2003)
239. Gupta, R.C., Gupta, P.L., Gupta, R.D.: Modelling failure time data with Lehmann alternative. Comm. Stat. Theor. Meth. **27**, 887–904 (1998)
240. Gupta, R.C., Gupta, R.D.: Proportional reversed hazards model and its applications. J. Stat. Plann. Infer. **137**, 3525–3536 (2007)
241. Gupta, R.C., Kirmani, S.N.U.A.: On order relationships between reliability measures. Comm. Stat. Stoch. Model. **3**, 149–156 (1987)
242. Gupta, R.C., Kirmani, S.N.U.A.: The role of weighted distributions in stochastic modelling. Comm. Stat. Theor. Meth. **19**, 3147–3162 (1990)
243. Gupta, R.C., Kirmani, S.N.U.A.: Residual coefficient of variation and some characterization results. J. Stat. Plann. Infer. **91**, 23–31 (2000)
244. Gupta, R.C., Kirmani, S.N.U.A.: Moments of residual life and some characterizations. J. Appl. Stat. Sci. **13**, 155–167 (2004)
245. Gupta, R.C., Kirmani, S.N.U.A.: Some characterizations of distributions by functions of failure rate and mean residual life. Comm. Stat. Theor. Meth. **33**, 3115–3131 (2004)
246. Gupta, R.C., Kirmani, S.N.U.A., Launer, R.L.: On life distributions having monotone residual variance. Probab. Eng. Inform. Sci. **1**, 299–307 (1987)
247. Gupta, R.C., Langford, E.S.: On the determination of a distribution by its median residual life function: a functional equation. J. Appl. Probab. **21**, 120–128 (1984)
248. Gupta, R.C., Lvin, S.: Monotonicity of failure rate and mean residual life of a gamma type model. Appl. Math. Comput. **165**, 623–633 (2005)
249. Gupta, R.C., Warren, R.: Determination of change points of non-monotonic failure rates. Comm. Stat. Theor. Meth. **30**, 1903–1920 (2001)
250. Gupta, R.D., Kundu, D.: Generalized exponential distribution. Aust. New Zeal. J. Stat. **41**, 173–178 (1999)
251. Gupta, R.D., Kundu, D.: Exponentiated exponential family: An alternative to Gamma and Weibull distributions. Biometrical J. **43**, 117–130 (2001)
252. Gupta, R.D., Kundu, D.: Discriminating between Weibull and generalized exponential distributions. Comput. Stat. Data Anal. **43**, 179–196 (2003)
253. Gupta, R.D., Kundu, D.: Generalized exponential distribution: Existing results and some recent developments. J. Stat. Plann. Infer. **137**, 3537–3547 (2007)
254. Gupta, R.D., Nanda, A.K.: Some results on reversed hazard rate ordering. Comm. Stat. Theor. Meth. **30**, 2447–2457 (2001)
255. Gurland, J., Sethuraman, J.: Reversal of increasing failure rates when pooling failure data. Technometrics **36**, 416–418 (1994)
256. Guttman, N.B.: The use of L-moments in the determination of regional precipitation climates. J. Clim. **6**, 2309–2325 (1993)
257. Hahn, G.J., Shapiro, S.S.: Statistical Models in Engineering. Wiley, New York (1967)
258. Haines, A.L., Singpurwalla, N.D.: Some contributions to the stochastic characterization of wear. In: Proschan, F., Serfling, R.J. (eds.) Reliability and Biometry: Statistical Analysis of Lifelength, pp. 47–80. Society for Industrial and Applied Mathematics, Philadelphia (1974)

259. Hankin, R.K.S., Lee, A.: A new family of non-negative distributions. Aust. New Zeal. J. Stat. **48**, 67–78 (2006)
260. Haritha, N.H., Nair, N.U., Nair, K.R.M.: Modelling incomes using generalized lambda distributions. J. Income Distrib. **17**, 37–51 (2008)
261. Harlow, W.V.: Asset allocation in a downside-risk frame work. Financ. Anal. J. **47**, 28–40 (1991)
262. Harrell, F.E., Davis, D.E.: A new distribution free quantile estimator. Biometrika **69**, 635–640 (1982)
263. Hartley, H.O., David, H.A.: Universal bounds for mean range and extreme observations. Ann. Math. Stat. **25**, 85–99 (1954)
264. Hastings, C., Mosteller, F., Tukey, J.W., Winsor, C.P.: Low moments for small samples: A comparative study of statistics. Ann. Math. Stat. **18**, 413–426 (1947)
265. Haupt, E., Schäbe, H.: A new model for lifetime distribution with bathtub shaped failure rate. Microelectron. Reliab. **32**, 633–639 (1992)
266. Haupt, E., Schäbe, H.: The TTT transformation and a new bathtub distribution model. J. Stat. Plann. Infer. **60**, 229–240 (1997)
267. Hemmati, F., Khorram, E., Rezekhah, S.: A new three parameter ageing distribution. J. Stat. Plann. Infer. **141**, 2266–2275 (2011)
268. Hendi, M.I.: On decreasing cumulative conditional class of life distributions. Pakistan J. Stat. **7**, 71–79 (1991)
269. Hendi, M.I., Mashhour, A.F., Montasser, M.A.: Closure of NBUC class under the formation of parallel system. J. Appl. Probab. **30**, 975–978 (1993)
270. Hesselager, O., Wang, S., Willmot, G.: Exponential scale mixtures and equilibrium distributions. Scand. Actuarial. J. **2**, 125–142 (1997)
271. Hinkley, D.V.: On power transformations to symmetry. Biometrika **62**, 101–111 (1975)
272. Hjorth, U.: A reliability distribution with increasing, decreasing, constant and bathtub shaped failure rates. Technometrics **22**, 99–107 (1980)
273. Hogben, D.: Some properties of Tukey's test for non-additivity. Ph.D. thesis, The State University of New Jersey, New Jersey (1963)
274. Hollander, M., Park, D.H., Proschan, F.: Testing whether new is better than used of specific age with randomly censored data. Can. J. Stat. **13**, 45–52 (1985)
275. Honfeng, Z., Yi, W.W.: The NBEFR and NWEFR classes of distributions. Microelectron. Reliab. **37**, 919–922 (1997)
276. Hosking, J.R.M.: L-moments: analysis and estimation of distribution using linear combination of order statistics. J. Roy. Stat. Soc. B **52**, 105–124 (1990)
277. Hosking, J.R.M.: Moments or L-moments? An example comparing two measures of distributional shape. The Am. Stat. **46**, 186–189 (1992)
278. Hosking, J.R.M.: The use of L-moments in the analysis of censored data. In: Balakrishnan, N. (ed.) Recent Advances in Life Testing and Reliability, pp. 545–564. CRC Press, Boca Raton (1995)
279. Hosking, J.R.M.: Some theoretical results concerning L-moments. Research Report, RC 14492. IBM Research Division, Yorktown Heights, New York (1996)
280. Hosking, J.R.M.: On the characterization of distributions by their L-moments. J. Stat. Plann. Infer. **136**, 193–198 (2006)
281. Hosking, J.R.M.: Some theory and practical uses of trimmed L-moments. J. Stat. Plann. Infer. **137**, 3024–3029 (2007)
282. Hosking, J.R.M., Wallis, J.R.: Regional Frequency Analysis: An Approach based on L-Moments. Cambridge University Press, Cambridge (1997)
283. Hougaard, P.: Life table methods for heterogeneous populations, distributions describing the heterogeneity. Biometrika **71**, 75–83 (1984)
284. Hu, C., Lin, G.D.: Characterization of the exponential distribution by stochastic ordering properties of the geometric compound. Ann. Inst. Stat. Math. **55**, 499–506 (2003)
285. Hu, T., He, P.: A note on comparisons of k-out-of n systems with respect to hazard and reversed hazard rate orders. Probab. Eng. Inform. Sci. **14**, 27–32 (2000)

286. Hu, T., Wei, Y.: Stochastic comparison of spacings from restricted families of distributions. Stat. Probab. Lett. **53**, 91–99 (2001)
287. Hu, T., Xie, H.: Proofs of closure properties of NBUC and NBU(2) under convolution. J. Appl. Probab. **39**, 224–227 (2002)
288. Hu, T., Zhu, Z., Wei, Y.: Likelihood ratio and mean residual life orders for order statistics of heterogeneous random variables. Probab. Eng. Inform. Sci. **15**, 259–272 (2001)
289. Huber-Carol, C., Balakrishnan, N., Nikulin, M.S., Mesbah, M. (eds.): Goodness-of-Fit Tests and Model Validity. Birkhäuser, Boston (2002)
290. Jain, K., Singh, H., Bagai, I.: Relation for reliability measures for weighted distributions. Comm. Stat. Theor. Meth. **18**, 4393–4412 (1989)
291. Jaisingh, L.R., Kolarik, W.J., Dey, D.K.: A flexible bathtub hazard model for nonrepairable systems with uncensored data. Microelectron. Reliab. **27**, 87–103 (1987)
292. Jiang, R., Ji, P., Xiao, X.: Ageing property of univariate failure rate models. Reliab. Eng. Syst. Saf. **79**, 113–116 (2003)
293. Jiang, R., Murthy, D.N.P.: Parametric study of competing risk model involving two Weibull distributions. Int. J. Reliab. Qual. Saf. Eng. **4**, 17–34 (1997)
294. Jiang, R., Murthy, D.N.P.: Two sectional models involving three Weibull distributions. Qual. Reliab. Eng. Int. **13**, 83–96 (1997)
295. Jiang, R., Murthy, D.N.P.: Mixture of Weibull distributions—Parametric characterization of failure rate function. Appl. Stoch. Model. Data Anal. **14**, 47–65 (1998)
296. Jiang, R., Murthy, D.N.P.: The exponentiated Weibull family: A graphical approach. IEEE Trans. Reliab. **48**, 68–72 (1999)
297. Jiang, R., Murthy, D.N.P., Ji, P.: Models involving two inverse Weibull distributions. Reliab. Eng. Syst. Saf. **73**, 73–81 (2001)
298. Joag-Dev, K., Kochar, S.C., Proschan, F.: A general comparison theorem and its applications to certain partial orderings of distributions. Stat. Probab. Lett. **22**, 111–119 (1995)
299. Joannes, D.L., Gill, C.A.: Comparing measures of sample skewness and kurtosis. The Statistician **47**, 183–189 (1998)
300. Joe, H.: Characterization of life distributions from residual lifetimes. Ann. Inst. Stat. Math. **37**, 165–172 (1985)
301. Joe, H., Proschan, F.: Percentile residual life functions. Oper. Res. **32**, 668–678 (1984)
302. Johnson, N.L., Kotz, S., Balakrishnan, N.: Continuous Univariate Distributions, vol. 2, 2nd edn. Wiley, New York (1995)
303. Johnson, N.L., Kotz, S., Kemp, A.W.: Univariate Discrete Distributions, 2nd edn. Wiley, New York (1992)
304. Joiner, B.L., Rosenblatt, J.R.: Some properties of the range of samples from Tukey's symmetric lambda distribution. J. Am. Stat. Assoc. **66**, 394–399 (1971)
305. Jones, M.C.: Estimating densities, quantiles, quantile densities and density quantiles. Ann. Inst. Stat. Math. **44**, 721–727 (1992)
306. Jones, M.C.: On some expressions for variance, covariance, skewness and L-moments. J. Stat. Plann. Infer. **126**, 97–108 (2004)
307. Jones, M.C.: On a class of distributions defined by the relationship between their density and distribution functions. Comm. Stat. Theor. Meth. **36**, 1835–1843 (2007)
308. Kaigh, W.D., Lachenbruch, P.A.: A generalized quantile estimator. Comm. Stat. Theor. Meth. **11**, 2217–2238 (1982)
309. Kalla, S.L., Al-Saqabi, B.N., Khajah, A.G.: A unified form of gamma type distributions. Appl. Math. Comput. **8**, 175–187 (2001)
310. Kao, J.H.K.: A graphical estimation of mixed Weibull parameters in life testing of electronic tubes. Technometrics **10**, 389–407 (1959)
311. Kaplan, E.L., Meier, P.: Nonparametric estimation from incomplete observations. J. Am. Stat. Assoc. **53**, 457–481 (1958)
312. Karvanen, J.: Estimation of quantile mixture via L-moments and trimmed L-moments. Comput. Stat. Data Anal. **51**, 947–959 (2006)

313. Karvanen, J., Nuutinen, A.: Characterizing the generalized lambda distribution by L-moments. Comput. Stat. Data Anal. **52**, 1971–1983 (2008)
314. Karian, A., Dudewicz, E.J.: Fitting Statistical Distributions, the Generalized Lambda Distribution and Generalized Bootstrap Methods. Chapman and Hall/CRC Press, Boca Raton (2000)
315. Karian, A., Dudewicz, E.J.: Comparison of GLD fitting methods, superiority of percentile fits to moments in L^2 norm. J. Iran. Stat. Soc. **2**, 171–187 (2003)
316. Karian, A., Dudewicz, E.J.: Computational issues in fitting statistical distributions to data. Am. J. Math. Manag. Sci. **27**, 319–349 (2007)
317. Karian, A., Dudewicz, E.J.: Handbook of Fitting Statistical Distributions with R. CRC Press, Boca Raton (2011)
318. Kayid, M.: A general family of NBU class of life distributions. Stat. Meth. **4**, 1895–1905 (2007)
319. Kayid, M., Ahmad, I.A.: On the mean inactivity time ordering with reliability applications. Probab. Eng. Inform. Sci. **18**, 395–409 (2004)
320. Kayid, M., El-Bassiouny, A.H., Al-Wasel, I.A.: On some new stochastic orders of interest in reliability theory. Int. J. Reliab. Appl. **8**, 95–109 (2007)
321. Kebir, Y.: Laplace transform characterizations of probabilistic orderings. Probab. Eng. Inform. Sci. **8**, 125–134 (1994)
322. Kececioglu, D.B.: Reliability and Lifetesting Handbook, vol. 1. DEStech Publications, Lancaster (2002)
323. Keilson, J., Sumita, U.: Uniform stochastic ordering and related inequalities. Can. J. Stat. **15**, 63–69 (1982)
324. Kiefer, J.: On Bahadur's representation of sample quantiles. Ann. Math. Stat. **38**, 1323–1342 (1967)
325. Kijima, M.: Hazard rate and reversed hazard rate monotonicities in continuous Markov chains. J. Appl. Probab. **35**, 545–556 (1998)
326. King, R.A.R., MacGillivray, H.L.: A starship estimation method for the generalized λ distributions. Aust. New Zeal. J. Stat. **41**, 353–374 (1999)
327. King, R.A.R., MacGillivray, H.L.: Fitting the generalized lambda distribution with location and scale-free shape functionals. Am. J. Math. Manag. Sci. **27**, 441–460 (2007)
328. Kirmani, S.N.U.A.: On sample spacings from IMRL distributions. Stat. Probab. Lett. **29**, 159–166 (1996)
329. Kirmani, S.N.U.A.: On sample spacings from IMRL distributions. Stat. Probab. Lett. **37**, 315 (1998)
330. Klar, B.: A note on $\mathscr{L}$-class of distributions. J. Appl. Probab. **39**, 11–19 (2002)
331. Klar, B., Muller, A.: Characterization of classes of life distribution generalizing the NBUE class. J. Appl. Probab. **40**, 20–32 (2003)
332. Klefsjö, B.: The HNBUE and HNWUE classes of life distributions. Nav. Res. Logist. Q. **29**, 331–344 (1982)
333. Klefsjö, B.: HNBUE and HNWUE classes of life distributions. Nav. Res. Logist. Q. **29**, 615–626 (1982)
334. Klefsjö, B.: On ageing properties and total time on test transforms. Scand. J. Stat. **9**, 37–41 (1982)
335. Klefsjö, B.: Some tests against ageing based on the total time on test transform. Comm. Stat. Theor. Meth. **12**, 907–927 (1983)
336. Klefsjö, B.: Testing exponentiality against HNBUE. Scand. J. Stat. **10**, 67–75 (1983)
337. Klefsjö, B.: A useful ageing property based on Laplace transforms. J. Appl. Probab. **20**, 615–626 (1983)
338. Klefsjö, B.: TTT-transforms: A useful tool when analysing different reliability problems. Reliab. Eng. **15**, 231–241 (1986)
339. Klefsjö, B.: TTT-plotting: A tool for both theoretical and practical problems. J. Stat. Plann. Infer. **29**, 99–110 (1991)

340. Klefsjö, B., Westberg, U.: TTT plotting and maintenance policies. Qual. Eng. **9**, 229–235 (1996–1997)
341. Kleiber, C., Kotz, S.: Statistical Size Distributions in Economics and Actuarial Sciences. Wiley, Hoboken (2003)
342. Kleinbaum, D.G.: Survival Analysis-A Self Learning Text. Springer, New York (1996)
343. Klutke, G., Kiessler, C., Wortman, M.A.: A critical look at the bathtub curve. IEEE Trans. Reliab. **52**, 125–129 (2003)
344. Knopik, L.: Some results on ageing class. Contr. Cybern. **34**, 1175–1180 (2005)
345. Knopik, L.: Characterization of a class of lifetime distributions. Contr. Cybern. **35**, 407–414 (2006)
346. Kochar, S.C.: Distribution free comparison of two probability distributions with reference to their hazard rates. Biometrika **66**, 437–441 (1979)
347. Kochar, S.C.: On extensions of DMRL and related partial orderings of life distributions. Comm. Stat. Stoch. Model. **5**, 235–245 (1989)
348. Kochar, S.C., Deshpande, J.V.: On exponential scores for testing against positive ageing. Stat. Probab. Lett. **3**, 71–73 (1985)
349. Kochar, S.C., Li, X., Shaked, M.: The total time on test transform and the excess wealth stochastic orders of distributions. Adv. Appl. Probab. **34**, 826–845 (2002)
350. Kochar, S.C., Wiens, D.: Partial orderings of life distributions with respect to their ageing properties. Nav. Res. Logist. **34**, 823–829 (1987)
351. Koicheva, M.: A characterization of gamma distribution in terms of conditional moments. Appl. Math. **38**, 19–22 (1993)
352. Korwar, R.: On stochastic orders for the lifetime of k-out-of-n system. Probab. Eng. Inform. Sci. **17**, 137–142 (2003)
353. Kotlyar, V.Y.: A class of ageing distributions. Cybern. Syst. Anal. **28**, 170–176 (1992)
354. Kottas, A., Gelfand, A.E.: Bayesian semiparametric median regression modelling. J. Am. Stat. Assoc. **96**, 1458–1468 (2001)
355. Kotz, S., Seier, E.: An analysis of quantile measures of kurtosis, center and tails. Stat. Paper. **50**, 553–568 (2009)
356. Kulasekera, K.B., Park, H.D.: The class of better mean residual life at age t_0. Microelectron. Reliab. **27**, 725–735 (1987)
357. Kumar, D., Westberg, U.: Maintenance scheduling under age replacement policy using proportional hazards model and TTT-plotting. Eur. J. Oper. Res. **99**, 507–515 (1997)
358. Kundu, D., Gupta, R.D., Manglick, A.: Discriminating between lognormal and generalized exponential distributions. J. Stat. Plann. Infer. **127**, 213–227 (2005)
359. Kundu, D., Kannan, N., Balakrishnan, N.: On the hazard function of the Birnbaum-Saunders distribution and associated inference. Comput. Stat. Data Anal. **52**, 2692–2702 (2008)
360. Kundu, D., Nanda, A.K.: Some reliability properties of the inactivity time. Comm. Stat. Theor. Meth. **39**, 899–911 (2010)
361. Kundu, D., Raqab, M.Z.: Generalized Rayleigh distribution. Comput. Stat. Data Anal. **49**, 187–200 (2005)
362. Kunitz, H., Pamme, H.: The mixed gamma ageing model in life data analysis. Stat. Paper. **34**, 303–318 (1993)
363. Kupka, J., Loo, S.: The hazard and vitality measures of ageing. J. Appl. Probab. **26**, 532–542 (1989)
364. Kus, C.: A new lifetime distribution. Comput. Stat. Data Anal. **51**, 4497–4509 (2007)
365. Kvaloy, J.T., Lindqvist, B.H.: TTT based tests for trend in repairable systems data. Reliab. Eng. Syst. Saf. **60**, 13–28 (1998)
366. Lai, C.D., Moore, T., Xie, M.: The beta integrated failure rate model. In: Proceedings of the International Workshop on Reliability Modelling Analysis – From Theory to Practice, pp. 153–159. National University of Singapore, Singapore (1998)
367. Lai, C.D., Mukherjee, S.P.: A note on a finite range distribution of failure times. Microelectron. Reliab. **26**, 183–189 (1986)

368. Lai, C.D., Xie, M.: Stochastic Ageing and Dependence for Reliability. Springer, New York (2006)
369. Lai, C.D., Xie, M., Murthy, D.N.P.: Bathtub shaped failure rate life distributions. In: Balakrishnan, N., Rao, C.R. (eds.) Handbook of Statistics, vol. 20. Advances in Reliability, pp. 69–104. North-Holland, Amsterdam (2001)
370. Lai, C.D., Xie, M., Murthy, D.N.P.: Modified Weibull model. IEEE Trans. Reliab. **52**, 33–37 (2003)
371. Lakhany, A., Mausser, H.: Estimating parameters of the generalized lambda distribution. Algo Res. Q. **3**, 47–58 (2000)
372. Lan, Y., Leemis, L.M.: Logistic exponential survival function. Nav. Res. Logist. **55**, 252–264 (2008)
373. Landwehr, J.M., Matalas, N.C.: Probability weighted moments compared with some traditional techniques in estimating Gumbel parameters and quantiles. Water Resour. Res. **15**, 1055–1064 (1979)
374. Lariviere, M.A.: A note on probability distributions with generalized failure rates. Oper. Res. **54**, 602–605 (2006)
375. Lariviere, M.A., Porteus, E.L.: Setting to a news vendor: An analysis of price-only contracts. Manuf. Serv. Oper. Manag. **3**, 293–305 (2001)
376. Launer, R.L.: Graphical techniques for analysing failure time data with percentile residual life. IEEE Trans. Reliab. **42**, 71–80 (1983)
377. Launer, R.L.: Inequalities for NBUE and NWUE life distributions. Oper. Res. **32**, 660–667 (1984)
378. Lawless, J.F.: Construction of tolerance bounds for the extreme-value and Weibull distributions. Technometrics **17**, 255–261 (1975)
379. Leemis, L.M.: Lifetime distribution identities. IEEE Trans. Reliab. **35**, 170–174 (1986)
380. Lefante, J.J., Jr.: The generalized single hit model. Math. Biosci. **83**, 167–177 (1987)
381. Lefevre, C., Utev, S.: Comparison of individual risk models. Insur. Math. Econ. **28**, 21–30 (2001)
382. Lehmann, E.L.: The power of rank tests. Ann. Math. Stat. **24**, 23–42 (1953)
383. Lehmann, E.L., Rojo, J.: Invariant directional orderings. Ann. Stat. **20**, 2100–2110 (1992)
384. Lemonte, A.J., Cribari-Neto, F., Vasconcellos, K.L.P.: Improved statistical inference for two parameter Birnbaum-Saunders distribution. Comput. Stat. Data Anal. **51**, 4656–4681 (2007)
385. Lewis, P.A.W., Sheldler, G.S.: Simulation of nonhomogeneous Poisson process with log linear rate function. Biometrika **61**, 501–505 (1976)
386. Leiva, V., Riquelme, N., Balakrishnan, N., Sanhueza, A.: Lifetime analysis based on the generalized Birnbaum-Saunders distribution. Comput. Stat. Data Anal. **52**, 2079–2097 (2008)
387. Lewis, P.A.W., Sheldler, G.S.: Simulation of nonhomogeneous Poisson process with degree two exponential polynomial rate function. Oper. Res. **27**, 1026–1041 (1979)
388. Li, H., Shaked, M.: A general family of univariate stochastic orders. J. Stat. Plann. Infer. **137**, 3601–3610 (2007)
389. Li, X., Kochar, S.C.: Some new results involving the NBU(2) class of life distributions. J. Appl. Probab. **35**, 242–247 (2001)
390. Li, X., Li, Z., Jing, B.: Some results about NBUC class of life distributions. Stat. Probab. Lett. **61**, 235–236 (2003)
391. Li, X., Qiu, G.: Some preservation results of NBUC ageing properties with applications. Stat. Paper. **48**, 581–594 (2007)
392. Li, X., Shaked, M.: The observed total time on test and the observed excess wealth. Stat. Probab. Lett. **68**, 247–258 (2004)
393. Li, X., Xu, M.: Reversed hazard rate order of equilibrium distributions and a related ageing notion. Stat. Paper. **49**, 749–767 (2007)
394. Li, X., Yam, R.C.M.: Reversed properties of some negative ageing properties with application. Stat. Paper. **46**, 65–68 (2005)
395. Li, X., Zuo, M.J.: Preservation of stochastic orders for random minima and maxima with applications. Nav. Res. Logist. **51**, 332–344 (2000)

396. Li, Y.: Closure of NBU(2) class under formation of parallel system. Stat. Probab. Lett. **67**, 57–63 (2004)
397. Li, Y.: Preservation of NBUC and NBU(2) classes under mixtures. Probab. Eng. Inform. Sci. **19**, 277–298 (2005)
398. Li, Z., Li, X.: $\{IFR * t_0\}$ and $\{NBU * t_0\}$ classes of life distributions. J. Stat. Plann. Infer. **70**, 191–200 (1998)
399. Lillo, R.E.: On the median residual lifetime and its aging properties: A characterization theorem and applications. Nav. Res. Logist. **52**, 370–380 (2005)
400. Lin, G.D.: Characterization of the $\mathscr{L}$-class of life distributions. Stat. Probab. Lett. **40**, 259–266 (1998)
401. Lin, G.D.: Characterization of the exponential distribution via residual life time. Sankhyā Ser. B **65**, 249–258 (2003)
402. Lin, G.D.: On the characterization of life distributions by percentile residual lifetimes. Sankhyā **71**, 64–72 (2009)
403. Lin, G.D., Hu, C.: A note on the $\mathscr{L}$-class of life distributions. Sankhyā Ser. A **62**, 267–272 (2000)
404. Loh, W.Y.: A new generalization of the class of NBU distributions. IEEE Trans. Reliab. **33**, 419–422 (1984)
405. Lynch, J.D.: On condition for mixtures of increasing failure rate distributions to have an increasing failure rate. Probab. Eng. Inform. Sci. **13**, 33–36 (1999)
406. Ma, C.: A note on stochastic ordering of order statistics. J. Appl. Probab. **34**, 785–789 (1997)
407. MacGillivray, H.L.: Skewness properties of asymmetric forms of Tukey-lambda distribution. Comm. Stat. Theor. Meth. **11**, 2239–2248 (1982)
408. Mahdy, M.: Characterization and preservations of the variance inactivity time ordering and increasing variance inactivity time class. J. Adv. Res. **3**, 29–34 (2012)
409. Mann, H.B., Whitney, D.R.: On a test of whether one of two random variables is stochastically larger than the other. Ann. Math. Stat. **18**, 50–60 (1947)
410. Mann, N.R., Fertig, K.W.: Efficient unbiased quantile estimators for moderate-size complete samples from extreme-value and Weibull distributions; confidence bounds and tolerance and prediction intervals. Technometrics **19**, 87–93 (1977)
411. Marshall, A.W., Olkin, I.: A new method of adding a parameter to a family of distributions with application to exponential and Weibull families. Biometrika **84**, 641–652 (1997)
412. Marshall, A.W., Olkin, I.: Life Distributions. Springer, New York (2007)
413. Marshall, A.W., Proschan, F.: Classes of distributions applicable in replacement with renewal theory implications. In: Proceedings of the Sixth Berkeley Symposium on Mathematical Statistics and Probability, pp. 395–415. California Press, Berkeley (1972)
414. McDonald, J.B., Richards, D.O.: Hazard rates and generalized beta distributions. IEEE Trans. Reliab. **36**, 463–466 (1987)
415. McDonald, J.B., Richards, D.O.: Model selection: Some generalized distributions. Comm. Stat. Theor. Meth. **16**, 1049–1057 (1987)
416. Mercy, J., Kumaran, M.: Estimation of the generalized lambda distribution from censored data. Braz. J. Probab. Stat. **24**, 42–56 (2010)
417. Misra, N., Gupta, N., Dhariyal, I.D.: Preservation of some ageing properties and stochastic orders by weighted distributions. Comm. Stat. Theor. Meth. **37**, 627–644 (2008)
418. Mitra, M., Basu, S.K.: On some properties of bathtub failure rate family of distributions. Microelectron. Reliab. **36**, 679–684 (1996)
419. Mitra, M., Basu, S.K.: Shock models leading to nonmonotonic ageing classes of life distributions. J. Stat. Plann. Infer. **55**, 131–138 (1996)
420. Moore, T., Lai, C.D.: The beta failure rate distribution. In: Proceedings of the 30th Annual Conference of Operational Research Society of New Zealand, pp. 339–344. Palmerston, New Zealand (1994)
421. Moors, J.J.A.: A quantile alternative for kurtosis. The Statistician **37**, 25–32 (1988)
422. Mudholkar, G.S., Asubonting, K.O., Hutson, A.D.: Transformation of the bathtub failure rate data in reliability using the Weibull distribution. Stat. Methodol. **6**, 622–633 (2009)

423. Mudholkar, G.S., Hutson, A.D.: The exponentiated Weibull family: Some properties and flood data applications. Comm. Stat. Theor. Meth. **25**, 3059–3083 (1996)
424. Mudholkar, G.S., Hutson, A.D.: Analogues of *L*-moments. J. Stat. Plann. Infer. **71**, 191–208 (1998)
425. Mudholkar, G.S., Kollia, G.D.: The isotones of the test of exponentiality. In: ASA Proceedings, Statistical Graphics, Alexandria (1990)
426. Mudholkar, G.S., Kollia, G.D.: Generalized Weibull family–a structural analysis. Comm. Stat. Theor. Meth. **23**, 1149–1171 (1994)
427. Mudholkar, G.S., Srivastava, D.K., Freimer, M.: The exponentiated Weibull family: A reanalysis of bus motor failure data. Technometrics **37**, 436–445 (1995)
428. Mudholkar, G.S., Srivastava, D.K., Kollia, G.D.: A generalization of the Weibull distribution with applications to the analysis of survival data. J. Am. Stat. Assoc. **91**, 1575–1583 (1996)
429. Mudholkar, G.S., Srivastava, D.K.: Exponentiated Weibull family for analysing bathtub failure data. IEEE Trans. Reliab. **42**, 299–302 (1993)
430. Mukherjee, S.P., Islam, A.: A finite range distribution of failure times. Nav. Res. Logist. Q. **30**, 487–491 (1983)
431. Mukherjee, S.P., Roy, D.: Some characterizations of the exponential and related life distributions. Calcutta Stat. Assoc. Bull. **35**, 189–197 (1986)
432. Muller, A., Stoyan, D.: Comparison Methods for Stochastic Models and Risks. Wiley, New York (2002)
433. Muralidharan, K., Lathika, P.: Analysis of instantaneous and early failures in Weibull distribution. Metrika **64**, 305–316 (2006)
434. Murthy, D.N.P., Xie, M., Jiang, R.: Weibull Models. Wiley, Hoboken (2003)
435. Murthy, V.K., Swartz, G., Yuen, K.: Realistic models for mortality rates and estimation, I and II. Technical Reports, University of California, Los Angeles (1973)
436. Muth, E.J.: Reliability models with positive memory derived from mean residual life function. In: Tsokos, C.P., Shimi, I.N. (eds.) Theory and Application of Reliability, pp. 401–435. Academic, Boston (1977)
437. Nadarajah, S.: Bathtub shaped failure rate functions. Qual. Quant. **43**, 855–863 (2009)
438. Nair, N.U.: A characteristic property of the Gumbel's bivariate exponential distribution. Calcutta Stat. Assoc. Bull. **36**, 181–184 (1987)
439. Nair, N.U., Asha, G.: Characterizations using failure and reversed failure rates. J. Indian Soc. Probab. Stat. **8**, 45–56 (2004)
440. Nair, N.U., Nair, K.R.M., Haritha, H.N.: Some properties of income gap ratio and truncated Gini coefficient. Calcutta Stat. Assoc. Bull. **60**, 239–254 (2008)
441. Nair, N.U., Preeth, M.: On some properties of equilibrium distributions of order *n*. Stat. Meth. Appl. **18**, 453–464 (2009)
442. Nair, N.U., Sankaran, P.G.: Characterization of the Pearson family of distributions. IEEE Trans. Reliab. **40**, 75–77 (1991)
443. Nair, N.U., Sankaran, P.G.: Characterization of multivariate life distributions. J. Multivariate Anal. **99**, 2096–2107 (2008)
444. Nair, N.U., Sankaran, P.G.: Quantile-based reliability analysis. Comm. Stat. Theor. Meth. **38**, 222–232 (2009)
445. Nair, N.U., Sankaran, P.G.: Properties of a mean residual life arising from renewal theory. Nav. Res. Logist. **57**, 373–379 (2010)
446. Nair, N.U., Sankaran, P.G.: Some results on an additive hazard model. Metrika **75**, 389–402 (2010)
447. Nair, N.U., Sankaran, P.G., Vineshkumar, B.: Total time on test transforms and their implications in reliability analysis. J. Appl. Probab. **45**, 1126–1139 (2008)
448. Nair, N.U., Sankaran, P.G., Vineshkumar, B.: Modelling lifetimes by quantile functions using Parzen's score function. Statistics **46**, 799–811 (2012)
449. Nair, N.U., Sudheesh, K.K.: Characterization of continuous distributions by variance bound and its implications to reliability modelling and catastrophe theory. Comm. Stat. Theor. Meth. **35**, 1189–1199 (2006)

450. Nair, N.U., Sudheesh, K.K.: Some results on lower variance bounds useful in reliability modelling and estimation. Ann. Inst. Stat. Math. **60**, 591–603 (2008)
451. Nair, N.U., Sudheesh, K.K.: Characterization of continuous distributions by properties of conditional variance. Stat. Methodol. **7**, 30–40 (2010)
452. Nair, N.U., Vineshkumar, B.: *L*-moments of residual life. J. Stat. Plann. Infer. **140**, 2618–2631 (2010)
453. Nair, N.U., Vineshkumar, B.: Reversed percentile residual life and related concepts. J. Kor. Stat. Soc. **40**, 85–92 (2010)
454. Nair, N.U., Vineshkumar, B.: Ageing concepts: An approach based on quantile functions. Stat. Probab. Lett. **81**, 2016–2025 (2011)
455. Nair, U.S.: The standard error of Gini's mean difference. Biometrika **34**, 151–155 (1936)
456. Najjar, D., Bigerelle, M., Lefevre, C., Iost, A.: A new approach to predict the pit depth extreme value of a localized corrosion process. ISIJ Int. **43**, 720–725 (2003)
457. Nanda, A.K.: Generalized ageing classes in terms of Laplace transforms. Sankhyā **62**, 258–266 (2000)
458. Nanda, A.K., Bhattacharjee, S., Alam, S.S.: Properties of proportional mean residual life model. Stat. Probab. Lett **76**, 880–890 (2006)
459. Nanda, A.K., Bhattacharjee, S., Balakrishnan, N.: Mean residual life function, associated orderings and properties. IEEE Trans. Reliab. **59**, 55–65 (2010)
460. Nanda, A.K., Jain, K., Singh, H.: On closure of some partial orderings under mixtures. J. Appl. Probab. **33**, 698–706 (1996)
461. Nanda, A.K., Shaked, M.: The hazard rate and RHR orders with applications to order statistics. Ann. Inst. Stat. Math. **53**, 853–864 (2001)
462. Nanda, A.K., Singh, H., Misra, N., Paul, P.: Reliability properties of reversed residual life time. Comm. Stat. Theor. Meth. **32**, 2031–2042 (2003)
463. Nassar, M., Eissa, F.H.: Bayesian estimation of the exponentiated Weibull model. Comm. Stat. Theor. Meth. **33**, 2343–2362 (2007)
464. Nassar, M.M., Eissa, F.H.: On the exponentiated Weibull distribution. Comm. Stat. Theor. Meth. **32**, 1317–1336 (2003)
465. Navarro, J., Franco, M., Ruiz, J.M.: Characterization through moments of residual life and conditional spacing. Sankhyā Ser. A **60**, 36–48 (1998)
466. Navarro, J., Hernandez, P.J.: How to obtain bathtub shaped failure rate models from normal mixtures. Probab. Eng. Inform. Sci. **18**, 511–531 (2004)
467. Navarro, J., Lai, C.D.: Ordering properties of systems with two dependent components. Comm. Stat. Theor. Meth. **36**, 645–655 (2007)
468. Neath, A.A., Samaniego, F.J.: On the total time on test transform of an IFRA distribution. Stat. Probab. Lett. **14**, 289–291 (1992)
469. Nelson, W.: Accelerated Testing, Statistical Methods, Test Plans and Data Analysis. Wiley, New York (1990)
470. Ng, H.K.T., Kundu, D., Balakrishnan, N.: Modified moment estimation for the two-parameter Birnbaum-Saunders distribution. Comput. Stat. Data Anal. **43**, 283–298 (2003)
471. Ng, H.K.T., Kundu, D., Balakrishnan, N.: Point and interval estimation for the two-parameter Birnbaum-Saunders distribution based on Type-II censored samples. Comput. Stat. Data Anal. **50**, 3222–3242 (2006)
472. Nikulin, M.S., Haghighi, F.: A chi-square test for the generalized power Weibull family for the head-and-neck cancer censored data. J. Math. Sci. **133**, 1333–1341 (2006)
473. Oluyede, B.: Some inequalities and bounds for weighted reliability measures. J. Inequal. Pure Appl. Math. **3** (2002). Article 60
474. Ortega, E.M.M.: A note on some functional relationships involving the mean inactivity time order. IEEE Trans. Reliab. **58**, 172–178 (2009)
475. Osaki, S., Li, X.: Characterizations of the gamma and negative binomial distributions. IEEE Trans. Reliab. **37**, 379–382 (1988)
476. Osturk, A., Dale, R.F.: A study of fitting the generalized lambda distribution to solar radiation data. J. Appl. Meteorol. **12**, 995–1004 (1982)

477. Osturk, A., Dale, R.F.: Least squares estimation of the parameters of the generalized lambda distribution. Technometrics **27**, 81–84 (1985)
478. Owen, W.J.: A new three parameter extension to the Birnbaum-Saunders distribution. IEEE Trans. Reliab. **55**, 475–479 (2006)
479. Padgett, W.J., Tsai, S.K.: Prediction intervals for future observations from the inverse Gaussian distribution. IEEE Trans. Reliab. **35**, 406–408 (1986)
480. Pamme, H., Kunitz, H.: Detection and modelling of ageing properties in lifetime data. In: Basu, A.P. (ed.) Advances in Reliability, pp. 291–302. North-Holland, Amsterdam (1993)
481. Paranjpe, S.A., Rajarshi, M.B.: Modelling non-parametric survivorship data with bathtub distributions. Ecology **67**, 1693–1695 (1986)
482. Paranjpe, S.A., Rajarshi, M.B., Gore, A.P.: On a model for failure rates. Biometrical J. **27**, 913–917 (1985)
483. Park, D.H.: Class of NBU-t_0 life distributions. In: Pham, H. (ed.) Handbook of Reliability Engineering. Springer, New York (2003)
484. Parzen, E.: Nonparametric statistical data modelling. J. Am. Stat. Assoc. **74**, 105–122 (1979)
485. Parzen, E.: Unifications of statistical methods for continuous and discrete data. In: Page, C., Lepage, R. (eds.) Proceedings of Computer Science-Statistics. INTERFACE 1990, pp. 235–242. Springer, New York (1991)
486. Parzen, E.: Concrete statistics. In: Ghosh, S., Schucany, W.R., Smith, W.B. (eds.) Statistics of Quality. Marcel Dekker, New York (1997)
487. Parzen, E.: Quality probability and statistical data modelling. Stat. Sci. **19**, 652–662 (2004)
488. Pearson, C.P.: Application of *L*-moments to maximum river flows. New Zeal. Stat. **28**, 2–10 (1993)
489. Pearson, K.: Tables of Incomplete Beta Function, 2nd edn. Cambridge University Press, Cambridge (1968)
490. Pellerey, F.: On the preservation of some orderings of risks under convolution. Insur. Math. Econ. **16**, 23–30 (1995)
491. Pereira, A.M.F., Lillo, R.E., Shaked, M.: The decreasing percentile residual life ageing notion. Statistics **46**, 1–17 (2011)
492. Perez-Ocon, R., Gamiz-Perez, M.L., Ruiz-Castro, J.E.: A study of different ageing classes via total time on test transform and Lorenz curves. Appl. Stoch. Model. Data Anal. **13**, 241–248 (1998)
493. Pham, T.G., Turkkan, M.: The Lorenz and the scaled total-time-on-test transform curves, A unified approach. IEEE Trans. Reliab. **43**, 76–84 (1994)
494. Pham, T.G., Turkkan, M.: Reliability of a standby system with beta distributed component lives. IEEE Trans. Reliab. **43**, 71–75 (1994)
495. Phani, K.K.: A new modified Weibull distribution function. Comm. Am. Ceram. Soc. **70**, 182–184 (1987)
496. Prakasa Rao, B.L.S.: On distributions with periodic failure rate and related inference problems. In: Panchapakesan, S., Balakrishanan, N. (eds.) Advances in Statistical Theory and Applications. Birkhauser, Boston (1997)
497. Pregibon, D.: Goodness of link tests for generalized linear models. Appl. Stat. **29**, 15–24 (1980)
498. Pundir, S., Arora, S., Jain, K.: Bonferroni curve and the related statistical inference. Stat. Probab. Lett. **75**, 140–150 (2005)
499. Quetelet, L.A.J.: Letters Addressed to HRH the Grand Duke of Saxe Coburg and Gotha in the Theory of Probability. Charles and Edwin Laton, London (1846). Translated by Olinthus Gregory Downs
500. Rajarshi, S., Rajarshi, M.B.: Bathtub distributions: A review. Comm. Stat. Theor. Meth. **17**, 2597–2621 (1988)
501. Ramberg, J.S.: A probability distribution with applications to Monte Carlo simulation studies. In: Patil, G.P., Kotz, S., Ord, J.K. (eds.) Model Building and Model Selection. Statistical Distributions in Scientific Work, vol. 2. D. Reidel, Dordrecht (1975)

502. Ramberg, J.S., Dudewicz, E., Tadikamalla, P., Mykytka, E.: A probability distribution and its uses in fitting data. Technometrics **21**, 210–214 (1979)
503. Ramberg, J.S., Schmeiser, B.W.: An approximate method for generating symmetric random variables. Comm. Assoc. Comput. Mach. **15**, 987–990 (1972)
504. Ramberg, J.S., Schmeiser, B.W.: An approximate method for generating asymmetric random variables. Comm. Assoc. Comput. Mach. **17**, 78–82 (1974)
505. Ramos-Fernandez, A., Paradela, A., Narajas, R., Albar, J.P.: Generalized method for probability based peptitude and protein identification from tandem mass spectrometry data and sequence data base searching. Mol. Cell. Proteomics **7**, 1745–1754 (2008)
506. Rao, C.R., Shanbhag, D.N.: Choquet-Deny Type Functional Equations with Applications to Stochastic Models. Wiley, Chichester (1994)
507. Rieck, J.R.: A moment generating function with applications to the Birnbaum-Saunders distribution. Comm. Stat. Theor. Meth. **28**, 2213–2222 (1999)
508. Robinson, L.W., Chan, R.R.: Scheduling doctor's appointment, optimal and empirically based heuristic policies. IIE Trans. **35**, 295–307 (2003)
509. Rohatgi, V.K., Saleh, A.K.Md.E.: A class of distributions connected to order statistics with nonintegral sample size. Comm. Stat. Theor. Meth. **17**, 2005–2012 (1988)
510. Rojo, J.: Nonparametric quantile estimators until order constraints. J. Nonparametr. Stat. **5**, 185–200 (1995)
511. Rojo, J.: Estimation of the quantile function of an IFRA distribution. Scand. J. Stat. **25**, 293–310 (1998)
512. Rolski, T.: Mean residual life. Bull. Int. Stat. Inst. **46**, 266–270 (1975)
513. Samaniego, F.J.: System Signatures and Their Applications in Engineering Reliability. Springer, New York (2007)
514. Samaniego, F.J., Balakrishnan, N., Navarro, J.: Dynamic signatures and their use in comparing the reliability of new and used systems. Nav. Res. Logist. **56**, 577–591 (2009)
515. Sankaran, P.G., Nair, N.U.: Nonparametric estimation of the hazard quantile function. J. Nonparametr. Stat. **21**, 757–767 (2009)
516. Sankaran, P.G., Nair, N.U., Sindhu, T.K.: A generalized Pearson system useful in reliability analysis. Stat. Paper. **44**, 125–130 (2003)
517. Sankarasubramonian, A., Sreenivasan, K.: Investigation and comparison of *L*-moments and conventional moments. J. Hydrol. **218**, 13–34 (1999)
518. Sarabia, J.M.: A general definition of Leimkuhler curves. J. Informetrics **2**, 156–163 (2008)
519. Sarabia, J.M., Prieto, F., Sarabia, M.: Revisiting a functional form for the Lorenz curve. Econ. Lett. **105**, 61–63 (2010)
520. Sarhan, A.M., Kundu, D.: Generalized linear failure rate distribution. Comm. Stat. Theor. Meth. **38**, 642–660 (2009)
521. Scarsini, M., Shaked, M.: Some conditions for stochastic equality. Nav. Res. Logist. **37**, 617–625 (1990)
522. Schäbe, H.: Constructing lifetime distributions with bathtub shaped failure rate from DFR distributions. Microelectron. Reliab. **34**, 1501–1508 (1994)
523. Schmittlein, D.C., Morrison, D.G.: The median residual lifetime–a characterization problem and an application. Oper. Res. **29**, 392–399 (1981)
524. Sen, A.: An ordinal approach to measurement. Econometrica **44**, 219–231 (1976)
525. Sen, A., Bhattacharyya, G.K.: Inference procedure for linear failure rate model. J. Stat. Plann. Infer. **44**, 59–76 (1995)
526. Sengupta, D., Deshpande, J.V.: Some results on the relative ageing of two life distributions. J. Appl. Probab. **31**, 991–1003 (1994)
527. Serfling, R.J.: Approximation Theorems of Mathematical Statistics. Wiley, New York (1980)
528. Seshadri, V.: The Inverse Gaussian Distribution: A Case Study in Exponential Families. Oxford University Press, New York (1994)
529. Shaked, M.: Statistical inference for a class of life distributions. Comm. Stat. Theor. Meth. **6**, 1323–1329 (1977)

530. Shaked, M.: Exponential life functions with NBU components. Ann. Probab. **11**, 752–759 (1983)
531. Shaked, M., Shanthikumar, J.G.: Stochastic Orders. Springer, New York (2007)
532. Shaked, M., Spizzichino, F.: Mixtures and monotonicity of failure rate functions. In: Balakrishnan, N., Rao, C.R. (eds.) Handbook of Statistics: Advances in Reliability, pp. 185–197. North-Holland, Amsterdam (2001)
533. Shaked, M., Wang, T.: Preservation of stochastic orderings under random mapping by point processes. Probab. Eng. Inform. Sci. **9**, 563–580 (1995)
534. Shanmughapriya, S., Lakshmi, S.: Exponentiated Weibull distribution for analysis of bathtub failure-rate data. Int. J. Appl. Math. Stat. **17**, 37–43 (2010)
535. Shapiro, S.S., Gross, A.J.: Statistical Modelling Techniques. Marcel Dekker, New York (1981)
536. Shapiro, S.S., Wilk, M.B.: An analysis of variance test for normality. Biometrika **52**, 591–611 (1965)
537. Sillitto, G.P.: Derivation of approximants to the inverse distribution function of a continuous univariate population from the order statistics of a sample. Biometrika **56**, 641–650 (1969)
538. Silva, G.O., Ortega, E.M.M., Cordeiro, G.M.: The beta modified Weibull distribution. Lifetime Data Anal. **16**, 409–430 (2010)
539. Silva, R.B., Barreto-Souza, W., Cordiero, G.M.: A new distribution with decreasing, increasing and upside down bathtub failure rate. Comput. Stat. Data Anal. **54**, 915–944 (2010)
540. Silver, E.A.: A safety factor approximation based upon Tukey's lambda distribution. Oper. Res. Q. **28**, 743–746 (1977)
541. Singh, H.: On partial orderings of life distributions. Nav. Res. Logist. **36**, 103–110 (1989)
542. Singh, H., Deshpande, J.V.: On some new ageing properties. Scand. J. Stat. **12**, 213–220 (1985)
543. Singh, U., Gupta, P.K., Upadhyay, S.K.: Estimation of parameters of exponentiated Weibull family. Comput. Stat. Data Anal. **48**, 509–523 (2005)
544. Smith, R.M., Bain, L.J.: An exponential power life-testing distribution. Comm. Stat. Theor. Meth. **4**, 449–481 (1975)
545. Song, J.K., Cho, G.Y.: A note on percentile residual life. Sankhyā **57**, 333–335 (1995)
546. Stein, W.E., Dattero, R.: Bondesson's functions in reliability theory. Appl. Stoch. Model. Bus. Ind. **15**, 103–109 (1999)
547. Stigler, S.M.: Fractional order statistics with applications. J. Am. Stat. Assoc. **72**, 544–550 (1977)
548. Stoyanov, J., Al-sadi, M.H.M.: Properties of a class of distributions based on conditional variance. J. Appl. Probab. **41**, 953–960 (2004)
549. Su, S.: A discretised approach to flexibly fit generalized lambda distributions to data. J. Mod. Appl. Stat. Meth. **4**, 408–424 (2005)
550. Su, S.: Numerical maximum log likelihood estimation for generalized lambda distributions. Comput. Stat. Data Anal. **51**, 3983–3998 (2007)
551. Su, S.: Fitting single and mixture of the generalized lambda distributions via discretized and maximum likelihood methods: GILDEX in (*r*). J. Stat. Software **21**, 1–22 (2010)
552. Suleswki, P.: On differently defined skewness. Comput. Meth. Sci. Technol. **14**, 39–46 (2008)
553. Sultan, K.S., Ismail, M.A., Al-Moisheer, A.S.: Mixture of two inverse Weibull distributions. Comput. Stat. Data Anal. **51**, 5377–5387 (2007)
554. Sunoj, S.M.: Characterization of some continuous distributions using partial moments. Metron **66**, 353–362 (2004)
555. Sunoj, S.M., Maya, S.S.: The role of lower partial moments in stochastic modelling. Metron **66**, 223–242 (2008)
556. Sweet, A.L.: On the hazard rate of the lognormal distribution. IEEE Trans. Reliab. **39**, 325–328 (1990)
557. Szekli, R.: Stochastic Ordering and Dependence in Applied Probability. Lecture Notes in Statistics, vol. 97. Springer, New York (1995)
558. Tabot, J.P.P.: The bathtub myth. Qual. Assur. **3**, 107–108 (1997)

559. Tajuddin, I.H.: A simple measure of skewness. Stat. Neerl. **50**, 362–366 (1996)
560. Tang, Y., Xie, M., Goh, T.N.: Statistical analysis of Weibull extension model. Comm. Stat. Theor. Meth. **32**, 913–928 (2003)
561. Tanguy, C.: Mean time to failure for periodic failure rate. R&RATA **2** (2009)
562. Tarsitano, A.: Fitting generalized lambda distribution to income data. In: COMPSTAT 2004 Symposium, pp. 1861–1867. Springer, New York (2004)
563. Tarsitano, A.: Estimation of the generalised lambda distributions parameter for grouped data. Comm. Stat. Theor. Meth. **34**, 1689–1709 (2005)
564. Tarsitano, A.: Comparing estimation methods for the FPLD. J. Probab. Stat. 1–16 (2010)
565. Tieling, Z., Xie, M.: Failure data analysis with extended Weibull distribution. Comm. Stat. Simul. Comput. **36**, 579–592 (2007)
566. Topp, C.W., Leone, P.C.: A family of j-shaped frequency function. J. Am. Stat. Assoc. **50**, 209–219 (1995)
567. Tukey, J.W.: The practical relationship between the common transformations of percentages of count and of amount. Technical Report 36, Princeton University, Princeton (1960)
568. Tukey, J.W.: The future of data analysis. Ann. Math. Stat. **33**, 1–67 (1962)
569. Tukey, J.W.: Exploratory Data Analysis. Addisson-Wesley, Reading (1977)
570. Usagaonkar, S.G.G., Maniappan, V.: Additive Weibull model for reliability analysis. Int. J. Perform. Eng. **5**, 243–250 (2009)
571. van Dyke, J.: Numerical investigation of the random variable $y = c(u^{\lambda} - (1-u)^{\lambda})$. Unpublished working paper. National Bureau of Standards, Statistical Engineering Laboratory (1961)
572. van Staden, P.J., Loots, M.T.: L-moment estimation for the generalized lambda distribution. In: Third Annual ASEARC Conference, New Castle, Australia (2009)
573. Vera, F., Lynch, J.: K-Mart stochastic modelling using iterated total time on test transforms. In: Modern Statistical and Mathematical Methods in Reliability, pp. 395–409. World Scientific, Singapore (2005)
574. Vogel, R.M., Fennessey, N.M.: L-moment diagrams should replace product moment diagrams. Water Resour. Res. **29**, 1745–1752 (1993)
575. Voinov, V., Nikulin, M.S., Balakrishnan, N.: Chi-Squared Goodness-of-Fit Tests with Applications. Academic, Boston (2013)
576. Wang, D., Hutson, A.D., Miecznikowski, J.C.: L-moment estimation for parametric survival models. Stat. Methodol. **7**, 655–667 (2010)
577. Wang, F.K.: A new model with bathtub-shaped hazard rate using an additive Burr XII distribution. Reliab. Eng. Syst. Saf. **70**, 305–312 (2000)
578. Westberg, U., Klefsjö, B.: TTT plotting for censored data based on piece-wise exponential estimator. Int. J. Reliab. Qual. Saf. Eng. **1**, 1–13 (1994)
579. Wie, X.: Test of exponentiality against a monotone hazards function alternative based on TTT transformations. Microelectron. Reliab. **32**, 607–610 (1992)
580. Wie, X.: Testing whether one distribution is more IFR than another. Microelectron. Reliab. **32**, 271–273 (1992)
581. Willmot, G.E., Cai, J.: On classes of life distributions with unknown age. Probab. Eng. Inform. Sci. **14**, 473–484 (2000)
582. Willmot, G.E., Drekic, S., Cai, J.: Equilibrium component distributions and stoploss moments. Scand. Actuarial J. **1**, 6–24 (2005)
583. Willmot, G.E., Lin, X.S.: Lindberg approximations for compound distributions with applications. Lecture Notes in Statistics, vol. 156. Springer, New York (2001)
584. Wondmagegnehu, E.T.: On the behaviour and shape of mixture failure rate from a family of IFR Weibull distributions. Nav. Res. Logist. **51**, 491–500 (2004)
585. Wondmagegnehu, E.T., Navarro, J., Hernandez, P.J.: Bathtub shaped failure rates from mixtures: A practical point of view. IEEE Trans. Reliab. **54**, 270–275 (2005)
586. Wong, K.L.: Unified (field) failure theory—demise of the bathtub curve. In: Proceedings of the Annual Reliability and Maintainability Symposium, Philadelphia, PA, pp. 402–407 (1981)

587. Wong, K.L.: The bathtub does not hold water any more. Qual. Reliab. Eng. Int. **4**, 279–282 (1988)
588. Wong, K.L.: Roller-coaster curve is in. Qual. Reliab. Eng. Int. **5**, 29–36 (1989)
589. Wong, K.L.: The physical basin for the roller-coaster hazard rate curve for electronics. Qual. Reliab. Eng. Int. **7**, 489–495 (1991)
590. Xiang, X.: A law of the logarithm for kernel quantile density estimators. Ann. Probab. **22**, 1078–1091 (1994)
591. Xie, F.L., Wei, B.C.: Diagnostic analysis for the log Birnbaum-Saunders regression models. Comput. Stat. Data Anal. **51**, 4692–4706 (2007)
592. Xie, M.: Testing constant failure rate against some partially monotone alternatives. Microelectron. Reliab. **27**, 557–565 (1987)
593. Xie, M.: Some total time on test quantities useful for testing constant against bathtub shaped failure rate distributions. Scand. J. Stat. **16**, 137–144 (1989)
594. Xie, M., Lai, C.D.: Reliability analysis using additive Weibull model with bathtub shaped failure rate function. Reliab. Eng. Syst. Saf. **52**, 87–93 (1995)
595. Xie, M., Tang, Y., Goh, T.N.: A modified Weibull extension with bathtub-shaped failure rate function. Reliab. Eng. Syst. Saf. **76**, 279–285 (2002)
596. Yitzhaki, S.: Gini's mean difference: A superior measure of variability for nonnormal distributions. Metron **61**, 285–316 (2003)
597. Yu, Y.: Stochastic orderings of exponential family of distributions and their mixtures. J. Appl. Probab. **46**, 244–254 (2009)
598. Yue, D., Cao, J.: The NBUL class of life distributions and replacement policy comparisons. Nav. Res. Logist. **48**, 578–591 (2001)
599. Zang, Z., Li, X.: Some new properties of stochastic ordering and ageing properties of coherent systems. IEEE Trans. Reliab. **59**, 718–724 (2010)
600. Zhang, T., Xie, M.: On the upper truncated Weibull distribution and its reliability implications. Reliab. Eng. Syst. Saf. **96**, 194–200 (2011)
601. Zhao, N., Song, Y.H., Lu, H.: Risk assessment strategies using total time on test transforms. IEEE (2006). doi: 10.1109/PES.2006.1709062
602. Zhao, P., Balakrishnan, N.: Mean residual life order of convolutions of heterogeneous exponential random variables. J. Multivariate Anal. **100**, 1792–1801 (2009)
603. Zhou, Y., Yip, P.S.F.: Nonparametric estimation of quantile density functions for truncated and censored data. J. Nonparametr. Stat. **12**, 17–39 (1999)
604. Zimmer, W., Keats, J.B., Wang, F.K.: The Burr XII distribution in reliability analysis. J. Qual. Technol. **20**, 386–394 (1998)

Index

N.U. Nair et al., *Quantile-Based Reliability Analysis*, Statistics for Industry and Technology, DOI 10.1007/978-0-8176-8361-0,

M

O

P

Q

R

S

T

Author Index

N.U. Nair et al., *Quantile-Based Reliability Analysis*, Statistics for Industry and Technology, DOI 10.1007/978-0-8176-8361-0,

Printed in the United States
By Bookmasters

Printed in the United States
By Bookmasters